Intelligent Systems Reference Library

Volume 100

Series editors

About this Series

The aim of this series is to publish a Reference Library, including novel advances and developments in all aspects of Intelligent Systems in an easily accessible and well structured form. The series includes reference works, handbooks, compendia, textbooks, well-structured monographs, dictionaries, and encyclopedias. It contains well integrated knowledge and current information in the field of Intelligent Systems. The series covers the theory, applications, and design methods of Intelligent Systems. Virtually all disciplines such as engineering, computer science, avionics, business, e-commerce, environment, healthcare, physics and life science are included.

More information about this series at http://www.springer.com/series/8578

Erik Cuevas · Daniel Zaldívar
Marco Perez-Cisneros

Applications of Evolutionary Computation in Image Processing and Pattern Recognition

Erik Cuevas
CUCEI
Universidad de Guadalajara
Guadalajara
Mexico

Marco Perez-Cisneros
Centro Universitario de Tonalá
Universidad de Guadalajara
Guadalajara
Mexico

Daniel Zaldívar
CUCEI
Universidad de Guadalajara
Guadalajara
Mexico

ISSN 1868-4394 ISSN 1868-4408 (electronic)
Intelligent Systems Reference Library
ISBN 978-3-319-26460-8 ISBN 978-3-319-26462-2 (eBook)
DOI 10.1007/978-3-319-26462-2

Library of Congress Control Number: 2015954589

Springer Cham Heidelberg New York Dordrecht London

Printed on acid-free paper

Springer International Publishing AG Switzerland is part of Springer Science+Business Media (www.springer.com)

Foreword

Combinations among evolutionary computation, image processing, and pattern recognition have made significant advances recently and have become trendy disciplines with increasing research projects in both academia and industry. Likewise, the demand for robust image processing and pattern recognition applications is growing at a remarkable rate despite an evident shortfall in appropriate approaches to meet such fast-growing demand and an increase on its associated challenges.

This book is a significant addition to this field and an excellent effort to address some key challenges and current trends. Erik Cuevas, Daniel Zaldivar, and Marco Perez have all investigated three recent related areas that have attracted a remarkable attention from the scientific community, such as image processing, pattern recognition, and evolutionary computation.

The motivation in compiling this book has strengthened from the fact that such paradigms have demonstrated an efficient management of image uncertainties that are typically difficult to eliminate. Such uncertainties include various types of information that are incomplete, noisy, imprecise, fragmentary, not fully reliable, vague, contradictory, or deficient. Commonly, such issues yield from the lack of the full and precise knowledge of the system, including the selection of evaluation criteria, alternatives, weights, assignment scores, and the final integrated decision result. The authors employ an interesting approach that capitalizes on an application-oriented method in order to find answers for such difficult issues. They have chosen an application-oriented scheme to explain the basic concepts and characteristics of the different evolutionary computation methods.

In particular, the authors have developed an interesting treatment that complements what is unknown for both research communities (evolutionary computation and image processing). Therefore, each chapter concentrates on a specific combination of evolutionary method with a complex image processing or pattern recognition problem; however, all chapters offer an interesting integral point of view from different paradigms.

The organization of each chapter builds on explaining the fundamentals of a given computer vision algorithm. At the same time, it includes the fundamentals and the analysis of an appropriate evolutionary method that can reinforce the performance of such artificial vision algorithm. It is expected that students will appreciate the ability to run the simulation exemplars in order to reinforce their understanding of the mathematical derivations. Adopting a simulation-based approach for learning about basic and advanced issues on evolutionary vision algorithms undoubtedly ensures that students will possess all concepts in an enjoyable and interactive way.

I believe that the authors have done a good job in addressing and studying successful cases for each chapter. I consider the book a good addition to the areas of image processing, pattern recognition, and evolutionary computation since it will deliver an effective support for students and researchers who aim to design better state-of-the-art image processing and pattern recognition applications, still delivering high performance and robustness with the use of evolutionary approaches.

September 2015

Raul Rojas
Freie Universität Berlin

Preface

Image processing and pattern recognition emerge as important components of decision support systems for industrial, medical, military applications, among others. Several methods have been presented in the literature to solve image processing and pattern recognition tasks. They require some method-specific parameters to be optimally tuned in order to achieve the best performance. Such requirement has naturally steered both methods toward being converted into optimization problems.

Optimization has a fundamental importance in solving many problems in image processing and pattern recognition. Such a fact is evident from a quick look at special issues, congresses, and specialized journals that focus on such areas, where a significant number of manuscripts report on the use of optimization techniques.

Classical optimization methods often face great difficulties while dealing with images or systems containing noise and distortions. Under such conditions, the use of evolutionary computation approaches has been recently extended to address challenging real-world image processing and pattern recognition problems.

Image processing and pattern recognition are both dynamic and fast moving fields of research. Each new approach that is developed by engineers, mathematicians, and computer scientists is quickly identified, understood, and assimilated in order to be applied to image processing and pattern recognition problems. In this book, we strive to bring some state-of-the-art techniques by using recent results in evolutionary computation after its application to challenging and significant problems in image processing and pattern recognition.

Evolutionary computation methods are vast and have many variants. There exist a rich amount of literature on the subject, including textbooks, tutorials, and journal papers that cover in detail practically every aspect of the field. The great amount of information available makes it difficult for no specialist to explore the literature and to find the right optimization technique for a specific image or pattern recognition application. Therefore, any attempt to present the whole area of evolutionary computation in detail would be a daunting task, probably doomed to failure. This task would be even more difficult if the goal is to understand the applications of evolutionary methods in the context of image processing and pattern recognition. For this reason, the best practice is to consider only a representative set of evolutionary approaches, just as it has been done in this book.

The aim of this book was to provide an overview of the different aspects of evolutionary methods in order to enable the reader in reaching a global understanding of the field and in conducting studies on specific evolutionary techniques that are related to applications in image processing and pattern recognition that for some reason attract his interest.

Our goal is to bridge the gap between recent evolutionary optimization techniques and novel image processing methods that profit on the convenient properties of evolutionary methods. To do this, at each chapter, we endeavor to explain basic ideas of the proposed applications in ways that can be understood by readers who may not possess the necessary backgrounds on either of the fields. Therefore, image processing and pattern recognition practitioners who are not evolutionary computation researchers will appreciate that the techniques discussed are beyond simple theoretical tools since they have been adapted to solve significant problems that commonly arise on such areas. On the other hand, members of the evolutionary computation community can learn the way in which image processing and pattern recognition problems can be translated into an optimization task.

This book has been structured so that each chapter can be read independently from the others. Chapter 1 describes evolutionary computation (EC). This chapter concentrates on elementary concepts of evolutionary algorithms. Readers that are familiar with EC may wish to skip this chapter.

In Chap. 2, an automatic image multi-threshold segmentation approach based on differential evolution (DE) is presented. The algorithm approximates the 1-D histogram of the image using a mixture of Gaussian functions whose parameters are calculated by using the differential evolution method. Each Gaussian function represents a pixel class and therefore a threshold point in the image.

Chapter 3 presents how the artificial bee colony (ABC) optimization can be used to reduce the number of search locations within a block matching process. In the algorithm, the computation of search locations is drastically reduced by considering a fitness calculation strategy which indicates when it is feasible to calculate or only estimate new search locations.

In Chap. 4, an effective technique for extracting multiple ellipses from an image is presented. The approach employs an evolutionary algorithm that mimics the way in which animals behave collectively, assuming the overall detection process as a multimodal optimization problem. In the algorithm, search agents emulate a group of animals that interact to each other by using simple biological rules that are modeled as evolutionary operators.

In Chap. 5, the template matching problem is solved by using an evolutionary approach based on the states of matter phenomenon. In the approach, individuals emulate molecules that experiment state transitions which represent different exploration–exploitation behaviors.

Chapter 6 presents a method for robustly estimating multiple view relations from point correspondences. The approach combines the RANSAC method and an evolutionary technique known as the clonal selection algorithm (CSA). Upon such combination, the method adopts a different sampling strategy in comparison with RANSAC in order to generate putative solutions. Under the new mechanism, new candidate solutions are iteratively built by considering the quality of models that have been generated by previous candidate solutions, rather than relying over a pure random selection as it is the case with classic RANSAC.

Chapter 7 exhibits an algorithm for the automatic detection of circular shapes from complicated and noisy images with no consideration of the conventional Hough transform principles. The detector is based on a newly developed evolutionary algorithm called the adaptive population with reduced evaluations (APRE). It reduces the number of function evaluations through the use of two mechanisms: (1) adapting dynamically the size of the population and (2) incorporating a fitness calculation strategy which decides whether the calculation or estimation of the newly generated individuals is feasible.

In Chap. 8, a segmentation algorithm based on the Harmony search (HS) algorithm is introduced. The approach combines the good search capabilities of HS with objective functions suggested by the popular segmentation methods of Otsu and Kapur.

Chapter 9 presents an algorithm for the automatic detection of white blood cells embedded into complicated and cluttered smear images that considers the complete process as a circle detection problem. The approach is based on a nature inspired technique known as the electromagnetism-like optimization (EMO) which is a heuristic method that follows electromagnetism principles for solving complex optimization problems.

Finally, Chap. 10 presents an algorithm for the automatic selection of pixel classes for image segmentation. The presented method combines a recent evolutionary method with the definition of a new objective function that appropriately evaluates the segmentation quality with respect to the number of classes. The evolutionary algorithm, called locust search (LS), is based on the behavior of swarms of locusts. Different to the most of existent evolutionary algorithms, it explicitly avoids the concentration of individuals in the best positions, avoiding critical flaws such as the premature convergence to sub-optimal solutions and the limited exploration–exploitation balance.

There are many people who are somehow involved in the writing process of this book. We thank Dr. Raul Rojas for supporting us to have it published. We express our gratitude to Prof. Lakhmi Jain, who so warmly sustained this project. Acknowledgements also go to Dr. Thomas Ditzinger, who so kindly agreed to its appearance. We also acknowledge that this work was supported by CONACYT under the grant number CB 181053.

Considering that writing this book has been a very enjoyable experience for the authors and that the overall topic of evolutionary techniques in image processing has become a fruitful subject, it has been tempting to introduce a large amount of new material and novel evolutionary methods. However, the usefulness and potential adoption of the book seems to be founded over a compact and appropriate presentation of successful algorithms, which in turn has driven the overall organization of the book that we hope may provide the clearest picture to the reader's eyes.

Guadalajara, Mexico
September 2015

Erik Cuevas
Daniel Zaldívar
Marco Perez-Cisneros

Contents

Chapter 1
Introduction

Abstract This chapter gives a conceptual overview of optimization techniques, describing their main characteristics. The goal of this introduction is to motivate the consideration of evolutionary methods for solving optimization problems. The study of the optimization methods is conducted in such a way that it is clear the necessity of using evolutionary optimization methods for the solution of image processing and pattern recognition problems.

1.1 Definition of an Optimization Problem

The vast majority of image processing and pattern recognition algorithms use some form of optimization, as they intend to find some solution which is "best" according to some criterion. From a general perspective, an optimization problem is a situation that requires to decide for a choice from a set of possible alternatives in order to reach a predefined/required benefit at minimal costs [1].

Consider a public transportation system of a city, for example. Here the system has to find the "best" route to a destination location. In order to rate alternative solutions and eventually find out which solution is "best," a suitable criterion has to be applied. A reasonable criterion could be the distance of the routes. We then would expect the optimization algorithm to select the route of shortest distance as a solution. Observe, however, that other criteria are possible, which might lead to different "optimal" solutions, e.g., number of transfers, ticket price or the time it takes to travel the route leading to the fastest route as a solution.

Mathematically speaking, optimization can be described as follows: Given a function $f : S \rightarrow \mathbb{R}$ which is called the objective function, find the argument which minimizes f:

$$x^* = \arg \min_{x \in S} f(x) \qquad (1.1)$$

E. Cuevas et al., *Applications of Evolutionary Computation in Image Processing and Pattern Recognition*, Intelligent Systems Reference Library 100,
DOI 10.1007/978-3-319-26462-2_1

S defines the so-called solution set, which is the set of all possible solutions for the optimization problem. Sometimes, the unknown(s) x are referred to design variables. The function f describes the optimization criterion, i.e., enables us to calculate a quantity which indicates the "quality" of a particular x.

In our example, S is composed by the subway trajectories and bus lines, etc., stored in the database of the system, x is the route the system has to find, and the optimization criterion $f(x)$ (which measures the quality of a possible solution) could calculate the ticket price or distance to the destination (or a combination of both), depending on our preferences.

Sometimes there also exist one or more additional constraints which the solution x^* has to satisfy. In that case we talk about constrained optimization (opposed to unconstrained optimization if no such constraint exists). As a summary, an optimization problem has the following components:

- One or more design variables x for which a solution has to be found
- An objective function $f(x)$ describing the optimization criterion
- A solution set S specifying the set of possible solutions x
- (optional) One or more constraints on x

In order to be of practical use, an optimization algorithm has to find a solution in a reasonable amount of time with reasonable accuracy. Apart from the performance of the algorithm employed, this also depends on the problem at hand itself. If we can hope for a numerical solution, we say that the problem is well-posed. For assessing whether an optimization problem is well-posed, the following conditions must be fulfilled:

1. A solution exists.
2. There is only one solution to the problem, i.e., the solution is unique.
3. The relationship between the solution and the initial conditions is such that small perturbations of the initial conditions result in only small variations of x^*.

1.2 Classical Optimization

Once a task has been transformed into an objective function minimization problem, the next step is to choose an appropriate optimizer. Optimization algorithms can be divided in two groups: derivative-based and derivative-free [2].

In general, $f(x)$ may have a nonlinear form respect to the adjustable parameter x. Due to the complexity of $f(\cdot)$, in classical methods, it is often used an iterative algorithm to explore the input space effectively. In iterative descent methods, the next point x_{k+1} is determined by a step down from the current point x_k in a direction vector $\mathbf{d}$:

$$x_{k+1} = x_k + \alpha \mathbf{d}, \tag{1.2}$$

where α is a positive step size regulating to what extent to proceed in that direction. When the direction d in Eq. 1.1 is determined on the basis of the gradient ($\mathbf{g}$) of the objective function $f(\cdot)$, such methods are known as gradient-based techniques.

The method of steepest descent is one of the oldest techniques for optimizing a given function. This technique represents the basis for many derivative-based methods. Under such a method, the Eq. 1.3 becomes the well-known gradient formula:

$$x_{k+1} = x_k - \alpha \mathbf{g}(f(x)), \tag{1.3}$$

However, classical derivative-based optimization can be effective as long the objective function fulfills two requirements:

- The objective function must be two-times differentiable.
- The objective function must be uni-modal, i.e., have a single minimum.

A simple example of a differentiable and uni-modal objective function is

$$f(x_1, x_2) = 10 - e^{-\left(x_1^2 + 3 \cdot x_2^2\right)} \tag{1.4}$$

Figure 1.1 shows the function defined in Eq. 1.4.

Unfortunately, under such circumstances, classical methods are only applicable for a few types of optimization problems. For combinatorial optimization, there is no definition of differentiation.

Furthermore, there are many reasons why an objective function might not be differentiable. For example, the “floor” operation in Eq. 1.5 quantizes the function in Eq. 1.4, transforming Fig. 1.1 into the stepped shape seen in Fig. 1.2. At each step’s edge, the objective function is non-differentiable:

$$f(x_1, x_2) = \text{floor}\left(10 - e^{-\left(x_1^2 + 3 \cdot x_2^2\right)}\right) \tag{1.5}$$

Even in differentiable objective functions, gradient-based methods might not work. Let us consider the minimization of the Griewank function as an example.

$$\begin{aligned} \text{minimize} \quad & f(x_1, x_2) = \frac{x_1^2 + x_2^2}{4000} - \cos(x_1)\cos\left(\frac{x_2}{\sqrt{2}}\right) + 1 \\ \text{subject to} \quad & -30 \le x_1 \le 30 \\ & -30 \le x_2 \le 30 \end{aligned} \tag{1.6}$$

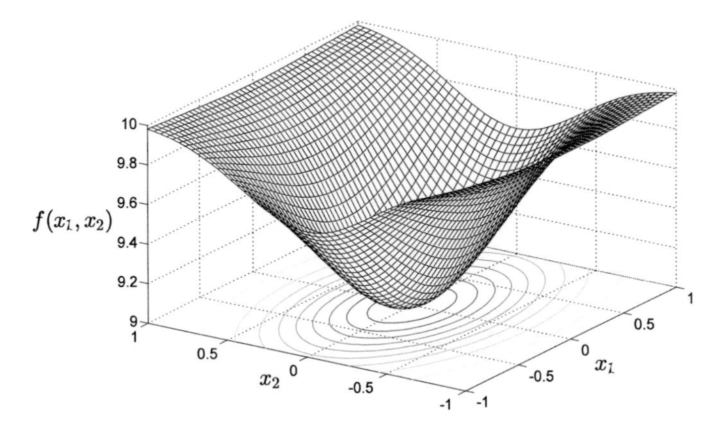

Fig. 1.1 Uni-modal objective function

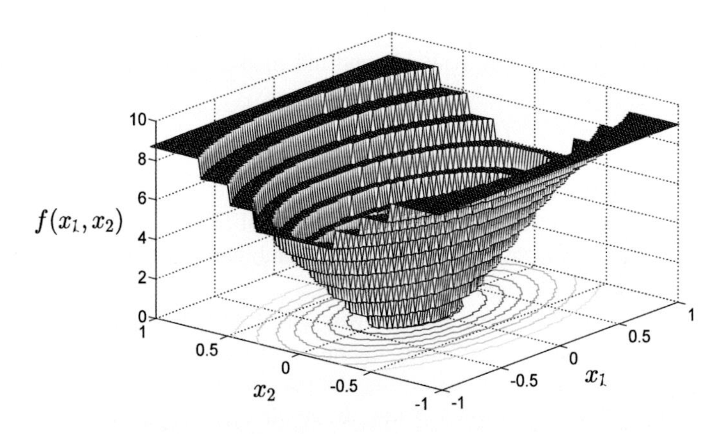

Fig. 1.2 A non-differentiable, quantized, uni-modal function

From the optimization problem formulated in Eq. 1.6, it is quite easy to understand that the global optimal solution is $x_1 = x_2 = 0$. Figure 1.3 visualizes the function defined in Eq. 1.6. According to Fig. 1.3, the objective function has many local optimal solutions (multimodal) so that the gradient methods with a randomly generated initial solution will converge to one of them with a large probability.

Considering the limitations of gradient-based methods, image processing and pattern recognition problems make difficult their integration with classical optimization methods. Instead, some other techniques which do not make assumptions and which can be applied to wide range of problems are required [3].

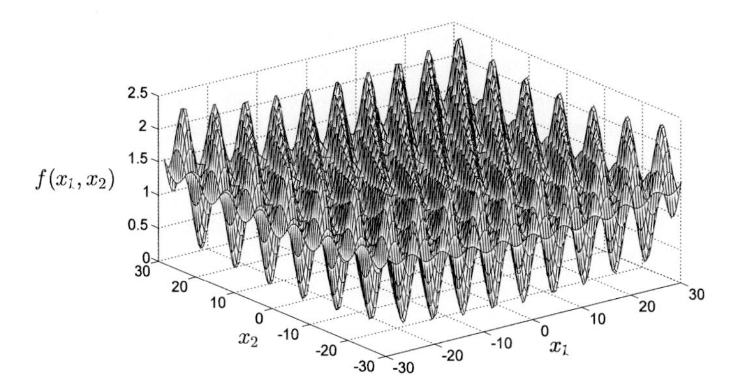

Fig. 1.3 The Griewank multi-modal function

1.3 Evolutionary Computation Methods

Evolutionary computation (EC) [4] methods are derivative-free procedures, which do not require that the objective function must be neither two-times differentiable nor uni-modal. Therefore, EC methods as global optimization algorithms can deal with non-convex, nonlinear, and multimodal problems subject to linear or nonlinear constraints with continuous or discrete decision variables.

The field of EC has a rich history. With the development of computational devices and demands of industrial processes, the necessity to solve some optimization problems arose despite the fact that there was not sufficient prior knowledge (hypotheses) on the optimization problem for the application of a classical method. In fact, in the majority of image processing and pattern recognition cases, the problems are highly nonlinear, or characterized by a noisy fitness, or without an explicit analytical expression as the objective function might be the result of an experimental or simulation process. In this context, the EC methods have been proposed as optimization alternatives.

An EC technique is a general method for solving optimization problems. It uses an objective function in an abstract and efficient manner, typically without utilizing deeper insights into its mathematical properties. EC methods do not require hypotheses on the optimization problem nor any kind of prior knowledge on the objective function. The treatment of objective functions as "black boxes" [5] is the most prominent and attractive feature of EC methods.

EC methods obtain knowledge about the structure of an optimization problem by utilizing information obtained from the possible solutions (i.e., candidate solutions) evaluated in the past. This knowledge is used to construct new candidate solutions which are likely to have a better quality.

Recently, several EC methods have been proposed with interesting results. Such approaches uses as inspiration our scientific understanding of biological, natural or social systems, which at some level of abstraction can be represented as optimization processes [6]. These methods include the social behavior of bird flocking and fish schooling such as the Particle Swarm Optimization (PSO) algorithm [7], the cooperative behavior of bee colonies such as the Artificial Bee Colony (ABC) technique [8], the improvisation process that occurs when a musician searches for a better state of harmony such as the Harmony Search (HS) [9], the emulation of the bat behavior such as the Bat Algorithm (BA) method [10], the mating behavior of firefly insects such as the Firefly (FF) method [11], the social-spider behavior such as the Social Spider Optimization (SSO) [12], the simulation of the animal behavior in a group such as the Collective Animal Behavior [13], the emulation of immunological systems as the clonal selection algorithm (CSA) [14], the simulation of the electromagnetism phenomenon as the electromagnetism-Like algorithm [15], and the emulation of the differential and conventional evolution in species such as the Differential Evolution (DE) [16] and Genetic Algorithms (GA) [17], respectively.

1.3.1 Structure of an Evolutionary Computation Algorithm

From a conventional point of view, an EC method is an algorithm that simulates at some level of abstraction a biological, natural or social system. To be more specific, a standard EC algorithm includes:

1. One or more populations of candidate solutions are considered.
2. These populations change dynamically due to the production of new solutions.
3. A fitness function reflects the ability of a solution to survive and reproduce.
4. Several operators are employed in order to explore an exploit appropriately the space of solutions.

The EC methodology suggest that, on average, candidate solutions improve their fitness over generations (i.e., their capability of solving the optimization problem). A simulation of the evolution process based on a set of candidate solutions whose fitness is properly correlated to the objective function to optimize will, on average, lead to an improvement of their fitness and thus steer the simulated population towards the global solution.

Most of the optimization methods have been designed to solve the problem of finding a global solution of a nonlinear optimization problem with box constraints in the following form:

$$\begin{aligned} &\text{maximize} \quad f(\mathbf{x}), \quad \mathbf{x} = (x_1, \ldots, x_d) \in \mathbb{R}^d \\ &\text{subject to} \quad \mathbf{x} \in \mathbf{X} \end{aligned} \tag{1.7}$$

where $f : \mathbb{R}^d \rightarrow \mathbb{R}$ is a nonlinear function whereas $\mathbf{X} = \{\mathbf{x} \in \mathbb{R}^d | l_i \leq x_i \leq u_i, i = 1, \ldots, d.\}$ is a bounded feasible search space, constrained by the lower (l_i) and upper (u_i) limits.

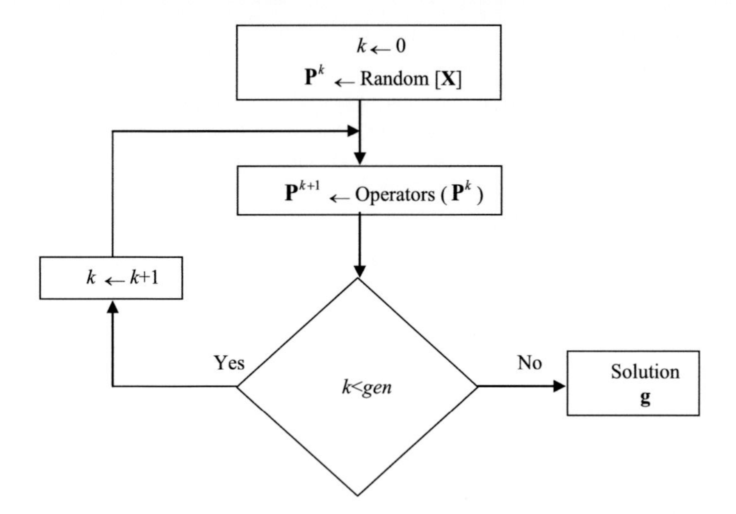

Fig. 1.4 The basic cycle of an EC method

In order to solve the problem formulated in Eq. 1.6, in an evolutionary computation method, a population $\mathbf{P}^k(\{\mathbf{p}_1^k, \mathbf{p}_2^k, \ldots, \mathbf{p}_N^k\})$ of N candidate solutions (individuals) evolves from the initial point ($k = 0$) to a total *gen* number iterations ($k = gen$). In its initial point, the algorithm begins by initializing the set of N candidate solutions with values that are randomly and uniformly distributed between the pre-specified lower (l_i) and upper (u_i) limits. In each iteration, a set of evolutionary operators are applied over the population $\mathbf{P}^k$ to build the new population $\mathbf{P}^{k+1}$. Each candidate solution $\mathbf{p}_i^k$ ($i \in [1, \ldots, N]$) represents a d-dimensional vector $\left\{p_{i,1}^k, p_{i,2}^k, \ldots, p_{i,d}^k\right\}$ where each dimension corresponds to a decision variable of the optimization problem at hand. The quality of each candidate solution $\mathbf{p}_i^k$ is evaluated by using an objective function $f\left(\mathbf{p}_i^k\right)$ whose final result represents the fitness value of $\mathbf{p}_i^k$. During the evolution process, the best candidate solution $\mathbf{g}$ ($g_1, g_2, \ldots g_d$) seen so-far is preserved considering that it represents the best available solution. Figure 1.4 presents a graphical representation of a basic cycle of an EC method.

References

1. Akay, B., Karaboga, D.: A survey on the applications of artificial bee colony in signal, image, and video processing. SIViP **9**(4), 967–990 (2015)
2. Yang, X-S.: Engineering Optimization. Wiley, Hoboken (2010)
3. Treiber, M.A.: Optimization for Computer Vision an Introduction to Core Concepts and Methods. Springer, Berlin (2013)
4. Simon, D.: Evolutionary Optimization Algorithms. Wiley, Berlin (2013)
5. Blum, C., Roli, A.: Metaheuristics in combinatorial optimization: overview and conceptual comparison. ACM Comput. Surv. (CSUR) **35**(3), 268–308 (2003). doi:10.1145/937503.937505
6. Nanda, S.J., Panda, G.: A survey on nature inspired metaheuristic algorithms for partitional clustering. Swarm Evol. Comput. **16**, (2014), 1-18
7. Kennedy, J., Eberhart, R.: Particle swarm optimization. In: Proceedings of the 1995 IEEE International Conference on Neural Networks, vol. 4, pp. 1942–1948, December 1995
8. Karaboga, D.: An idea based on honey bee swarm for numerical optimization. In: Technical Report-TR06. Engineering Faculty, Computer Engineering Department, Erciyes University (2005)
9. Geem, Z.W., Kim, J.H., Loganathan, G.V.: A new heuristic optimization algorithm: harmony search. Simulations **76**, 60–68 (2001)
10. Yang, X.S.: A new metaheuristic bat-inspired algorithm. In: Cruz, C., González, J., Krasnogor, G.T.N., Pelta, D.A. (eds.) Nature Inspired Cooperative Strategies for Optimization (NISCO 2010), Studies in Computational Intelligence, vol. 284, pp. 65–74. Springer, Berlin (2010)
11. Yang, X.S.: Firefly algorithms for multimodal optimization. In: Stochastic Algorithms: Foundations and Applications, SAGA 2009, Lecture Notes in Computer Sciences, vol. 5792, pp. 169–178 (2009)
12. Cuevas, E., Cienfuegos, M., Zaldívar, D., Pérez-Cisneros, M.: A swarm optimization algorithm inspired in the behavior of the social-spider. Expert Syst. Appl. **40**(16), 6374–6384 (2013)
13. Cuevas, E., González, M., Zaldivar, D., Pérez-Cisneros, M., García, G.: An algorithm for global optimization inspired by collective animal behaviour. Discrete Dyn. Nat. Soc. art. no. 638275 (2012)
14. de Castro, L.N., von Zuben, F.J.: Learning and optimization using the clonal selection principle. IEEE Trans. Evol. Comput. **6**(3), 239–251 (2002)
15. Birbil, Ş.I., Fang, S.C.: An electromagnetism-like mechanism for global optimization. J. Glob. Optim. **25**(1), 263–282 (2003)
16. Storn, R., Price, K.: Differential evolution -a simple and efficient adaptive scheme for global optimisation over continuous spaces. In: Technical Report TR-95–012, ICSI, Berkeley, CA (1995)
17. Goldberg, D.E.: Genetic Algorithm in Search Optimization and Machine Learning. Addison-Wesley, Boston (1989)

Chapter 2
Image Segmentation Based on Differential Evolution Optimization

Abstract Threshold selection is a critical preprocessing step for image analysis, pattern recognition and computer vision. On the other hand Differential Evolution (DE) is a heuristic method for solving complex optimization problems, yielding promising results. DE is easy to use, keeps a simple structure and holds acceptable convergence properties and robustness. In this chapter, an automatic image multi-threshold approach based on differential evolution optimization is presented. Hereby the segmentation process is considered to be similar to an optimization problem. First, the algorithm approximates the 1-D histogram of the image using a mixture of Gaussian functions whose parameters are calculated using the differential evolution method. Each Gaussian function approximating the histogram represents a pixel class and therefore a threshold point. The resultant approach is not only computationally efficient but also does not require prior assumptions whatsoever about the image. The method is likely to be most useful for applications considering different and perhaps initially unknown image classes. Experimental results demonstrate the algorithm's ability to perform automatic threshold selection while preserving main features from the original image.

2.1 Introduction

Several image processing applications aim to detect and mark remarkable features which in turn might be used to perform high-level tasks. In particular, image segmentation seeks to group pixels within meaningful regions. Commonly, gray levels belonging to the object are substantially different from the gray levels featuring the background. Thresholding is thus a simple but effective tool to isolate objects of interest from the background. Its applications include several classics such as document image analysis, whose goal is to extract printed characters [1, 2], logos, graphical content, or musical scores; also it is used for map processing which aims to locate lines, legends, and characters [3]. It is also used for scene processing, aiming for object detection and marking [4]; Similarly, it has been employed to quality inspection for materials [5, 6], discarding defective parts.

E. Cuevas et al., *Applications of Evolutionary Computation in Image Processing and Pattern Recognition*, Intelligent Systems Reference Library 100,
DOI 10.1007/978-3-319-26462-2_2

Thresholding selection techniques can be classified into two categories: bi-level and multi-level. In bi-level thresholding, one limit value is chosen to segment an image into two classes: one represents the object and the other represents the background. When an image is composed of several distinct objects, multiple threshold values have to be selected for proper segmentation. This is called multilevel thresholding.

A variety of thresholding approaches have been proposed for image segmentation, including conventional methods [7–10] and intelligent techniques such as in [11, 12]. Extending the algorithm to a multilevel approach may arise some inconveniences: (i) they may have no systematic and analytic solution when the number of classes to be detected increases and (ii) the number of classes is either difficult to be predicted or must be pre-defined. However, this parameter is unknown for many real applications.

In order to solve these problems, an alternative approach using a optimization algorithm based on differential evolution for multilevel thresholding is presented in this chapter. In the traditional multilevel optimal thresholding, the intensity distributions belonging to the object or to the background pixels are assumed to follow some Gaussian probability function; therefore a combination of probability density functions is usually adopted to model these functions. The parameters in the combination function are unknown and the parameter estimation is typically assumed to be a nonlinear optimization problem [13]. The unknown parameters that give the best fit to the processed histogram are determined by using the so-called differential evolution optimization algorithm. DE is a relatively new population-based evolutionary computational model [14]. An important difference among other evolutionary computational techniques, such as genetic algorithms (GA), is that the GA relies on the crossover operator which provides the exchange of information required to build better solutions. DE algorithm fully relies on the mutation operation as its central procedure. DE is well known for its ability to efficiently and adaptively explore large search spaces and has two advantages: (1) DE has shown a faster convergence rate than other evolutionary algorithms on some problems [15], and (2) it has very few parameters to adjust, which makes it particularly easy to implement, as shown by several cases to which DE has been successfully applied to a diverse set of optimization problems [16–18].

2.2 Gaussian Approximation

Assuming an image has L gray levels $[0,\ldots,L-1]$, following a gray level distribution which can be displayed in the form of the histogram $h(g)$. In order to simplify the description, the histogram is normalized and is considered as a probability distribution function:

$$h(g) = \frac{n_g}{N}, \quad h(g) \geq 0, \quad N = \sum_{g=0}^{L-1} n_g, \quad \text{and} \quad \sum_{g=0}^{L-1} h(g) = 1, \tag{2.1}$$

assuming that n_g denotes the number of pixels with gray level g while N is the total number of pixels in the image. The histogram function can be contained into a mix of Gaussian probability functions, yielding:

$$p(x) = \sum_{i=1}^{K} P_i \cdot p_i(x) = \sum_{i=1}^{K} \frac{P_i}{\sqrt{2\pi}\sigma_i} \exp\left[\frac{-(x-\mu_i)^2}{2\sigma_i^2}\right], \tag{2.2}$$

considering that Pi is the a priori probability of class i, $p_i(x)$ is the probability distribution function of gray-level random variable x in class i, μ_i and σ_i are the mean and standard deviation of the ith probability distribution function, and K is the number of classes within the image. In addition, the constraint $\sum_{i=1}^{K} P_i = 1$ must be satisfied. The typical mean square error consideration is used to estimate the $3K$ parameters P_i, μ_i and σ_i, $i = 1, \ldots, K$. For example, the mean square error between the composite Gaussian function $p(x_i)$ and the experimental histogram function $h(x_i)$ is defined as follows:

$$E = \frac{1}{n}\sum_{j=1}^{n} \left[p(x_j) - h(x_j)\right]^2 + \omega \cdot \left|\left(\sum_{i=1}^{K} P_i\right) - 1\right| \tag{2.3}$$

Assuming an n-point histogram as in [13] and ω being the penalty associated with the constrain $\sum_{i=1}^{K} P_i = 1$.

In general, the determination of parameters that minimize the square error is not a simple problem. A straightforward method to decrease the partial derivatives of the error function to zero considers a set of simultaneous transcendental equations [13]. An analytical solution is not available due to the non-linear nature of the equations. The algorithm therefore makes use of a numerical procedure following an iterative approach based on the gradient information. However, considering that the gradient descent method may easily get stuck within local minima, the final solution is highly dependent on the initial values. According to our previous experience, the evolution-based approaches actually provide a satisfactory performance in case of image processing problems [18–20]. The DE algorithm is therefore adopted in order to find the parameters and their corresponding threshold values.

2.3 Differential Evolution Algorithms

In recent years, there has been a growing interest in evolutionary algorithms for diverse fields of Science and Engineering. The Differential Evolution algorithm (DE) [14, 15, 21], is a relatively novel optimization technique to solve numerical-optimization problems. The algorithm has successfully been applied to several sorts of problems as it has claimed a wider acceptance and popularity, following its simplicity, robustness, and good convergence properties [16–18].

The DE algorithm is population based including a simple and direct searching algorithm for globally optimizing multi-modal functions. Just like the Genetic Algorithms (GA), it employs crossover and mutation operators as selection mechanisms. As previously mentioned, an important difference among other evolutionary computational techniques, such as genetic algorithms (GA), is that the GA relies on the crossover operator which provides the exchange of information required to build better solutions. DE algorithm fully relies on the mutation operation as its central procedure.

The DE algorithm also employs a non-uniform crossover, taking child vector parameters from one parent more frequently than from others. The non-uniform crossover operator efficiently shuffles information about successful combinations. This enables the searching to be focused on the most promising area of the solution space. The DE algorithm introduces a novel mutation operation which is not only simple, but also significantly effective. The mutation operation is based on the differences of randomly sampled pairs of solutions within the population. Besides of being simple and capable of globally optimize a multi-modal search spaces, the DE algorithm shows other benefits: it is fast, easy to use, very easily to adapt in case of integer or discrete optimizations. It is quite effective for nonlinear constraint optimization, including penalty functions.

The version of DE algorithm used in this chapter is known as the DE/best/l/exp or "DE1" (see [14, 15]). Classic DE algorithm begins by initializing a population of N_p and an D-dimensional vectors with parameter values which are randomly and uniformly distributed between the pre-specified lower initial parameter bound $x_{j,\text{low}}$ and the upper initial parameter bound $x_{j,\text{high}}$

$$\begin{gathered} x_{j,i,t} = x_{j,\text{low}} + \text{rand}(0,1) \cdot (x_{j,\text{high}} - x_{j,\text{low}}); \\ j = 1,2,\ldots,D;\ i = 1,2,\ldots,N_p;\ t = 0. \end{gathered} \tag{2.4}$$

The subscript t is the generation index, while j and i are the parameter and population indexes respectively. Hence, $x_{j,i,t}$ is the jth parameter of the ith population vector in generation t.

To generate a trial solution, DE algorithm first mutates the best solution vector from the current population by adding to the scaled difference of two other vectors from the current population

$$\begin{gathered} \mathbf{v}_{i,t} = \mathbf{x}_{best,t} + F \cdot (\mathbf{x}_{r_1,t} - \mathbf{x}_{r_2,t}); \\ r_1, r_2 \in \{1,2,\ldots,N_p\}, \end{gathered} \tag{2.5}$$

with $\mathbf{v}_{i,t}$ being the mutant vector. Vector indexes r_1 and r_2 are randomly selected considering that both are distinct and different from the population index i (i.e., $r_1 \neq r_2 \neq i$). The mutation scale factor F is a positive real number typically less than 1.

Next step considers one or more parameter values of the mutant vector $\mathbf{v}_{i,t}$ to be exponentially crossed to those belonging to the ith population vector $\mathbf{x}_{i,t}$. The result is the trial vector $\mathbf{u}_{i,t}$

$$u_{j,i,t} = \begin{cases} v_{j,i,t}, & \text{if rand(0,1)} \leq Cr \text{ or } j = j_{\text{rand}}, \\ x_{j,i,t}, & \text{otherwise.} \end{cases} \tag{2.6}$$

with $j_{\text{rand}} \in \{1, 2, \ldots, D\}$.

The crossover constant $(0.0 \leq Cr \leq 1.0)$ controls the section of parameters belonging to the mutant vector which contributes to the trial vector. In addition, the trial vector always inherits the mutant vector parameter with a random index j_{rand} to ensure that the trial vector differs by at least one parameter from the vector to be compared $(\mathbf{x}_{i,t})$.

Finally, a selection operation is used to improve the solutions. If the cost function of the trial vector is less or equal to the target vector, then the trial vector replaces the target vector on the next generation. Otherwise, the target vector remains in the population for at least one new generation:

$$\mathbf{x}_{i,t+1} = \begin{cases} \mathbf{u}_{i,t}, & \text{if } f(\mathbf{u}_{i,t}) \leq f(\mathbf{x}_{i,t}), \\ \mathbf{x}_{i,t}, & \text{otherwise.} \end{cases} \tag{2.7}$$

here, f represents the cost funcion. These steps are repeated until a termination criterion is attained or a predetermined generation number is reached.

2.4 Determination of Thresholding Values

The next step is to determine the optimal threshold values. Considering that the data classes are organized such that $\mu_1 < \mu_2 < \cdots < \mu_K$, the threshold values are obtained by computing the overall probability error for two adjacent Gaussian functions, following:

$$E(T_i) = P_{i+1} \cdot E_1(T_i) + P_i \cdot E_2(T_i), \quad i = 1, 2, \ldots, K-1 \tag{2.8}$$

considering

$$E_1(T_i) = \int_{-\infty}^{T_i} p_{i+1}(x)dx, \tag{2.9}$$

and

$$E_2(T_i) = \int_{T_i}^{\infty} p_i(x)dx, \tag{2.10}$$

$E_1(T_i)$ is the probability of mistakenly classifying the pixels in the $(i + 1)$th class to the ith class, while $E_2(T_i)$ is the probability of erroneously classifying the pixels in the ith class to the $(i + 1)$th class. P_j's are the a priori probabilities within the combined probability density function, and T_i is the threshold value between the ith and the $(i + 1)$th classes. One T_i value is chosen such as the error $E(T_i)$ is minimized. By differentiating $E(T_i)$ with respect to T_i and equating the result to zero, it is possible to use the following equation to define the optimum threshold value T_i:

$$AT_i^2 + BT_i + C = 0 \tag{2.11}$$

considering

$$\begin{aligned} A &= \sigma_i^2 - \sigma_{i+1}^2 \\ B &= 2 \cdot (\mu_i \sigma_{i+1}^2 - \mu_{i+1} \sigma_i^2) \\ C &= (\sigma_i \mu_{i+1})^2 - (\sigma_{i+1} \mu_i)^2 + 2 \cdot (\sigma_i \sigma_{i+1})^2 \cdot \ln\left(\frac{\sigma_{i+1} P_i}{\sigma_i P_{i+1}}\right) \end{aligned} \tag{2.12}$$

Although the above quadratic equation has two possible solutions, only one of them is feasible (positive and inside the interval).

2.5 Experimental Results

In this section, two experiments are tested to evaluate the performance of the presented algorithm. The first one considers the well-known image of the "The Cameraman" shown in Fig. 2.1a. The corresponding histogram is presented by Fig. 2.1b. The goal is to segment the image in 3 different pixel classes. According to Eq. 2.2, during learning, the DE algorithm adjusts 9 parameters in this test, following the minimization procedure sketched by Eq. 2.3. In this experiment, a population of 90 (N_p) individuals is considered. Each candidate holds 9 dimensions, yielding:

$$I_N = \{P_1^N, \sigma_1^N, \mu_1^N, P_2^N, \sigma_2^N, \mu_2^N, P_3^N, \sigma_3^N, \mu_3^N\}, \tag{2.13}$$

with N representing the individual's number. The parameters (P, σ, μ) are randomly initialized, but assuming some restrictions to each parameter (for example μ must fall between 0 and 255). Before optimizing the elements of DE, some parameters have to be defined as follows: $F = 0.25$; $Cr = 0.8$, and $\omega = 1.5$.

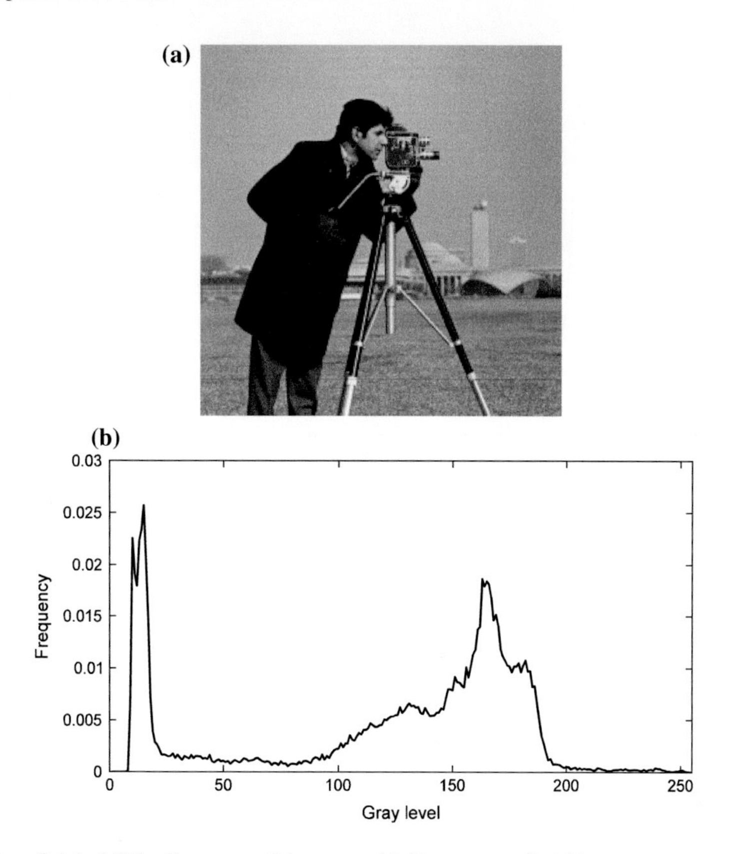

Fig. 2.1 **a** Original "The Cameraman" image, and **b** its correspondent histogram

In theory, the search for optimal values using DE covers all the search space. However, some sets of parameters chosen within the search space might not provide plausible solutions to the problem. Restricting the search space to a feasible region might also be difficult because there are not simple constraints [22]. In order to overcome this inconvenience, a penalty strategy is implemented in the DE algorithm [22, 23]. If a candidate parameter set is not a physically plausible solution, then an exaggerated cost function value is returned. Considering that such value is uncommonly large in comparison to usual cost function values, these "negative" offspring are usually eliminated in the next single generation.

The experiments suggest that after 200 iterations, the DE algorithm has converged to the global minimum. Figure 2.2a shows the obtained Gaussian functions (pixel classes), while Fig. 2.2b shows the combined graph. The layout in Fig. 2.2b suggest an easy combination of the Gaussian functions which approaches to shape of the graph shown in Fig. 2.1b which represents the histogram of the original image. Figure 2.3 shows the segmented image whose threshold values are calculated according to Eqs. 2.11 and 2.12.

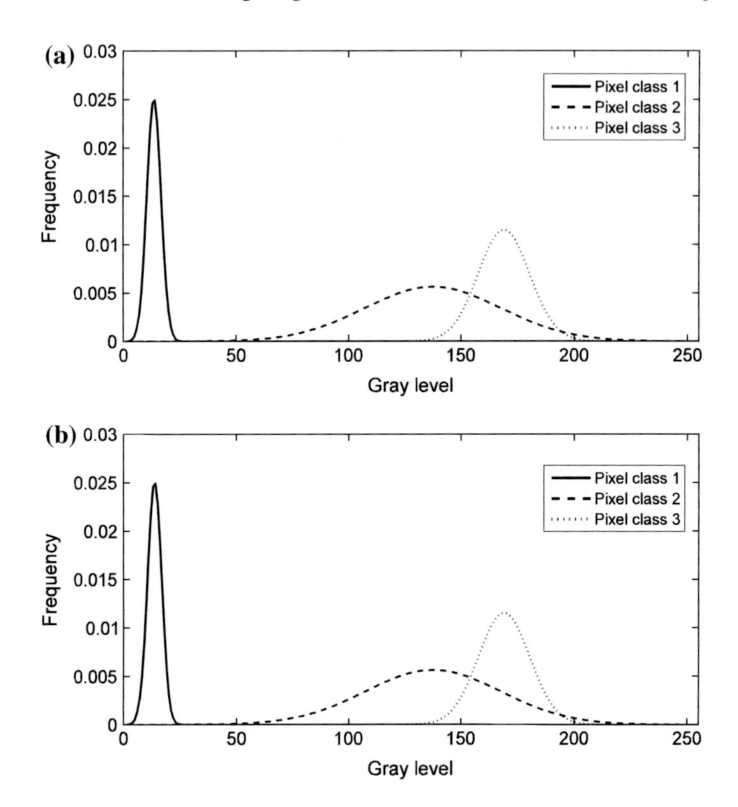

Fig. 2.2 Applying the DE algorithm for 3 classes and its results: **a** Gaussian functions for each class, **b** Mixed Gaussian functions (approach to the original histogram)

An important experiment to test the algorithm's ability to apply the segmentation algorithm with no previous knowledge is now presented. It also considers a higher number of classes by an increase to four instead of three segments. The algorithm follows the minimization presented by Eq. 2.3. Twelve parameters are now considered corresponding to the values of the four Gaussian functions. One population of 120 individuals and 12 dimensions are assumed for this test. Other parameters of the DE algorithm have kept the same values that in the experiment for three classes. Figure 2.4a shows the Gaussian functions after 200 iterations. The combined graph is presented in Fig. 2.4b. It can be seen in Fig. 2.4b the improvement of the new procedure in comparison to the original histogram shown in Fig. 2.1b. Finally, Fig. 2.5 shows the computed threshold values.

The second experiment considers a new image known as "The scene" shown by Fig. 2.6a. The histogram is presented in Fig. 2.6b. Following the first experiment, the image is segmented considering 3 pixel classes. The optimization is performed by the DE algorithm which results in the Gaussian functions shown by Fig. 2.7a. Figure 2.7b presents the combined graph as it results from the addition of the

Fig. 2.3 The image after the segmentation is applied, considering three classes

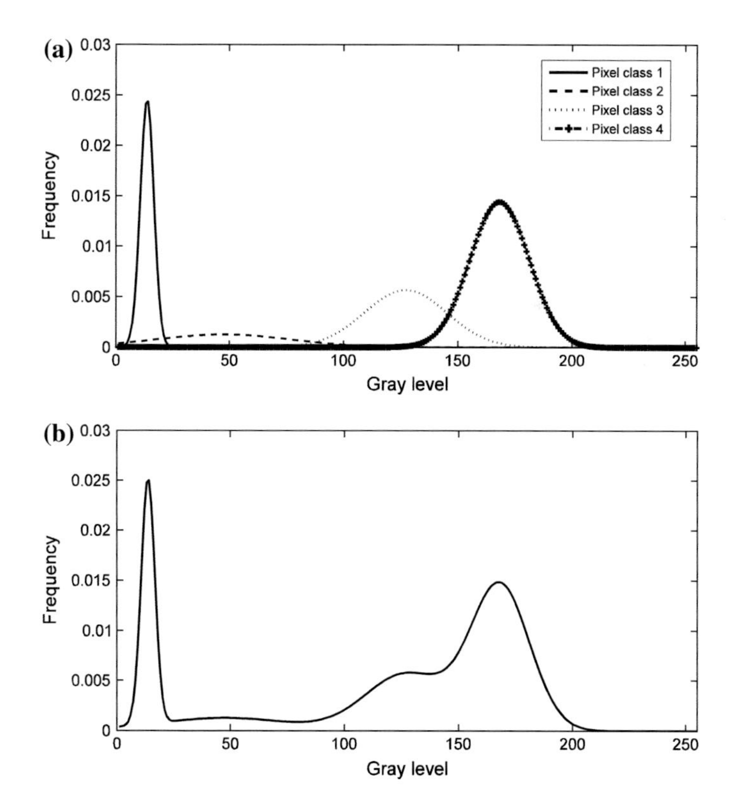

Fig. 2.4 Segmentation of the test image as it was obtained by the DE algorithm considering 4 classes: **a** Gaussian functions for each class. **b** Mixed Gaussian functions approaching the look of the original histogram

Fig. 2.5 The segmented image as it was obtained from considering 4 classes

(a)

(b)

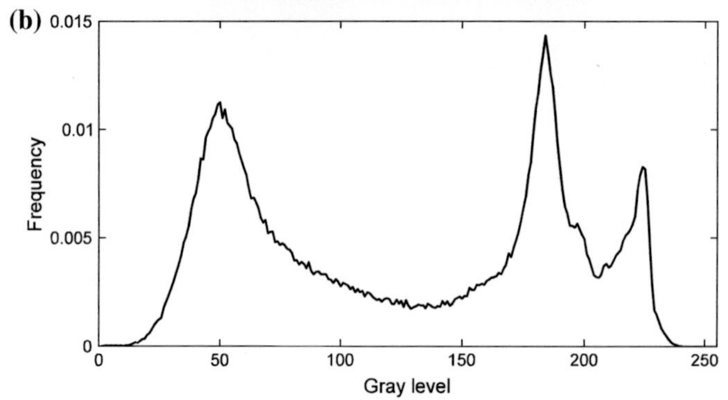

Fig. 2.6 Second experiment data, **a** the original image "The scene", and **b** its histogram

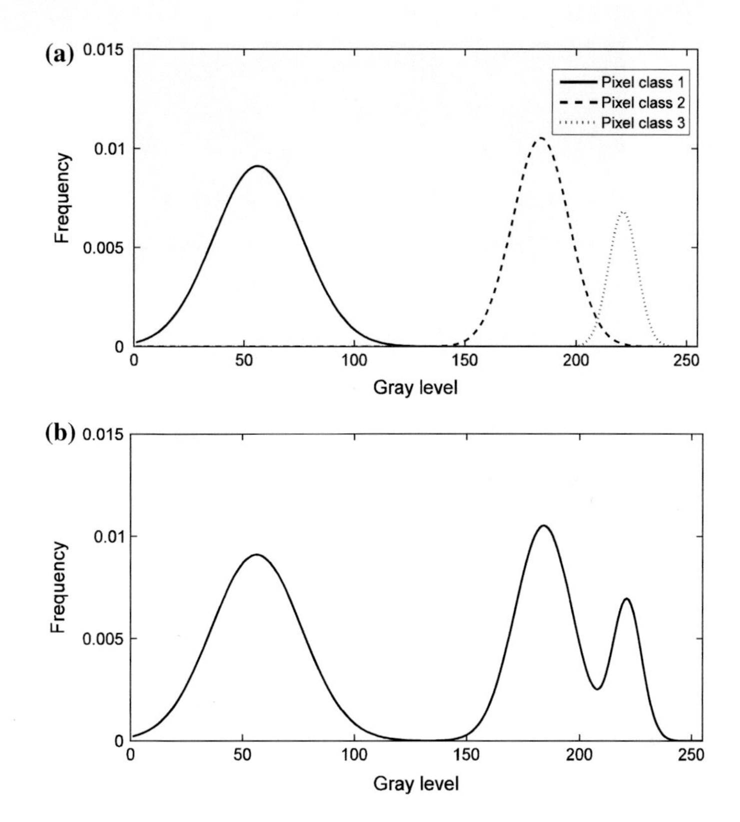

Fig. 2.7 Results obtained by the DE algorithm for 3 classes: **a** Gaussian functions at each class, **b** Combining the Gaussian functions, it approaches the original histogram

Gaussian functions. The experiment considered the parameters used in the DE algorithm for the first experiment.

While Fig. 2.8 shows the segmentation of the image considering 3 classes, the next experiment considers the same image "The scene", being segmented under 4 classes. After the optimization by the DE algorithm, the combined layout of the 4 Gaussian functions is obtained and shown by Fig. 2.9a. It is also evident that the resulting function approaches the original histogram. Figure 2.9b shows the results from the segmentation algorithm.

Fig. 2.8 "The scene" image considering 3 classes

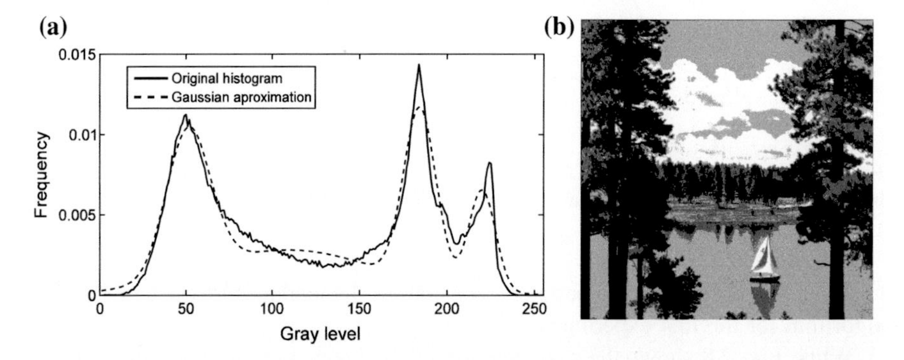

Fig. 2.9 Segmentation of "The Scene" considering the four classes for the DE algorithm: **a** comparison between the original histogram and the Gaussian approach, **b** the segmented image

2.6 Conclusions

In this chapter, an automatic image multi-threshold approach based on differential evolution optimization has been presented. The overall method can also be considered as an optimization algorithm based on differential evolution. The objects and background components within the image are assumed to fit into Gaussian distributions exhibiting non-equal mean and standard deviation. Therefore the histogram can thus be approximated by a mixture of Gaussian functions. The algorithm DE is used to estimate the parameters for the mixture density function as it seeks to get a minimum square error between the density function and the original histogram. Experimental results show that the presented approach can produce satisfactory results. Further chapter for extending the approach is under development, including a formal comparison to other state-of-the-art image segmentation techniques.

References

1. Abak, T., Baris, U., Sankur, B.: The performance of thresholding algorithms for optical character recognition. In: Proceedings of International Conference on Document Analytical Recognition 1997, pp. 697–700 (1997)
2. Kamel, M., Zhao, A.: Extraction of binary character/graphics images from grayscale document images. Graph. Models Image Process. **55**(3), 203–217 (1993)
3. Trier, O.D., Jain, A.K.: Goal-directed evaluation of binarization methods. IEEE Trans. Pattern Anal. Mach. Intell. **17**(12), 1191–1201 (1995)
4. Bhanu, B.: Automatic target recognition: state ofthe art survey. IEEE Trans. Aerosp. Electron. Syst. **22**, 364–379 (1986)
5. Sezgin, M., Sankur, B.: Comparison of thresholding methods for non-destructive testing applications. In: IEEE International Conference on Image Processing 2001, pp. 764–767 (2001)
6. Sezgin, M., Tasaltin, R.: A new dichotomization technique to multilevel thresholding devoted to inspection applications. Pattern Recogn. Lett. **21**(2), 151–161 (2000)
7. Guo, R., Pandit, S.M.: Automatic threshold selection based on histogram modes and discriminant criterion. Mach. Vis. Appl. **10**, 331–338 (1998)
8. Pal, N.R., Pal, S.K.: A review on image segmentation techniques. Pattern Recogn. **26**, 1277–1294 (1993)
9. Shaoo, P.K., Soltani, S., Wong, A.K.C., Chen, Y.C.: Survey: a survey of thresholding techniques. Comput. Vis. Graph. Image Process. **41**, 233–260 (1988)
10. Snyder, W., Bilbro, G., Logenthiran, A., Rajala, S.: Optimal thresholding: a new approach. Pattern Recogn. Lett. **11**, 803–810 (1990)
11. Chen, S., Wang, M.: Seeking multi-thresholds directly from support vectors for image segmentation. Neurocomputing **67**(4), 335–344 (2005)
12. Chih-Chih, L.: A novel image segmentation approach based on particle swarm optimization. IEICE Trans. Fundam. **89**(1), 324–327 (2006)
13. Gonzalez, R.C., Woods, R.E.: Digital image processing. Addison Wesley, Reading, MA (1992)
14. Storn, R.M., Price, K.V.: Differential evolution: a simple and efficient heuristic for global optimization over continuous spaces. J. Global Optim. **11**, 341–359 (1997)
15. Price, K.V., Storn, R.M., Lampinen, J.: Differential evolution: a practical approach to global optimization. Springer, Berlin (2005)
16. Chang, W.D.: Parameter identification of rossler's chaotic system by an evolutionary algorithm. Chaos, Solutions Fractals **29**(5), 1047–1053 (2006)
17. Liu, B., Wang, L., Jin, Y.H., Huang, D.X., Tang, F.: Control and synchronization of chaotic systems by differential evolution algorithm. Chaos, Solutions Fractals **34**(2), 412–419 (2007)
18. Baştürk, A., Günay, E.: Efficient edge detection in digital images using a cellular neural network optimized by differential evolution algorithm. Expert Syst. Appl. **36**(8), 2645–2650 (2009)
19. Lai, C.-C., Tseng, D.-C.: An optimal L-filter for reducing blocking artifacts using genetic algorithms. Signal Process. **81**(7), 1525–1535 (2001)

20. Tseng, D.-C., Lai, C.-C.: A genetic algorithm for MRF-based segmentation of multispectral textured images. Pattern Recogn. Lett. **20**(14), 1499–1510 (1999)
21. Price, K.V.: Differential evolution: a fast and simple numerical optimizer. In: IEEE Biennial Conference of the North American Fuzzy Information Processing Society, NAFIPS, Berkeley, CA, pp. 524–527 (1996)
22. Franco, G., Betti, R., Lus, H.: Identification of structural systems using an evolutionary strategy. Eng. Mech. **130**(10), 1125–1139 (2004)
23. Koziel, S., Michalewicz, Z.: Evolutionary algorithms, homomorphous mappings, and constrained parameter optimization. Evol. Comput. **7**(1), 19–44 (1999)

Chapter 3
Motion Estimation Based on Artificial Bee Colony (ABC)

Abstract Block matching (BM) motion estimation plays a very important role in video coding. In a BM approach, image frames in a video sequence are divided into blocks. For each block in the current frame, the best matching block is identified inside a region of the previous frame, aiming to minimize the sum of absolute differences (SAD). Unfortunately, the SAD evaluation is computationally expensive and represents the most consuming operation in the BM process. Therefore, BM motion estimation can be approached as an optimization problem, where the goal is to find the best matching block within a search space. The simplest available BM method is the full search algorithm (FSA) which finds the most accurate motion vector through an exhaustive computation of SAD values for all elements of the search window. Recently, several fast BM algorithms have been proposed to reduce the number of SAD operations by calculating only a fixed subset of search locations at the price of poor accuracy. In this chapter, an algorithm based on Artificial Bee Colony (ABC) optimization is presented to reduce the number of search locations in the BM process. In the algorithm, the computation of search locations is drastically reduced by considering a fitness calculation strategy which indicates when it is feasible to calculate or only estimate new search locations. Since the resultant algorithm does not consider any fixed search pattern or any other movement assumption as most of other BM approaches do, a high probability for finding the true minimum (accurate motion vector) is expected. Conducted simulations show that the method achieves the best balance over other fast BM algorithms, in terms of both estimation accuracy and computational cost.

3.1 Introduction

Video coding is currently employed for a considerable number of applications including fixed and mobile telephony, real-time video conferencing, DVD and high-definition digital television. Motion Estimation (ME) is an important part of any video coding system, since it can achieve significant compression by exploiting

E. Cuevas et al., *Applications of Evolutionary Computation in Image Processing and Pattern Recognition*, Intelligent Systems Reference Library 100,
DOI 10.1007/978-3-319-26462-2_3

the temporal redundancy that commonly exists in a video sequence. Several ME methods have been studied seeking for a complexity reduction at video coding such as block matching (BM) algorithms, parametric-based models [1], optical flow [2] and pel-recursive techniques [3]. Among such methods, BM seems to be the most popular technique due to its effectiveness and simplicity for both software and hardware implementations [4]. In order to reduce the computational complexity in ME, many BM algorithms have been proposed and employed at implementations for several video compression standards such as MPEG-4 [5] and H.264 [6].

In BM algorithms, the video frames are partitioned into non-overlapping blocks of pixels. Each block is predicted from a block of equal size in the previous frame. In particular, for each block at the current frame, the algorithm aims for the best matching block within a search window from the previous frame, while minimizing a certain matching metric. The most used matching measure is the sum of absolute differences (SAD) which is computationally expensive and represents the most consuming operation in the BM process. The best matching block thus represents the predicted block, whose displacement from the previous block is represented by a transitional motion vector (MV). Therefore, BM is essentially an optimization problem whose goal is to find the best matching block within a search space.

The full search algorithm (FSA) [7] is the simplest block-matching algorithm that can deliver the optimal estimation solution with respect to a minimal matching error as it checks all candidates one at a time. However, such exhaustive search and the full-matching error calculation at each checking point, yields an extremely computational expensive BM method that seriously constraints its use for real-time video applications.

In order to decrease the computational complexity of the BM process, several BM algorithms have been proposed considering the following three techniques: (1) using a fixed pattern: the search operation is conducted over a fixed subset of the total search window. The Three Step Search (TSS) [8], the New Three Step Search (NTSS) [9], the Simple and Efficient TSS (SES) [10], the Four Step Search (4SS) [11] and the Diamond Search (DS) [12], all represent some of its well-known examples. Although such approaches have been algorithmically considered as the fastest, they are not able to eventually match the dynamic motion-content, sometimes delivering false motion vectors (image distortions). (2) Reducing the search points: the algorithm chooses as search points only those locations that iteratively minimize the error-function (SAD values). This category includes the Adaptive Rood Pattern Search (ARPS) [13], the Fast Block Matching Using Prediction (FBMAUPR) [14], the Block-based Gradient Descent Search (BBGD) [15] and the Neighbourhood Elimination algorithm (NE) [16]. Such approaches assume that the error-function behaves monotonically, holding well for slow-moving sequences but failing for other kind of movements in video sequences [17], making the algorithm prone to get trapped into local minima. (3) Decreasing the computational overhead for every search point: the matching cost (SAD operation) is replaced by a partial or a simplified version that features less complexity. The New pixel-Decimation (ND) [18], the Efficient Block Matching Using Multilevel Intra, the Inter-Sub-blocks [9] and the Successive Elimination Algorithm [19], all assume that all pixels within each block, move by the same finite distance and a good estimate of the motion can be obtained through only a fraction of the pixel pool. However, since only a fraction of pixels enters into the matching computation, the use of such regular sub-sampling techniques can seriously affect the accuracy of the detection of motion vectors due to noise or illumination changes.

Another popular group of BM algorithms employ spatio-temporal correlation by using neighboring blocks in the spatial and temporal domain in order to predict MVs. The main advantage of such algorithms is that they alleviate the local minimum problem to some extent as the new initial or predicted search center is usually closer to the global minimum and therefore the chance of getting trapped in a local minimum decreases. This idea has been incorporated by many fast-block motion estimation algorithms such as the Enhanced Predictive Zonal Search (EPZS) [20] and the UMHexagonS [21]. However, the information delivered by the neighboring blocks occasionally conduces to false initial search points producing distorted motion vectors. Such problem is typically caused by the movement of very small objects contained in the image sequences [22].

Alternatively, evolutionary approaches such as genetic algorithms (GA) [23] and particle swarm optimization (PSO) [24] are well known for delivering the location of the global optimum in complex optimization problems. Despite of such fact, only few evolutionary approaches have specifically addressed the problem of BM, such as the Light-weight Genetic Block Matching (LWG) [25], the Genetic Four-step Search (GFSS) [26] and the PSO-BM [27]. Although these methods support an accurate identification of the motion vector, their spending times are very long in comparison to other BM techniques.

Karaboga has recently presented one bee-swarm algorithm for solving numerical optimization problems which is known as the Artificial Bee Colony (ABC) method [28]. Inspired by the intelligent foraging behaviour of honeybee swarm, the ABC algorithm consists of three essential components: food source positions, nectar amount and different honey bee classes. Each food source position represents a feasible solution for the problem under consideration and the nectar amount of a food source represents the quality of such solution corresponding to its fitness value. Each class of bee symbolizes one particular operation for generating new candidate food source positions (candidate solutions). The ABC algorithm starts by producing a randomly distributed initial population (food source locations). After initialization, an objective function evaluates whether such candidates represent an acceptable solution (nectar amount) or not. Guided by the values of such objective function, the candidate solutions are evolved through different ABC operations (honey bee types). When the fitness function (nectar amount) cannot be further improved after a maximum number of cycles is reached, its related food source is assumed to be abandoned and replaced by a new randomly chosen food source location. The performance of ABC algorithm has been compared to other optimization methods such as GA, Differential Evolution (DE) and Particle Swarm Optimization (PSO) [29, 30]. The results showed that ABC can produce optimal solutions and thus is more effective than other methods in several optimization problems. Such characteristics have motivated the use of ABC to solve different sorts of engineering problems such as signal processing [31, 32], flow shop scheduling [33], structural inverse analysis [34], clustering [35, 36], vehicle path planning [37] and image processing [38, 39].

One particular difficulty in applying ABC to real-world applications is about its demand for a large number of fitness evaluations before delivering a satisfying result. However, fitness evaluations are not always straightforward in many real-world applications as either an explicit fitness function does not exist or the fitness evaluation is computationally very expensive. Furthermore, since random numbers are involved in the calculation of new individuals, they may encounter same positions (repetition) that have been visited by other individuals at previous iterations, particularly when individuals are confined to a small area.

The problem of considering expensive fitness evaluations has already been faced in the field of evolutionary algorithms (EA) and is better known as fitness approximation [40]. In such approach, the idea is to estimate the fitness value of so many individuals as it is possible instead of evaluating the complete set. Such estimations are based on an approximate model of the fitness landscape. Thus, the individuals to be evaluated and those to be estimated are determined following some fixed criteria which depend on the specific properties of the approximate model [41]. The models involved at the estimation can be built during the actual EA run, since EA repeatedly samples the search space at different points [42]. There are many possible approximation models and several have already been used in combination with EA (e.g. polynomials [43], the kriging model [44], the feed-forward neural networks that includes multi-layer Perceptrons [45] and radial basis-function networks [46]). These models can be either global, which make use of all available data or local which make use of only a small set of data around the point where the function is to be approximated. Local models, however, have a number of advantages [42]: they are well-known and suitably established techniques with relatively fast speeds. Moreover, they employ the most remarkable information for the estimation of newer points: the closest neighbors.

In this chapter, an algorithm based on ABC is presented to reduce the number of search locations in the BM process. The algorithm uses a simple fitness calculation approach which is based on the Nearest Neighbor Interpolation (NNI) algorithm in order to estimate the fitness value (SAD operation) for several candidate solutions (search locations). As a result, the approach can substantially reduce the number of function evaluations yet preserving the good search capabilities of ABC. The method achieves the best balance over other fast BM algorithms, in terms of both estimation accuracy and computational cost.

3.2 Artificial Bee Colony (ABC) Algorithm

The ABC algorithm assumes the existence of a set of operations that may resemble some features from the honey bee behavior. For instance, each solution within the search space includes a parameter set representing food source locations. The "fitness value" refers to the food source quality which is strongly linked to the food's location. The process mimics the bee's search for valuable food sources yielding an analogous process for finding the optimal solution.

3.2.1 Biological Bee Profile

The minimal model for a honey bee colony consists of three classes: employed bees, onlooker bees and scout bees. The employed bees will be responsible for investigating about food sources and sharing the information to recruit onlooker bees, which in turn, will make a decision on choosing food sources considering such information. The food source holding a higher quality will have a larger chance to be selected by onlooker bees than one showing a lower quality. An employed bee, whose food source is rejected as low quality by employed and onlooker bees, will change to a scout bee to randomly search for new food sources. Therefore, the exploitation is driven by employed and onlooker bees while the exploration is maintained by scout bees. The implementation details of such bee-like operations in the ABC algorithm are described in the next sub-section.

3.2.2 Description of the ABC Algorithm

Resembling other swarm based approaches, the ABC algorithm is an iterative process. It starts with a population of randomly generated solutions or food sources. The following three operations are applied until a termination criterion is met [30]:

1. Send the employed bees.
2. Select the food sources by the onlooker bees.
3. Determine the scout bees.

3.2.3 Initializing the Population

The algorithm begins by initializing N_p food sources. Each food source is a D-dimensional vector containing the parameter values to be optimized which are randomly and uniformly distributed between the pre-specified lower initial parameter bound x_j^{low} and the upper initial parameter bound x_j^{high}.

$$\begin{aligned} x_{j,i} =& x_j^{low} + \text{rand}(0,1) \cdot (x_j^{high} - x_j^{low}); \\ j =& 1,2,\ldots,D; \quad i = 1,2,\ldots,N_p. \end{aligned} \tag{3.1}$$

with j and i being the parameter and individual indexes respectively. Hence, $x_{j,i}$ is the jth parameter of the i-th individual.

3.2.4 Send Employed Bees

The number of employed bees is equal to the number of food sources. At this stage, each employed bee generates a new food source in the neighborhood of its present position as follows:

$$\begin{aligned} & v_{j,i} = x_{j,i} + \phi_{j,i}(x_{j,i} - x_{j,k}); \\ & k \in \{1,2,\ldots,N_p\}; \quad j \in \{1,2,\ldots,D\} \end{aligned} \tag{3.2}$$

$x_{j,i}$ is a randomly chosen j parameter of the ith individual and k is one of the N_p food sources, satisfying the conditional $i \neq k$. If a given parameter of the candidate solution v_i exceeds its predetermined boundaries, that parameter should be adjusted in order to fit the appropriate range. The scale factor $\phi_{j,i}$ is a random number between $[-1,1]$. Once a new solution is generated, a fitness value representing the profitability associated to a particular solution is calculated. The fitness value for a minimization problem can be assigned to each solution v_i by the following expression

$$fit_i = \begin{cases} \frac{1}{1+J_i} & \text{if } J_i \geq 0 \\ 1 + abs(J_i) & \text{if } J_i < 0 \end{cases} \tag{3.3}$$

where J_i is the objective function to be minimized. A greedy selection process is thus applied between v_i and x_i. If the nectar amount (fitness) of v_i is better, then the solution x_i is replaced by v_i, otherwise x_i remains.

3.2.5 *Select the Food Sources by the Onlooker Bees*

Each onlooker bee selects one of the proposed food sources depending on their fitness value which has been recently defined by employed bees. The probability that a food source will be selected can be obtained from the equation below:

$$Prob_i = \frac{fit_i}{\sum_{i=1}^{N_p} fit_i}, \tag{3.4}$$

where fit_i is the fitness value of the food source i, which is related to the objective function value (J_i) corresponding to the food source i. The probability of a food source being selected by the onlooker bees increases as the fitness value of a food source increases. After the food source is selected, onlooker bees will go to the selected food source and select a new candidate food source position inside the neighborhood of the selected food source. T he new candidate food source can be expressed and calculated by Eq. (3.2). In case that the nectar amount i.e. fitness of the new solution, is better than before, such position is held, otherwise the old solution remains.

3.2.6 *Determine the Scout Bees*

If a food source i (candidate solution) cannot be further improved through a predetermined number "*limit*" of trials, the food source is assumed to be abandoned, and the corresponding employed or onlooker bee becomes a scout. A scout bee explores the searching space with no previous information i.e. the new solution is generated randomly as in Eq. (3.1). In order to verify if a candidate solution has reached the predetermined "*limit*", a counter A_i is assigned to each food source i. Such counter is incremented as a consequence of a bee-operation failing to improve the food source's fitness.

3.3 Fitness Approximation Method

Evolutionary algorithms that use fitness approximation aim to find the global minimum of a given function by considering only a small number of function evaluations and a large number of estimations. Such algorithms commonly employ alternative models of the function landscape in order to approximate the actual fitness function. The application of this method requires that the objective function fulfils two conditions: a heavy computational overhead and a small number of dimensions (up to five) [47].

Recently, several fitness estimators have been reported in the literature [43–46] in which the number of function evaluations is considerably reduced to hundreds, dozens or even less. However, most of these methods produce complex algorithms whose performance is conditioned to the quality of the training phase and the learning algorithm in the construction of the approximation model.

In this chapter, we explore the use of a local approximation scheme based on the nearest-neighbor-Interpolation (NNI) for reducing the function evaluation number. The model estimates fitness values based on previously evaluated neighboring individuals which have been stored during the evolution process. At each generation, some individuals of the population are evaluated through the accurate (actual) fitness function while other remaining individuals are only estimated. The positions to be accurately evaluated are determined either by their proximity to the best individual or regarding their uncertain fitness value.

3.3.1 Updating the Individual Database

In a fitness approximation method, every evaluation or estimation of an individual produces one data point (individual position and fitness value) that is potentially considered for building the approximation model during the evolution process. In our presented approach, all seen-so-far evaluations are kept in a history array **T** which is employed to select the closest neighbor and to estimate the fitness value of a newer individual. Since all data are preserved and potentially available for their use, the model construction is faster because only the most relevant data points are actually used by the approach.

3.3.2 Fitness Calculation Strategy

This section discusses details about the strategy to decide which individuals are to be evaluated or estimated. The presented fitness calculation scheme estimates most of fitness values to reduce the computational overhead at each generation. In the model, those individuals lying nearer to the best fitness value holder, currently registered in the array **T** (rule 1), are evaluated by using the actual fitness function. Such individuals are relevant as they possess a stronger influence on the evolution process than others. On the other hand, evaluation is also compulsory for those individuals lying in a region of the search space which has been unexplored so far (rule 2). The fitness values for such individuals are uncertain since there is no close reference (close points contained in **T**) to calculate their estimates. The rest of the individuals, lying in a region of the search space that contains enough previously calculated points, must be estimated using the NNI (rule 3). This rule indicates that the fitness value for such individuals must be estimated by assigning the fitness value from the nearest individual stored in **T**.

Therefore, the fitness calculation model follows three important rules to evaluate or estimate fitness values:

1. ***Exploitation rule (evaluation)***. If a new individual (search position) P is located closer than a distance d with respect to the nearest individual $L_q(q = 1, 2, 3, \ldots, m$; where m is the number of elements contained in **T**) with a fitness value F_{L_q} that corresponds to the best fitness value seen-so-far, then the fitness value of P is evaluated using the actual fitness function. Figure 3.1b draws this rule procedure.
2. ***Exploration rule (evaluation)***. If a new individual P is located further away than a distance d with respect to the nearest individual L_q, then its fitness value is evaluated by using the actual fitness function. Figure 3.1c outlines the rule procedure.
3. ***NNI rule (estimation)***. If a new individual P is located closer than a distance d with respect to the nearest individual L_q, whose fitness value F_{L_q} does not correspond to the best fitness value, then its fitness value is estimated by assigning the same fitness that $L_q(F_P = F_{L_q})$. Figure 3.1d sketches the rule procedure.

The d value controls the trade-off between the evaluation and the estimation of search locations. Typical values of d range from 2 to 4. Figure 3.1 illustrates the procedure of fitness computation for a new solution (point P). In the problem (Fig. 3.1a), it is considered the fitness function f with respect to two parameters (x_1, x_2), where the individuals database array **T** contains five different elements $(L_1, L_2, L_3, L_4, L_5)$ and their corresponding fitness values $(F_{L_1}, F_{L_2}, F_{L_3}, F_{L_4}, F_{L_5})$. Figure 3.1b, c show the fitness evaluation $(f(x_1, x_2))$ of the new solution P, following the rule 1 and rule 2 respectively, whereas Fig. 3.1d presents the fitness estimation of P using the NNI approach which has been laid by rule 3.

The presented fitness calculation strategy, seen from an optimization perspective, favors the exploitation and exploration in the search process. For the exploration, the method evaluates the fitness function of new search locations which have been located far away from previously calculated positions. Additionally, it also estimates those which are closer. For the exploitation, the presented method evaluates the actual fitness function of those new individuals which are located nearby the position that holds the minimum fitness value seen-so-far, aiming to improve its minimum. After several simulations, the value of $d = 3$ has shown the best balance between the exploration and exploitation inside the search space (in the context of a BM application); thus it has been used in this chapter.

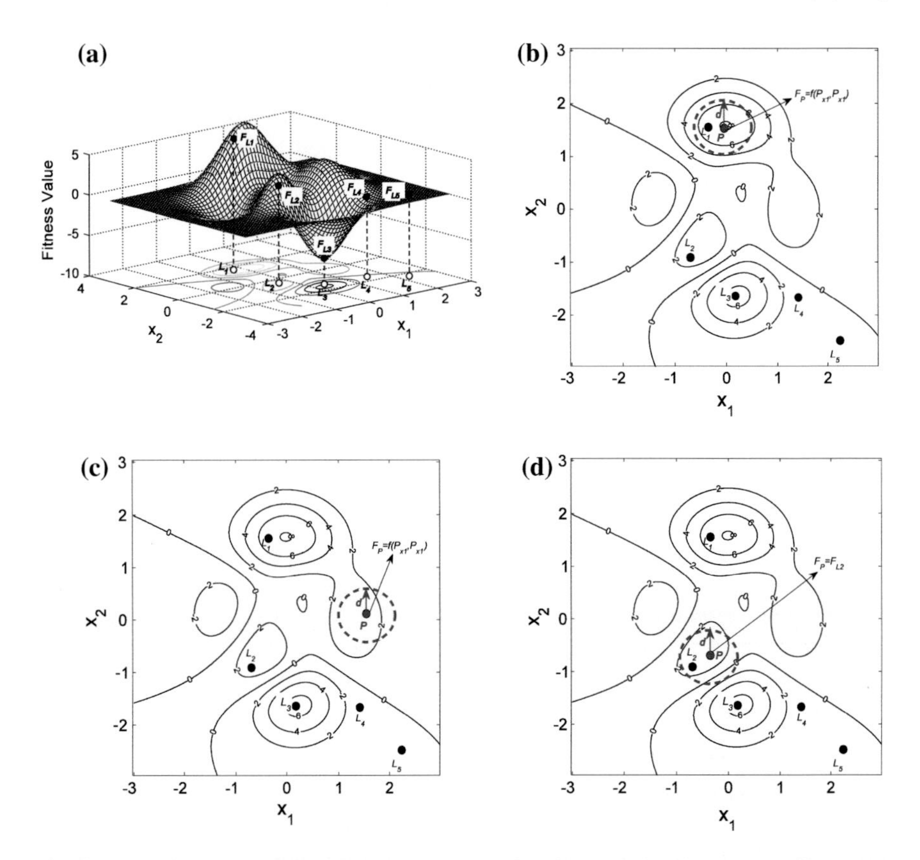

Fig. 3.1 The fitness calculation strategy. **a** Fitness function and the history array **T** content. **b** According to the rule 1, the individual (search position) P is evaluated as it is located closer than a distance d with respect to the best individual L_1. **c** According to the rule 2, the search point P is evaluated because there is no reference within its neighborhood. **d** According to rule 3, the fitness value of P is estimated by the NNI-estimator, assigning $F_P = F_{L_2}$

3.3.3 Presented ABC Optimization Method

In this section, the incorporation of the fitness calculation strategy to the ABC algorithm is presented. Only the fitness calculation scheme shows the difference between the conventional ABC and the enhanced approach. In the modified ABC, only some individuals are actually evaluated (rules 1 and 2) at each generation. The fitness values for the rest are estimated using the NNI-approach (rule 3). The estimation is executed using individuals that have been already stored in the array **T**.

Figure 3.2 shows the difference between the conventional ABC and its modified version. In Fig. 3.2a, it is clear that the way in which the fitness value is calculated represents the only difference between both methods. In the original ABC, each individual is evaluated according to traditional evolutionary algorithms by using the objective function. On the other hand, the modified ABC, the presented fitness calculation strategy for obtaining the fitness value has been employed. Figure 3.2b shows the components of the fitness calculation strategy: the fitness evaluation, the fitness estimation and the updating of the individual database. As a result, the ABC approach can substantially reduce the number of function evaluations yet preserving the good search capabilities of ABC.

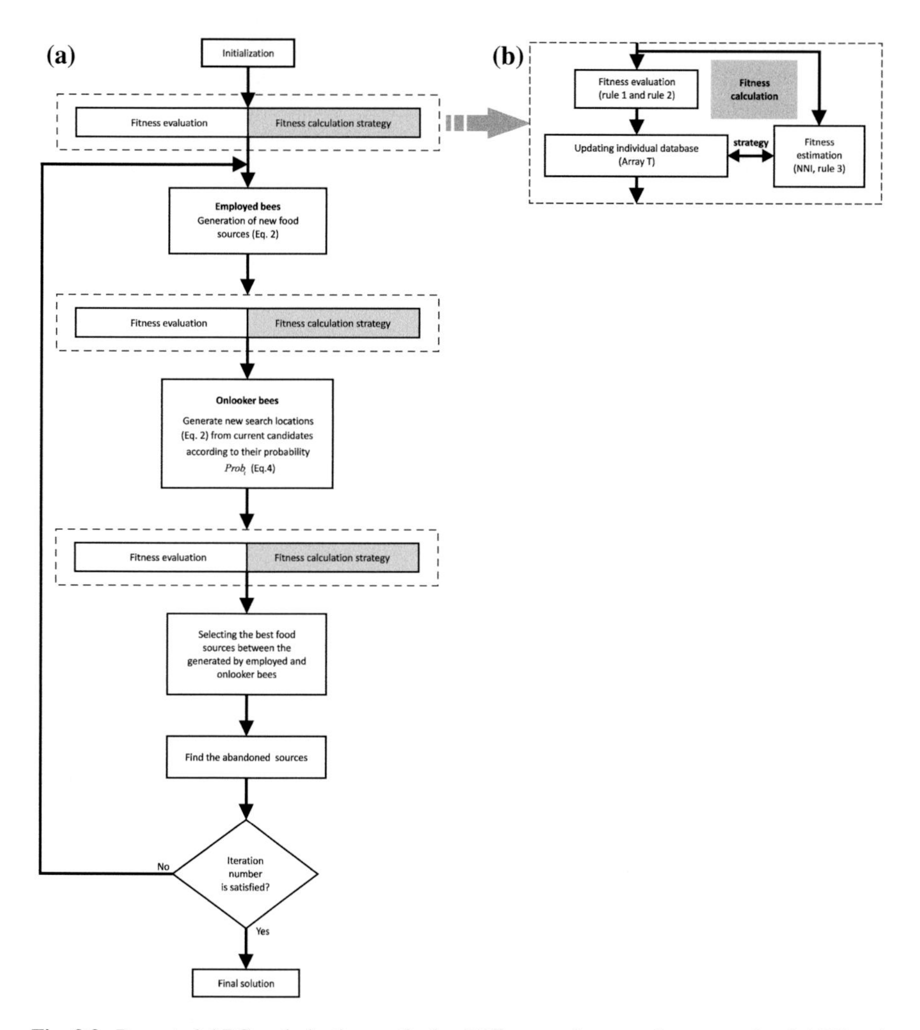

Fig. 3.2 Presented ABC optimization method. **a** Differences between the conventional ABC and the modified ABC. **b** The presented fitness calculation strategy

3.4 Motion Estimation and Block Matching

For motion estimation through a BM algorithm, the current frame of an image sequence I_t is divided into non-overlapping blocks of $N \times N$ pixels. For each template block in the current frame, the best matched block within a search window (S) of size $(2W + 1) \times (2W + 1)$ in the previous frame I_{t-1} is determined, where W is the maximum allowed displacement. The position difference between a template block in the current frame and the best matched block in the previous frame is called the motion vector (MV) (see Fig. 3.3). Under such perspective, BM can be approached as an optimization problem aiming for finding the best MV within a search space. The most well-known criterion for BM algorithms is the sum of absolute differences (SAD). It is defined in Eq. (3.5) considering a template block at position (x, y) in the current frame and the candidate block at position $(x + \hat{u}, y + \hat{v})$ in the previous frame I_{t-1}:

$$\mathrm{SAD}(\hat{u}, \hat{v}) = \sum_{j=0}^{N-1}\sum_{i=0}^{N-1} |g_t(x+i, y+j) - g_{t-1}(x+\hat{u}+i, y+\hat{v}+j)|, \tag{3.5}$$

where $g_t(\cdot)$ is the gray value of a pixel in the current frame I_t and $g_{t-1}(\cdot)$ is the gray level of a pixel in the previous frame I_{t-1}. Therefore, the MV in (u, v) is defined as follows:

$$(u, v) = \arg \min_{(u,v)\in S} \mathrm{SAD}(\hat{u}, \hat{v}), \tag{3.6}$$

where $S = \{(\hat{u}, \hat{v}) | -W \leq \hat{u}, \hat{v} \leq W \text{ and } (x + \hat{u}, y + \hat{v}) \text{ is a valid pixel position } I_{t-1}\}$.

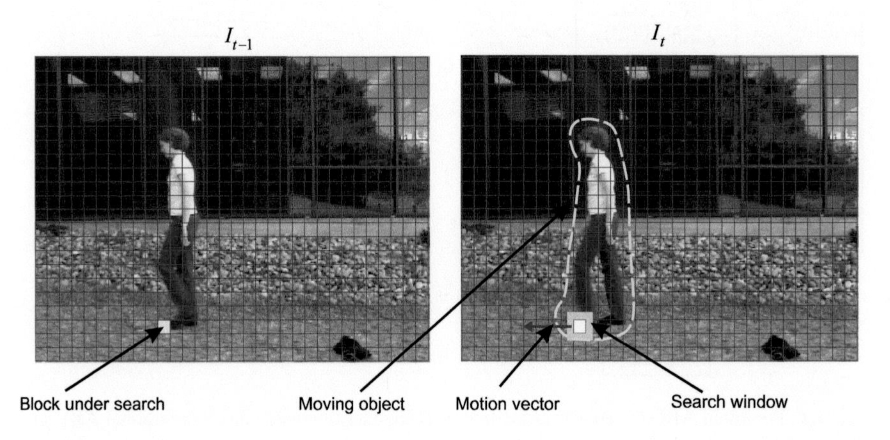

Fig. 3.3 Block matching procedure

In the context of BM algorithms, the FSA is the most robust and accurate method to find the MV. It tests all possible candidate blocks from I_{t-1} within the search area to find the block with the minimum SAD. For the maximum displacement of W, the FSA requires $(2W+1)^2$ search points. For instance, if the maximum displacement W is ∓ 7, the total search points are 225. Each SAD calculation requires $2N^2$ additions and the total number of additions for the FSA to match a 16×16 block is 130,560. Such computational requirement makes the application of FSA difficult for real time tasks.

3.5 BM Algorithm Based on ABC with the Estimation Strategy

FSA finds the global minimum (the accurate MV), considering all locations within the search space S. Nevertheless, the approach has a high computational cost for practical use. In order to overcome such a problem, many fast algorithms have been developed yielding only a poorer precision than the FSA. A better BM algorithm should spend less computational time on the search and get more accurate motion vectors (MVs).

The BM algorithm presented in this chapter is comparable to the fastest algorithm yet delivering a similar precision to the FSA approach. Since most of fast algorithms use a regular search pattern or assume a characteristic error function (uni-modal) for searching about the motion vector, they may get trapped into local minima considering that, for many cases (i.e. complex motion sequences), an uni-modal error is no longer valid. Figure 3.4 shows a typical error surface which has been computed around the search window for a fast-moving sequence. On the other hand, the presented BM algorithm uses a non-uniform search pattern for locating global minimum distortion. Under the effect of the ABC operators, the

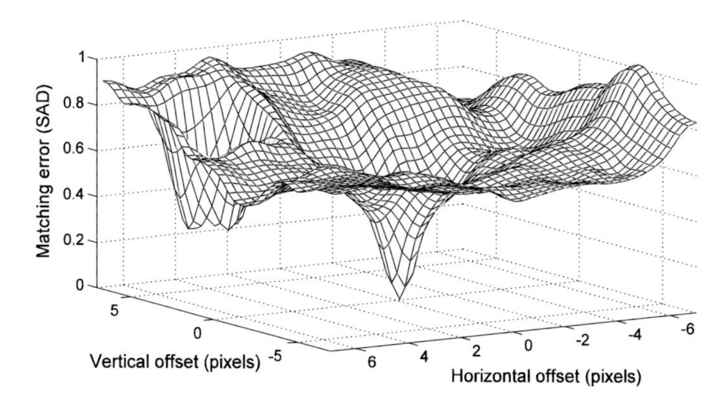

Fig. 3.4 Common non-uni-modal error surface with multiple local minimum error points

search locations vary from generation to generation, avoiding to get trapped into a local minimum. Besides, since the presented algorithm uses a fitness calculation strategy for reducing the evaluation of the SAD values, it requires fewer search positions.

In the algorithm, the search space S consists of a set of 2-D motion vectors $\hat{u}$ and $\hat{v}$ representing the x and y components of the motion vector, respectively. The particle is thus defined as:

$$P_i = \{\hat{u}_i, \hat{v}_i | -W \leq \hat{u}_i, \hat{v}_i \leq W\}, \tag{3.7}$$

where each particle i represents a possible motion vector. In this chapter, all search windows that are considered for the simulations are set to ±8 and ±16 pixels. Both configurations have been selected in order to compare their results to other approaches presented in the literature.

3.5.1 Initial Population

The first step in ABC optimization is to generate an initial group of individuals. The standard literature of evolutionary algorithms generally suggests the use of random solutions as initial population, considering the absence of knowledge about the problem [48]. However, Li [49] and Xiao [50] demonstrated that the use of solutions generated through some domain knowledge to set the initial population (i.e., non-random solutions) can significantly improve its performance. In order to obtain appropriate initial solutions (based on knowledge), an analysis over the motion vector distribution has been conducted. After considering several sequences (see Table 3.1 and Fig. 3.8), it can be seen that 98 % of the MVs are found to lie at the origin of the search window for a slow-moving sequence such as the one at *Container*, whereas complex motion sequences, such as the *Carphone* and the *Foreman* examples, have only 53.5 and 46.7 % of their MVs in the central search region. The *Stefan* sequence, showing the most complex motion content, has only 36.9 %. Figure 3.5 shows the surface of the MV distribution for the *Foreman* and the *Stefan*. On the other hand, although it is less evident, the MV distribution of

Table 3.1 Test sequences used in the comparison test

Sequence	Format	Total frames	Motion type
Container	QCIF(176 × 144)	299	Low
Carphone	QCIF(176 × 144)	381	Medium
Foreman	QCIF(352 × 288)	398	Medium
Akiyo	QCIF(352 × 288)	211	Medium
Stefan	CIF(352 × 288)	89	High
Football	SIF(352 × 240)	300	High

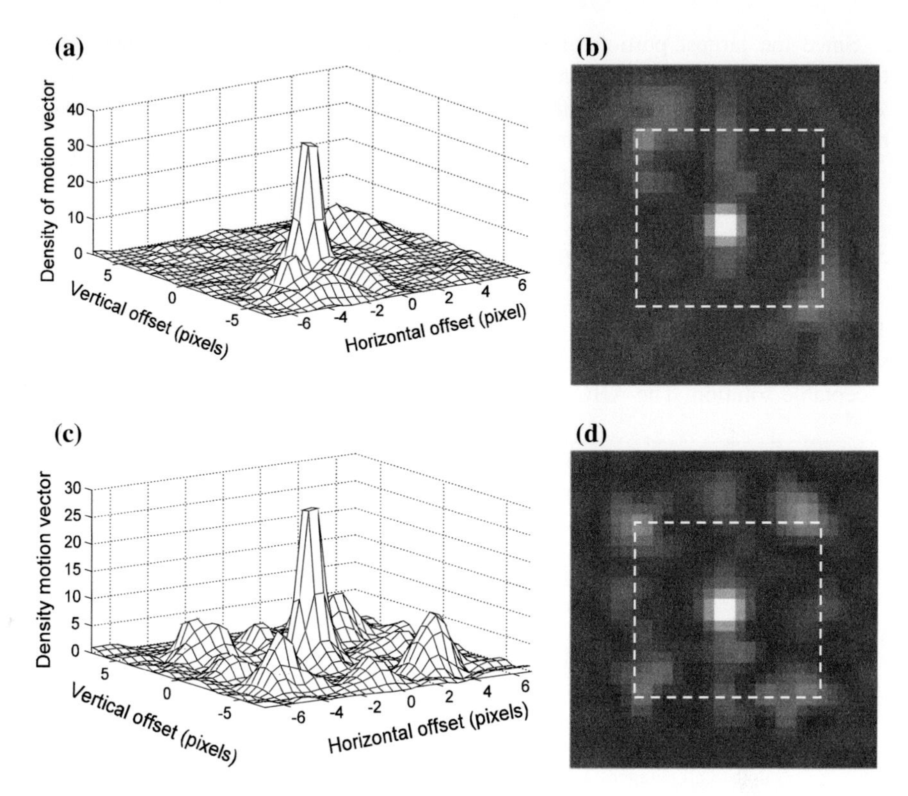

Fig. 3.5 Motion vector distribution for *Foreman* and Stefan sequences. **a–b** MV distribution for the *Foreman* sequence. **c–d** MV distribution for the *Stefan* sequence

several sequences shows small peaks at some locations lying away from the center as they are contained inside a rectangle which is shown in Fig. 3.5b, d by a white overlay. Real-world moving sequences concentrate most of the MVs under a limit due to the motion continuity principle [16]. Therefore, in this chapter, initial solutions are selected from five fixed locations which represent points showing the higher concentration in the MV distribution, just as it is shown by Fig. 3.6.

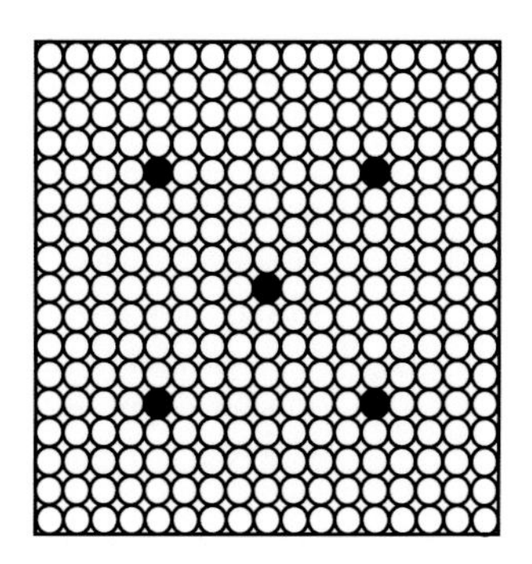

Fig. 3.6 Fixed pattern of five elements in the search window of ±8 which are to be used as initial solutions

Since the largest portion of the BM translations between neighboring video frames tend to behave as it is shown by the pattern of Fig. 3.6, such initial solutions are used in order to accelerate the ABC method. This consideration is taken regardless of the employed search window size (±8 or ±16).

3.5.2 *The ABC-BM Algorithm*

The goal of our BM-approach is to reduce the number of evaluations of the SAD values (actual fitness function) avoiding any performance loss and achieving an acceptable solution. The ABC-BM method is listed below:

Step 1:	Set the ABC parameters (*limit* = 10)
Step 2:	Initialize the population of 5 individuals $\mathbf{P} = \{P_1, \ldots, P_5\}$ using the pattern that has been shown in Fig. 3.6 and the individual database array **T**, as an empty array. Clear all counters $A_i (i \in 1, \ldots, 5)$ too
Step 3:	Compute the fitness values for each individual according to the fitness calculation strategy presented in Sect. 3.3. Since all individuals of the initial population fulfill rule 2 conditions, they are evaluated through the actual fitness function by calculating the actual SAD values
Step 4:	Update new evaluations in the individual database array **T**
Step 5:	Modify each element of the current population **P** (search locations) as stated by Eq. 3.2, in order to produce a new solution vector $\mathbf{E} = \{E_1, \ldots, E_5\}$. Likewise, update all counters A_i
Step 6:	Compute fitness values for each individual by using the fitness calculation strategy presented in Sect. 3.3
Step 7:	Update new evaluations in the individual database array **T**
Step 8:	Calculate the probability value $Prob_i$ for each candidate solution that is to be used as a preference index by onlooker bees
Step 9:	Generate new search locations (using the Eq. 3.2) from **T** according to their probability $Prob_i$ (send onlooker bees to their selected food source), in order to produce the solution vector $\mathbf{O} = \{O_1, \ldots, O_5\}$. Likewise, update counters A_i
Step 10:	Compute fitness values for each individual by using the fitness calculation strategy presented in Sect. 3.3
Step 11:	Update new evaluations in the individual database array **T**
Step 12:	Generate the new population **P**. In case that the fitness value (evaluated or approximated) of the new solution O_i, is better than the solution E_i, such position is selected as an element of **P**, otherwise the solution E_i is chosen
Step 13:	If the solution counter A_i exceeds the number "*limit*" of trials, the solution i is re-started and generated randomly using Eq. 3.1. Likewise, the counter A_i is cleared
Step 14:	Determine the best individual of the current new population. If the new fitness (SAD) value is better than the old one, then update $\hat{u}_{best}$, $\hat{v}_{best}$
Step 15:	If the number of target iterations has been reached (four in the case of a search window size of ±8 and eight for ±16), then the MV is $\hat{u}_{best}$, $\hat{v}_{best}$; otherwise go back to Step 5

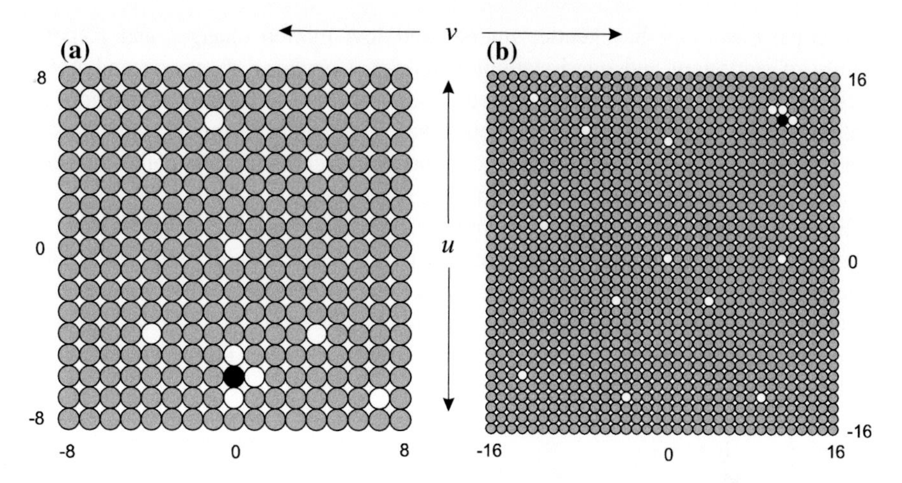

Fig. 3.7 Search-patterns generated by the ABC-BM algorithm. **a** Search window pattern ±8 with solution $\hat{u}^1_{best} = 0$ and $\hat{v}^1_{best} = -6$. **b** Search window pattern ±16 with solution $\hat{u}^2_{best} = 11$ and $\hat{v}^2_{best} = 12$

The presented ABC-BM algorithm considers different search locations during the complete optimization process with 45 in the case of a search window of ±8 and 85 for ±16, with 4 and 8 different iterations respectively. However, only a few of search locations are to be evaluated using the actual fitness function (between 5 and 14 in the case of a search window of ±8 and between 7 and 22 for the case of ±16) while all other remaining positions are just estimated. Figure 3.7 shows two search-pattern examples that have been generated by the ABC-BM approach. Such patterns exhibit the evaluated search-locations (rule 1 and 2) in white-cells, whereas the minimum location is marked in black. Grey-cells represent those that have been estimated (rule 3) or not visited at all, during the optimization process.

3.6 Experimental Results

3.6.1 ABC-BM Results

This section presents the results of comparing the presented ABC-BM algorithm with other existing fast BM algorithms. Simulations have been performed over the luminance component of popular video sequences that are listed in Table 3.1. Such sequences consist of different degrees and types of motion including QCIF (176 × 144), CIF (352 × 288) and SIF (352 × 240) respectively. The first four sequences are *Container*, *Carphone*, *Foreman* and *Akiyo* in QCIF format. The next two sequences are *Stefan* in CIF format and *Football* in SIF format. Among such sequences, *Container* has gentle, smooth and low motion changes and consists mainly of stationary and quasi-stationary blocks. *Carphone*, *Foreman* and *Akiyo* have moderately complex motion getting a "medium" category regarding its motion content. Rigorous motion which is based on camera panning with translation and complex motion content can be found in sequences of *Stefan* and *Football*. Figure 3.8 shows a sample frame from each sequence.

Each picture frame is partitioned into macro-blocks with the sizes of 16 × 16 pixels for motion estimation, where the maximum displacement within the search space W is ±8 pixels in both horizontal and vertical directions for the sequences *Container*, *Carphone*, *Foreman*, *Akiyo* and *Stefan*. The sequence *Football* has been simulated with a window size $W = \pm 16$ and requires a greater number of iterations (8 iterations) by the ABC-BM method.

In order to compare the performance of the ABC-BM approach, different search algorithms such as FSA, TSS [8], 4SS [11], NTSS [9], BBGD [15], DS [12], NE [16], ND [18], LWG [25], GFSS [26] and PSO-BM [27] have been all implemented in our simulations. For comparison purposes, all six video sequences in Fig. 3.8 have been all used. All simulations are performed on a Pentium IV 3.2 GHz PC with 1 GB of memory.

In the comparison, two relevant performance indexes have been considered: the distortion performance and the search efficiency.

Fig. 3.8 Test video sequences

3.6.1.1 Distortion Performance

First, all algorithms are compared in terms of their distortion performance which is characterized by the Peak-Signal-to-Noise-Ratio (PSNR) value. Such value indicates the reconstruction quality when motion vectors, which are computed through a BM approach, are used. In PSNR, the signal comes from original data frames whereas the noise is the error introduced by the calculated motion vectors. The PSNR is thus defined as:

$$\mathrm{PSNR} = 10 \cdot \log_{10}\left(\frac{255^2}{MSE}\right), \tag{3.8}$$

where *MSE* is the mean square between the original frames and those compensated by the motion vectors. Additionally, as an alternative performance index, the PSNR degradation ratio (D_{PSNR}) is used in the comparison. This ratio expresses in percentage (%) the level of mismatch between the PSNR of a BM approach and the PSNR of the FSA which is considered as a reference. Thus the D_{PSNR} is defined as:

$$D_{\mathrm{PSNR}} = -\left(\frac{\mathrm{PSNR}_{\mathrm{FSA}} - \mathrm{PSNR}_{\mathrm{BM}}}{\mathrm{PSNR}_{\mathrm{FSA}}}\right) \cdot 100\,\% \tag{3.9}$$

Table 3.2 shows the comparison of PSNR values and PSNR degradation ratios (D_{PSNR}) among BM algorithms. The experiment considers all six image sequences presented in Fig. 3.8. It is evident at the slow-moving sequence *Container*, that the PSNR values (the D_{PSNR} ratios) of all BM algorithms are similar. For the medium motion content sequences such as *Carphone*, *Foreman* and *Akiyo*, the approaches consistent of fixed patterns (TSS, 4SS and NTSS) exhibit the worst PSNR value (high D_{PSNR} ratio) except for the DS algorithm. On the other hand, BM methods that use evolutionary algorithms (LWG, GFSS, PSO-BM and ABC-BM) present the lowest D_{PSNR} ratio, only one step under the FSA approach which is considered as a reference. Finally, approaches based on the error-function minimization (BBGD and NE) and pixel-decimation (ND), show an acceptable performance. For the high motion sequence of *Stefan*, since the motion content of such sequences is complex and produces error surfaces with more than one minimum, the performance, in general, becomes worst for most of the algorithms in particular for those using fixed patterns. In the sequence *Football,* which has been simulated with a window size of ±16, the methods based on the evolutionary algorithms present the best PNSR values. Such performance is related to the fact that evolutionary methods adapt better to complex optimization problems with a bigger search area and a higher number of local minima. As a summary of the distortion performance, the last column of Table 3.2 presents the average PSNR degradation ratio (D_{PSNR}) from all sequences. According to such values, the presented ABC-BM method is superior to any other approach. Due to its computation complexity, the FSA is considered just as a reference. Best entries are bold-cased in Table 3.2.

Table 3.2 PSNR values and D_{PSNR} comparison of the BM methods

	Container $W = \pm 8$		Carphone $W = \pm 8$		Foreman $W = \pm 8$		Akiyo $W = \pm 8$		Stefan $W = \pm 8$		Football $W = \pm 16$		Total Average (D_{PSNR})
	PSNR	D_{PSNR}	PSNR	D_{PSNR}	PSNR	D_{PSNR}	PSNR	D_{PSNR}	PSNR	D_{PSNR}	PSNR	D_{PSNR}	
FSA	43.18	0	31.51	0	31.69	0	29.07	0	25.95	0	23.07	0	0
TSS	43.10	−0.20	30.27	−3.92	29.37	−7.32	26.21	−9.84	21.14	−18.52	20.03	−13.17	−8.82
4SS	43.12	−0.15	30.24	−4.01	29.34	−7.44	26.21	−9.84	21.41	−17.48	20.10	−12.87	−8.63
NTSS	43.12	−0.15	30.35	−3.67	30.56	−3.57	27.12	−6.71	22.52	−13.20	20.21	−12.39	−6.61
BBGD	43.14	−0.11	31.30	−0.67	31.00	−2.19	28.10	−3.33	25.17	−3.01	22.03	−4.33	−2.27
DS	43.13	−0.13	31.26	−0.79	31.19	−1.59	28.00	−3.70	24.98	−3.73	22.35	−3.12	−2.17
NE	43.15	−0.08	31.36	−0.47	31.23	−1.47	28.53	−2.69	25.22	−2.81	22.66	−1.77	−1.54
ND	43.15	−0.08	31.35	−0.50	31.20	−1.54	28.32	−2.56	25.21	−2.86	22.60	−2.03	−1.59
LWG	43.16	−0.06	31.40	−0.36	31.31	−1.21	28.71	−1.22	25.41	−2.09	22.90	−0.73	−0.95
GFSS	43.15	−0.06	31.38	−0.40	31.29	−1.26	28.69	−1.28	25.34	−2.36	22.92	−0.65	−1.01
PSO-BM	43.15	−0.07	31.39	−0.38	31.27	−1.34	28.65	1.42	25.39	−2.15	22.88	−0.82	−1.03
ABC-BM	43.17	**−0.02**	31.50	**−0.02**	31.62	**−0.22**	29.02	**−0.17**	25.90	**−0.19**	23.02	**−0.21**	**−0.17**

3.6.1.2 Search Efficiency

The search efficiency is used in this section as a measurement of computational complexity. The search efficiency is calculated by counting the average number of search points (or the average number of SAD computations) for the MV estimation. In Table 3.3, the search efficiency is compared and the best results are bold-cased. Just above FSA, some evolutionary algorithms such as LWG, GFSS and PSO-BM hold the highest number of search points per block. In contrast, the presented ABC-BM algorithm can be considered as a fast approach as it maintains a similar performance to DS. From data shown in Table 3.3, the average number of search locations that correspond to the ABC-BM method represents the number of SAD evaluations (the number of SAD estimations is not considered whatsoever). Additionally, the last two columns of Table 3.3 present the number of search locations that have been averaged (over the six sequences) and their performance rank. According to these values, the presented ABC-BM method is ranked at the first place. The average number of search points visited by the ABC-BM algorithm ranges from 9.0 to 16.3, representing the 4 % and the 7.4 % respectively in comparison to the FSA method. Such results demonstrate that our approach can significantly reduce the number of search points. Hence, the ABC-BM algorithm presented at this chapter is comparable to the fastest algorithms and delivers a similar precision to the FSA approach.

3.6.2 *Results on H.264*

In order to evaluate the performance of the presented algorithm, a set of experiments has been developed for JM-12.2 [51] of H.264/AVC reference software. Considering such encoder profile, the test conditions have been set as follows: For each test sequence only the first frame has been coded as “I frame” while remaining frames are coded as “P frames”. Only one reference frame has been used. The sum of absolute differences (SAD) distortion function is employed as the block distortion measure. The simulation platform in our experiments is a PC with Intel Pentium IV 2.66 GHz CPU.

Test sequences used in the experiments are *Container*, *Akiyo* and *Football*. Such sequences exhibit a variety of motion that is generally encountered in real video. A search window of ±8 is selected for sequences *Container* and *Akiyo* while the *Football* sequence is processed through a search window of ±16. Image formats used by the sequences are QCIF, CIF and SIF, testing at 30 fps over 200 different frames.

The group of experiments has been performed over test sequences at four different quantization parameters (QP = 28, 32, 36, 40) in order to test the algorithm at different bit rates. In the simulations, we have compared DS [12], EPZS [20] and the presented ABC-BM algorithm against the FSA, in terms of coding efficiency and computational complexity. The FSA is used as the basis of the image quality.

Table 3.3 Averaged number of visited search points per block for all ten BM methods

Algorithm	Container $W = \pm 8$	Carphone $W = \pm 8$	Foreman $W = \pm 8$	Akiyo $W = \pm 8$	Stefan $W = \pm 8$	Football $W = \pm 16$	Total Average	Rank
FSA	289	289	289	289	289	1089	422.3	12
TSS	25	25	25	25	25	25	25	8
4SS	19	25.5	24.8	27.3	25.3	25.6	24.58	7
NTSS	17.2	21.8	22.1	23.5	25.4	26.5	22.75	6
BBGD	9.1	14.5	14.5	13.2	17.2	22.3	15.13	3
DS	7.5	12.5	13.4	11.8	15.2	17.8	13.15	2
NE	11.7	13.8	14.2	14.5	19.2	24.2	16.36	5
ND	10.8	13.4	13.8	14.1	18.4	25.1	16.01	4
LWG	75	75	75	75	75	75	75	11
GFSS	60	60	60	60	60	60	60	10
PSO-BM	32.5	48.5	48.1	48.5	52.2	52.2	47	9
ABC-BM	**9.0**	**11.2**	**10.2**	**12.5**	16.1	**16.3**	**12.14**	**1**

Table 3.4 Performance comparison of DS, EPZS and ABC-BM for three different sequences in JM-12.2

Sequence	DS			EPZS			ABC-BM		
	BDBR (%)	BDPSNR (dB)	SUR	BDBR (%)	BDPSNR (dB)	SUR	BDBR (%)	BDPSNR (dB)	SUR
Container W = ±8	+4.22	−0.48	50.33	+0.33	−0.02	28.12	+0.35	−0.02	65.87
Akiyo W = ±8	+6.15	−0.54	44.87	+0.64	−0.07	24.79	+0.67	−0.08	58.14
Football W = ±16	+7.11	−0.56	48.78	+0.81	−0.11	25.12	+0.88	−0.12	60.47
Average	+5.82	−0.52	48.00	+0.59	−0.06	26.01	+0.63	−0.07	61.49

For the evaluation of coding efficiency, the Bjontegaard Delta PSNR (BDPSNR) and the Bjontegaard Delta Bit-Rate (BDBR) are used as performance indexes. Such indexes objectively express the average differences of PSNR and Bit-Rate when two methods are compared [52]. In order to calculate the computational complexity, the speed up ratio (SUR) is employed as computational efficiency index. SUR specifies the averaged speed up ratio between a given algorithm's motion estimation time and the time demanded by FSA. Table 3.4 shows the comparative results of DS, EPZS and the presented ABC-BM algorithm against the FSA. The sign (−) in BDPSNR and the sign (+) for BDBR indicate a loss in the coding performance. It is observed from experimental results that the presented ABC-BM algorithm is effective for reducing the computational complexity of the motion estimation. The speed up ratio (SUR) is about 60 times faster than FSA and it is about 3 times faster than EPZS, preserving a similar computation load in comparison to DS. Likewise, the coding performance of the ABC-BM algorithm is efficient because the loss in terms of BDBR and BDPSNR is low with only 0.63 % and 0.07 dB, respectively. Such coding performance is similar to the one produced by the EPZS method whereas it is better than the obtained by the DS algorithm.

3.6.3 Experiments with High Definition Sequences

This section presents yet more experiments in the JM-12.2 encoder profile in order to evaluate the performance of the presented algorithm over high definition sequences. Three testing video sequences are taken into consideration in the experiments. As it is shown in Table 3.5, such testing sequences include video data of various resolutions and different motion activities. Figure 3.9 presents a sample frame from each sequence.

Table 3.5 High definition test sequences used in the comparison

Sequence	Resolution	Total frames	Motion type
Blue-sky	1024 × 786	200	Low
Pedestrian-Area	1024 × 786	200	Medium
Rush-field	640 × 480	200	High

Blue-sky *Pedestrian-Area* *Rush-field*

Fig. 3.9 High definition test video sequences

Table 3.6 Performance comparison of DS, EPZS and ABC-BM for three different high definition sequences in JM-12.2

Sequence	DS			EPZS			ABC-BM		
	BDBR (%)	BDPSNR (dB)	SUR	BDBR (%)	BDPSNR (dB)	SUR	BDBR (%)	BDPSNR (dB)	SUR
Blue-sky	+7.05	−0.61	60.91	+0.21	−0.01	26.11	+0.22	−0.02	64.22
Pedestrian-Area.	+8.24	−0.72	57.14	+0.72	−0.04	24.12	+0.73	−0.05	60.04
Rush-field	+9.03	−0.79	51.92	+0.62	−0.08	19.17	+0.66	−0.10	54.21
Average	+8.10	−0.70	56.65	+0.51	−0.04	23.13	+0.53	−0.05	59.50

For encoding purposes, the JM-12.2 encoder profile has been set by using the first frame which has been coded as "I frame" and the remaining frames which have been coded as "P frames". Similarly to previous experiments, only one reference frame has been employed and the sum of absolute differences (SAD) distortion function is used as the block distortion measure. All sequences are tested at 30 fps considering 200 different frames. The search range for the sequences is ±32 whereas four QP (28, 32, 36, 40) values are used for calculating the BDPSNR and the BDBR. Table 3.6 shows the performance comparison of the presented ABC-BM algorithm with DS [12] and EPZS [20] while using the FSA method as the basis of the image quality. As it is shown in Table 3.6, the DS algorithm possesses the worst coding efficiency with 8.1 % and 0.7 dB respectively. However, it also presents a competitive SUR of 56.65. The EPZS method has the best coding performance (0.51 % and 0.04 dB), holding the worst speed up ratio at 21.13. On the other hand, the presented ABC-BM algorithm maintains the best tradeoff between coding efficiency (0.53 % and 0.05 dB) and computational complexity (59.5).

3.7 Conclusions

In this chapter, an algorithm based on Artificial Bee Colony (ABC) is presented to reduce the number of search locations in the BM process. The algorithm uses a simple fitness calculation approach which is based on the NNI algorithm. The method is able to save computational time by identifying which fitness value can be just estimated or must be calculated instead. As a result, the approach can substantially reduce the number of function evaluations, yet preserving the good search capabilities of ABC.

Since the algorithm does not consider any fixed search pattern or any other movement assumption, a high probability for finding the true minimum (accurate motion vector) is expected regardless of the movement complexity contained in the sequence. Therefore, the chance of being trapped into a local minimum is reduced in comparison to other BM algorithms.

The performance of ABC-BM has been compared to other existing BM algorithms by considering different sequences which present a great variety of formats and movement types. Experimental results demonstrate the high performance of the presented method in terms of coding efficiency and computational complexity.

Although the experimental results indicate that the ABC-BM method can yield better results on complicated sequences, it should be noticed that the aim of our chapter is not intended to beat all the BM methods which have been proposed earlier, but to show that the fitness approximation can effectively serve as an attractive alternative to evolutionary algorithms for solving complex optimization problems, yet demanding fewer function evaluations.

References

1. Tzovaras, D., Kompatsiaris, I., Strintzis, M.G.: 3D object articulation and motion estimation in model-based stereoscopic videoconference image sequence analysis and coding. Sig. Process. Image Commun. **14**(10), 817–840 (1999)
2. Barron, J.L., Fleet, D.J., Beauchemin, S.S.: Performance of optical flow techniques. Int. J. Comput. Vis. **12**(1), 43–77 (1994)
3. Skowronski, J.: Pel recursive motion estimation and compensation in subbands. Sig. Process. Image Commun. **14**, 389–396 (1999)
4. Huang, T., Chen, C., Tsai, C., Shen, C., Chen, L.: Survey on block matching motion estimation algorithms and architectures with new results. J. VLSI Sig. Proc. **42**, 297–320 (2006)
5. MPEG4. Information Technology Coding of Audio Visual Objects Part 2: Visual. JTC1/SC29/WG11, ISO/IEC14469-2(MPEG-4Visual) (2000)
6. H.264. Joint Video Team (JVT) of ITU-T and ISO/IEC JTC1, Geneva, JVT ofISO/IEC MPEG and ITU-T VCEG, JVT-g050r1, Draft ITU-TRec and Final Draft International Standard of Joint Video Specification (ITU-T Rec.H.264-ISO/IEC14496-10AVC) (2003)
7. Jain, J.R., Jain, A.K.: Displacement measurement and its application in inter-frame image coding. IEEE Trans. Commun. **COM-29**, 1799–1808 (1981)
8. Jong, H.-M., Chen, L.-G., Chiueh, T.-D.: Accuracy improvement and cost reduction of 3-step search block matching algorithm for video coding. IEEE Trans. Circ. Syst. Video Technol. **4**, 88–90 (1994)
9. Li, R., Zeng, B., Liou, M.L.: A new three-step search algorithm for block motion estimation. IEEE Trans. Circ. Syst. Video Technol. **4**(4), 438–442 (1994)
10. Jianhua, L., Liou, M.L.: A simple and efficient search algorithm for block-matching motion estimation. IEEE Trans. Circu. Syst. Video Technol. **7**(2), 429–433 (1997)
11. Po, L.-M., Ma, W.-C.: A novel four-step search algorithm for fast block motion estimation. IEEE Trans. Circ. Syst. Video Technol. **6**(3), 313–317 (1996)
12. Zhu, S., Ma, K.-K.: A new diamond search algorithm for fast block-matching motion estimation. IEEE Trans. Image Process. **9**(2), 287–290 (2000)
13. Nie, Y., Ma, K.-K.: Adaptive rood pattern search for fast block-matching motion estimation. IEEE Trans. Image Process. **11**(12), 1442–1448 (2002)
14. Yi-Ching, L., Jim, L., Zuu-Chang, H.: Fast block matching using prediction and rejection criteria. Sig. Process. **89**, 1115–1120 (2009)
15. Liu, L., Feig, E.: A block-based gradient descent search algorithm for block motion estimation in video coding. IEEE Trans. Circ. Syst. Video Technol. **6**(4), 419–422 (1996)
16. Saha, A., Mukherjee, J., Sural, S.: A neighborhood elimination approach for block matching in motion estimation. Sig. Process Image Commun. **26**, 8–9, 438–454 (2011)
17. Chow, K.H.K., Liou, M.L.: Generic motion search algorithm for video compression. IEEE Trans. Circ. Syst. Video Technol. **3**, 440–445 (1993)
18. Saha, A., Mukherjee, J., Sural, S.: New pixel-decimation patterns for block matching in motion estimation. Sig. Process. Image Commun. **23**, 725–738 (2008)
19. Song, Y., Ikenaga, T., Goto, S.: Lossy strict multilevel successive elimination algorithm for fast motion estimation. IEICE Trans. Fundam. **E90**(4), 764–770 (2007)
20. Tourapis, A.M.: Enhanced predictive zonal search for single and multiple frame motion estimation. In: Proceedings of Visual Communications and Image Processing, pp. 1069–1079. California, USA, January 2002
21. Chen, Z., Zhou, P., He, Y., Chen, Y.: Fast Integer Pel and Fractional Pel Motion Estimation for JVT, ITU-T.Doc.#JVT-F-017, December 2002
22. Nisar, H., Malik, A.S., Choi, T.-S.: Content adaptive fast motion estimation based on spatio-temporal homogeneity analysis and motion classification. Pattern Recogn. Lett. **33**, 52–61 (2012)

23. Holland, J.H.: Adaptation in Natural and Artificial Systems. University of Michigan Press, Ann Arbor, MI (1975)
24. Kennedy, J., Eberhart, R.C.: Particle swarm optimization. In: Proceedings of the 1995 IEEE International Conference on Neural Networks, vol. 4, pp. 1942–1948 (1995)
25. Chun-Hung, L., Ja-Ling, W.: A lightweight genetic block-matching algorithm for video coding. IEEE Trans. Circ. Syst. Video Technol. **8**(4), 386–392 (1998)
26. Wu, A., So, S.: VLSI implementation of genetic four-step search for block matching algorithm. IEEE Trans. Consum. Electron. **49**(4), 1474–1481 (2003)
27. Yuan, X., Shen, X. Block matching algorithm based on particle swarm optimization. International In: Conference on Embedded Software and Systems (ICESS 2008), Sichuan, China (2008)
28. Karaboga, D.: An idea based on honey bee swarm for numerical optimization, technical report-TR06. Erciyes University, Engineering Faculty, Computer Engineering Department (2005)
29. Karaboga, D., Basturk, B.: On the performance of artificial bee colony (ABC) algorithm. Appl. Soft Comput. **8**(1), 687–697 (2008)
30. Karaboga, D., Akay, B.: A comparative study of artificial bee colony algorithm. Appl. Math. Comput. **214**, 108–132 (2009)
31. Karaboga, N.: A new design method based on artificial bee colony algorithm for digital IIR filters. J. Franklin Inst. **346**, 328–348 (2009)
32. Sabat, S.L., Udgata, S.K., Abraham, A.: Artificial bee colony algorithm for small signal model parameter extraction of MESFET. Eng. Appl. Artif. Intell. **23**, 689–694 (2010)
33. Pan, Q.-K., Tasgetiren, M.F., Suganthan, P.N., Chua, T.J.: A discrete artificial bee colony algorithm for the lot-streaming flow shop scheduling problem. Inf. Sci. (2010). doi:10.1016/j.ins.2009.12.025
34. Kang, F., Li, J., Xu, Q.: Structural inverse analysis by hybrid simplex artificial bee colony algorithms. Comput. Struct. **87**, 861–870 (2009)
35. Zhang, C., Ouyang, D., Ning, J.: An artificial bee colony approach for clustering. Expert Syst. Appl. **37**, 4761–4767 (2010)
36. Karaboga, D., Ozturk, C.: A novel clustering approach: artificial bee colony (ABC) algorithm. Appl. Soft Comput. **11**, 652–657 (2011)
37. Xu, C., Duan, H., Liu, F.: Chaotic artificial bee colony approach to uninhabited combat air vehicle (UCAV) path planning. Aerosp. Sci. Technol. **14**, 535–541 (2010)
38. Cuevas, E., Sección-Echauri, F., Zaldivar, D., Pérez-Cisneros, M.: Multi-circle detection on images using artificial bee colony (ABC) optimization. Soft Comput. (2011). doi:10.1007/s00500-011-0741-0
39. Horng, M.-H.: Multilevel thresholding selection based on the artificial bee colony algorithm for image segmentation. Expert Syst. Appl. **38**(11), 13785–13791 (2011)
40. Jin, Y.: Comprehensive survey of fitness approximation in evolutionary computation. Soft. Comput. **9**, 3–12 (2005)
41. Jin, Yaochu: Surrogate-assisted evolutionary computation: recent advances and future challenges. Swarm Evol. Comput. **1**, 61–70 (2011)
42. Branke, J., Schmidt, C.: Faster convergence by means of fitness estimation. Soft. Comput. **9**, 13–20 (2005)
43. Zhou, Z., Ong, Y., Nguyen, M., Lim, D.: A study on polynomial regression and gaussian process global surrogate model in hierarchical surrogate-assisted evolutionary algorithm. In: IEEE Congress on Evolutionary Computation (ECiDUE'05), Edinburgh, United Kingdom, 2–5 Sept 2005
44. Ratle, A.: Kriging as a surrogate fitness landscape in evolutionary optimization. Artif. Intell. Eng. Des. Anal. Manuf. **15**, 37–49 (2001)
45. Lim, D., Jin, Y., Ong, Y., Sendhoff, B.: Generalizing surrogate-assisted evolutionary computation. IEEE Trans. Evol. Comput. **14**(3), 329–355 (2010)

46. Ong, Y., Lum, K., Nair, P.: Evolutionary algorithm with hermite radial basis function interpolants for computationally expensive adjoint solvers. Comput. Optim. Appl. **39**(1), 97–119 (2008)
47. Luoa, C., Shao-Liang, Z., Wanga, C., Jiang, Z.: A metamodel-assisted evolutionary algorithm for expensive optimization. J. Comput. Appl. Math. (2011). doi:10.1016/j.cam.2011.05.047
48. Goldberg, D.E.: Genetic Algorithms in Search, Optimization And Machine Learning. Addison-Wesley Professional, Menlo Park, CA (1989)
49. Li, X., Xiao, N., Claramunt, C., Lin, H.: Initialization strategies to enhancing the performance of genetic algorithms for the p-median problem. Comput. Ind. Eng. (2011). doi:10.1016/j.cie.2011.06.015
50. Xiao, N.: A unified conceptual framework for geographical optimization using evolutionary algorithms. Ann. Assoc. Am. Geogr. **98**, 795–817 (2008)
51. Joint Video Team Reference Software. Version 12.2 (JM12.2). http://iphome.hhi.de/suehring/tml/download/ (2007)
52. Bjontegaard, G.: Calculation of average PSNR differences between RD-Curves, ITU SG16 Doc.VCEG-M33 (2001)

Chapter 4
Ellipse Detection on Images Inspired by the Collective Animal Behavior

Abstract This chapter presents an effective technique for extracting multiple ellipses from an image. The approach employs an evolutionary algorithm that mimic the way in which animals behave collectively assuming the overall detection process as a multi-modal optimization problem. In the algorithm, searcher agents emulate a group of animals that interact to each other using simple biological rules which are modeled as evolutionary operators. In turn, such operators are applied to each agent considering that the complete group has a memory to store optimal solutions (ellipses) seen so-far by applying a competition principle. The detector uses a combination of five edge points as parameters to determine ellipse candidates (possible solutions) while a matching function determines if such ellipse candidates are actually present in the image. Guided by the values of such matching functions, the set of encoded candidate ellipses are evolved through the evolutionary algorithm so that the best candidates can be fitted into the actual ellipses within the image. Just after the optimization process ends, an analysis over the embedded memory is executed in order to find the best obtained solution (the best ellipse) and significant local minima (remaining ellipses). Experimental results over several complex synthetic and natural images have validated the efficiency of the resultant technique regarding accuracy, speed and robustness.

4.1 Introduction

Ellipse is one of the most commonly occurring geometric shapes in real images. Even perfect circles in 3D space are projected into elliptical shapes in the image. Therefore, the extraction of ellipses from images is a key problem in computer vision and pattern recognition. Applications include, among others, human face detection [1], iris recognition [2], driving assistance [3], industrial applications [4] and traffic sign detection [5].

E. Cuevas et al., *Applications of Evolutionary Computation in Image Processing and Pattern Recognition*, Intelligent Systems Reference Library 100,
DOI 10.1007/978-3-319-26462-2_4

Ellipse detection in real images is an open research problem since long time ago. Several approaches have been proposed which traditionally fall under three categories: Symmetry-based, Hough transform-based (HT) and Random sampling.

In symmetry-based detection [5–7], the ellipse geometry is taken into account. The most common elements used in ellipse geometry are the ellipse center and axis. Using these elements and edges in the image, the ellipse parameters can be found. Ellipse detection in digital images is commonly solved through the Hough Transform [8]. It works by representing the geometric shape by its set of parameters, then accumulating bins in the quantized parameter space. Peaks in the bins provide the indication of where ellipses may be. Obviously, since the parameters are quantized into discrete bins, the intervals of the bins directly affect the accuracy of the results and the computational effort. Therefore, for fine quantization of the space, the algorithm returns more accurate results, while suffering from large memory loads and expensive computation. In order to overcome such a problem, some other researchers have proposed other ellipse detectors following the Hough transform principles by using random sampling. In random sampling-based approaches [9, 10], a bin represents a candidate shape rather than a set of quantized parameters, as in the HT. However, like the HT, random sampling approaches go through an accumulation process for the bins. The bin with the highest score represents the best approximation of an actual ellipse in the target image. McLaughlin's work [11] shows that a random sampling-based approach produces improvements in accuracy and computational complexity, as well as a reduction in the number of false positives (nonexistent ellipses), when compared to the original HT and the number of its improved variants.

As an alternative to traditional techniques, the problem of ellipse detection has also been handled through optimization methods. In general, they have demonstrated to give better results than those based on the HT and random sampling with respect to accuracy, speed and robustness [12]. Such approaches have produced several robust ellipse detectors using different optimization algorithms such as Genetic algorithms (GA) [13, 14] and Particle Swarm Optimization (PSO) [15]. In the detection, the employed optimization algorithms perform well in locating a single optimum (only one shape) but fail to provide multiple solutions. Since extracting multiple ellipses primitives falls into the category of multi-modal optimization, they need to be applied several times in order to extract all the primitives. This entails the removal of detected primitives from the image, one at a time, while iterating, until there are no more candidates left in the image which results in highly time consuming algorithms.

Multimodal optimization is used to locate all the optima within the searching space, rather than one and only one optimum, and has been extensively studied by many researchers [16]. Many algorithms based on a large variety of different techniques have been proposed in the literature. Among them, 'niches and species', and a fitness sharing method [17] have been introduced to overcome the weakness of traditional evolutionary algorithms for multimodal optimization. Here, a new optimization algorithm based on the collective animal behavior is presented to solve multimodal problems, and then put into use in the application of multi-ellipse detection.

Many studies have been inspired by animal behavior phenomena in order to develop optimization techniques such as the Particle swarm optimization (PSO) algorithm which models the social behavior of bird flocking or fish schooling [18]. In recent years, there have been several attempts to apply the PSO to multi-modal function optimization problems [19, 20]. However, the performance of such approaches presents several flaws when it is compared to the other multi-modal metaheuristic counterparts [21].

Recently, the concept of individual-organization [22, 23] has been widely used to understand collective behavior of animals. The central principle of individual-organization is that simple repeated interactions between individuals can produce complex behavioral patterns at group level [22, 24, 25]. Such inspiration comes from behavioral patterns seen in several animal groups, such as ant pheromone trail networks, aggregation of cockroaches and the migration of fish schools, which can be accurately described in terms of individuals following simple sets of rules [26]. Some examples of these rules [25, 27] include keeping current position (or location) for best individuals, local attraction or repulsion, random movements and competition for the space inside of a determined distance. On the other hand, new studies have also shown the existence of collective memory in animal groups [28–30]. The presence of such memory establishes that the previous history, of group structure, influences the collective behavior exhibited in future stages. Therefore, according to these new developments, it is possible to model complex collective behaviors by using simple individual rules and configuring a general memory.

This chapter presents an algorithm for automatic detection of multiple ellipse shapes that considers the overall process as a multi-modal optimization problem. In the detection, the approach employs an evolutionary algorithm based on the way in which animals behave collectively. In such algorithm, searcher agents emulate a group of animals that interact to each other by simple biological rules which are modeled as evolution operators. Such operators are applied to each agent considering that the complete group has a memory which stores the optimal solutions seen so-far by applying a competition principle. The detector uses a combination of five edge points to determine ellipse candidates (possible solutions). A matching function determines if such ellipse candidates are actually present in the image. Guided by the values of such matching functions, the set of encoded candidate ellipses are evolved through the evolutionary algorithm so that the best candidates can be fitted into the actual ellipse within the image. Following the optimization process, the embedded memory is analyzed in order to find the best ellipse and significant local minima (remaining ellipses). Experimental results over several complex synthetic and natural images have validated the efficiency of the resultant technique regarding accuracy, speed and robustness.

4.2 Collective Animal Behavior Algorithm (CAB)

The CAB algorithm assumes the existence of a set of operations that resembles the interaction rules that model the collective animal behavior. In the approach, each solution within the search space represents an animal position. The "fitness value" refers to the animal dominance with respect to the group. The complete process mimics the collective animal behavior.

The approach in this chapter implements a memory for storing best solutions (animal positions) mimicking the aforementioned biologic process. Such memory is divided into two different elements, one for maintaining the best locations at each generation ($\mathbf{M}_g$) and the other for storing the best historical positions during the complete evolutionary process ($\mathbf{M}_h$).

4.2.1 Description of the CAB Algorithm

Following other metaheuristic approaches, the CAB algorithm is an iterative process that starts by initializing the population randomly (generated random solutions or animal positions). Then, the following four operations are applied until a termination criterion is met (i.e. the iteration number *NI*):

1. Keep the position of the best individuals.
2. Move from or to nearby neighbors (local attraction and repulsion).
3. Move randomly.
4. Compete for the space within a determined distance (update the memory).

4.2.1.1 Initializing the Population

The algorithm begins by initializing a set $\mathbf{A}$ of N_p animal positions ($\mathbf{A} = \{\mathbf{a}_1, \mathbf{a}_2, \ldots, \mathbf{a}_{N_p}\}$). Each animal position $\mathbf{a}_i$ is a *D*-dimensional vector contain ing parameter values to be optimized. Such values are randomly and uniformly distributed between the pre-specified lower initial parameter bound a_j^{low} and the upper initial parameter bound a_j^{high}.

$$a_{j,i} = a_j^{low} + \text{rand}(0,1) \cdot (a_j^{high} - a_j^{low}); \\ j = 1,2,\ldots,D; \quad i = 1,2,\ldots,N_p. \tag{4.1}$$

with j and i being the parameter and individual indexes respectively. Hence, $a_{j,i}$ is the jth parameter of the i-th individual.

All the initial positions $\mathbf{A}$ are sorted according to the fitness function (dominance) to form a new individual set $\mathbf{X} = \{\mathbf{x}_1, \mathbf{x}_2, \ldots, \mathbf{x}_{N_p}\}$, so that we can choose the best B positions and store them in the memory $\mathbf{M}_g$ and $\mathbf{M}_h$. The fact that both memories share the same information is only allowed at this initial stage.

4.2.1.2 Keep the Position of the Best Individuals

Analogous to the biological metaphor, this behavioral rule, typical from animal groups, is implemented as an evolutionary operation in our approach. In this operation, the first B elements ($\{\mathbf{a}_1, \mathbf{a}_2, \ldots, \mathbf{a}_B\}$), of the new animal position set $\mathbf{A}$, are generated. Such positions are computed by the values contained inside the historical memory $\mathbf{M}_h$, considering a slight random perturbation around them. This operation can be modeled as follows:

$$\mathbf{a}_l = \mathbf{m}_h^l + \mathbf{v} \tag{4.2}$$

where $l \in \{1, 2, \ldots, B\}$ while $\mathbf{m}_h^l$ represents the l-element of the historical memory $\mathbf{M}_h$. $\mathbf{v}$ is a random vector with a small enough length.

4.2.1.3 Move from or to Nearby Neighbors

From the biological inspiration, animals experiment a random local attraction or repulsion according to an internal motivation. Therefore, we have implemented new evolutionary operators that mimic such biological pattern. For this operation, a uniform random number r_m is generated within the range [0, 1]. If r_m is less than a threshold H, a determined individual position is moved (attracted or repelled) considering the nearest best historical position within the group (i.e. the nearest position in $\mathbf{M}_h$); otherwise, it goes to the nearest best location within the group for the current generation (i.e. the nearest position in $\mathbf{M}_g$). Therefore such operation can be modeled as follows:

$$\mathbf{a}_i = \begin{cases} \mathbf{x}_i \pm r \cdot (\mathbf{m}_h^{nearest} - \mathbf{x}_i) & \text{with probability } H \\ \mathbf{x}_i \pm r \cdot (\mathbf{m}_g^{nearest} - \mathbf{x}_i) & \text{with probability } (1-H) \end{cases} \tag{4.3}$$

where $i \in \{B+1, B+2, \ldots, N_p\}$, $\mathbf{m}_h^{nearest}$ and $\mathbf{m}_g^{nearest}$ represent the nearest elements of $\mathbf{M}_h$ and $\mathbf{M}_g$ to $\mathbf{x}_i$, while r is a random number.

4.2.1.4 Move Randomly

Following the biological model, under some probability P, one animal randomly changes its position. Such behavioral rule is implemented considering the next expression:

$$\mathbf{a}_i = \begin{cases} \mathbf{r} & \text{with probability } P \\ \mathbf{x}_i & \text{with probability } (1-P) \end{cases} \tag{4.4}$$

being $i \in \{B+1, B+2, \ldots, N_p\}$ and $\mathbf{r}$ a random vector defined in the search space. This operator is similar to re-initialize the particle in a random position, as it is done by Eq. (4.1).

4.2.1.5 Compete for the Space Within of a Determined Distance (Update the Memory)

Once the operations to keep the position of the best individuals, such as moving from or to nearby neighbors and moving randomly, have all been applied to the all N_p animal positions, generating N_p new positions, it is necessary to update the memory $\mathbf{M}_h$.

The concept of dominance is used to update the memory $\mathbf{M}_h$. Animals that interact within a group maintain a minimum distance among them. Such distance ρ depends on how aggressive an animal behaves [31, 32]. Hence, when two animals confront each other inside such distance, the most dominant individual prevails meanwhile the other withdraws. Figure 4.1 depicts the process.

In the presented algorithm, the historical memory $\mathbf{M}_h$ is updated considering the following procedure:

1. The elements of $\mathbf{M}_h$ and $\mathbf{M}_h$ are merged into $\mathbf{M}_U$ ($\mathbf{M}_U = \mathbf{M}_h \cup \mathbf{M}_g$).
2. Each element $\mathbf{m}_U^i$ of the memory $\mathbf{M}_h$ is compared pair-wise to remaining memory elements ($\{\mathbf{m}_U^1, \mathbf{m}_U^2, \ldots, \mathbf{m}_U^{2B-1}\}$). If the distance between both elements is less than ρ, the element getting a better performance in the fitness function prevails meanwhile the other is removed.
3. From the resulting elements of $\mathbf{M}_U$ (from step 2), it is selected the B best value to build the new $\mathbf{M}_h$.

Unsuitable values of ρ yield a lower convergence rate, a longer computational time, a larger function evaluation number, the convergence to a local maximum or

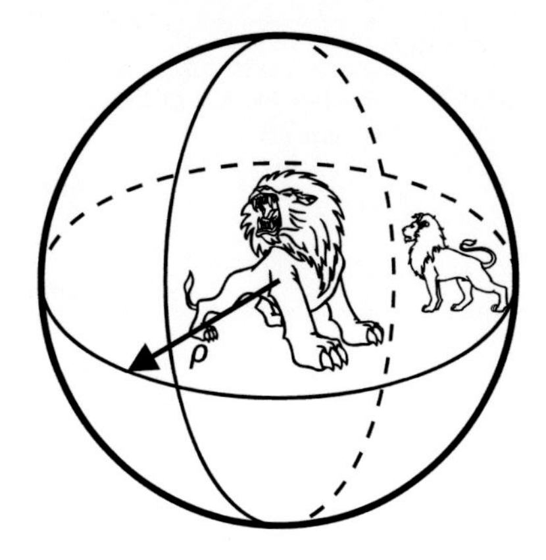

Fig. 4.1 Dominance concept as it is presented when two animals confront each other inside of a ρ distance

to an unreliable solution. The ρ value is computed considering the following equation:

$$\rho = \frac{\prod_{j=1}^{D} (a_j^{high} - a_j^{low})}{10 \cdot D}, \tag{4.5}$$

where a_j^{low} and a_j^{high} represent the pre-specified lower and upper bound of the j-parameter respectively, in an D-dimensional space.

4.2.1.6 Computational Procedure

The computational procedure for the presented algorithm can be summarized as follows:

Step 1	Set the parameters N_p, B, H, P and NI
Step 2	Generate randomly the position set $\mathbf{A} = \{\mathbf{a}_1, \mathbf{a}_2, \ldots, \mathbf{a}_{N_p}\}$ using Eq. 4.1
Step 3	Sort $\mathbf{A}$ according to the objective function (dominance) to build $\mathbf{X} = \{\mathbf{x}_1, \mathbf{x}_2, \ldots, \mathbf{x}_{N_p}\}$.
Step 4	Choose the first B positions of $\mathbf{X}$ and store them into the memory $\mathbf{M}_g$
Step 5	Update $\mathbf{M}_h$ according to Sect. 4.2.1.5 (during the first iteration: $\mathbf{M}_h = \mathbf{M}_g$)
Step 6	Generate the first B positions of the new solution set $\mathbf{A}$ ($\{\mathbf{a}_1, \mathbf{a}_2, \ldots, \mathbf{a}_B\}$. Such positions correspond to the elements of $\mathbf{M}_h$ making a slight random perturbation around them $\mathbf{a}_l = \mathbf{m}_h^l + \mathbf{v}$; being $\mathbf{v}$ a random vector of a small enough length

(continued)

(continued)

Step 7	Generate the rest of the **A** elements using the attraction, repulsion and random movements
	for i=B+1: N_p
	if (r_1<P) then
	attraction and repulsion movement
	{ if (r_2<H) then
	$\mathbf{a}_i = \mathbf{x}_i \pm r \cdot (\mathbf{m}_h^{nearest} - \mathbf{x}_i)$
	else if
	$\mathbf{a}_i = \mathbf{x}_i \pm r \cdot (\mathbf{m}_g^{nearest} - \mathbf{x}_i)$
	}
	else if
	random movement
	{
	$\mathbf{a}_i = \mathbf{r}$
	}
	end for where $r_1, r_2, r \in \text{rand}(0,1)$

Step 8	If NI is completed, the process is finished; otherwise go back to step 3 The best value in $\mathbf{M}_h$ represents the global solution for the optimization problem.

4.3 Ellipse Detection Using CAB

4.3.1 Data Preprocessing

In order to detect ellipse shapes, candidate images must be preprocessed first by the well-known Canny algorithm which yields a single-pixel edge-only image. Then, the (x_i, y_i) coordinates for each edge pixel p_i are stored inside the edge vector $P = \{p_1, p_2, \ldots, p_{N_p}\}$, with N_p being the total number of edge pixels.

4.3.2 Individual Representation

Each candidate solution E (ellipse candidate) uses five edge points. Under such representation, edge points are selected following a random positional index within the edge array P. This procedure will encode a candidate solution as the ellipse that passes through five points p_i, p_j, p_k, p_l and p_m ($E = p_i, p_j, p_k, p_l, p_m$). Thus, by substituting the coordinates of each point of E into Eq. 4.6, we gather a set of five simultaneous equations which are linear in the five unknown parameters a', b', c', f' and g'.

$$a'x^2 + 2h'xy + b'y^2 + 2g'x + 2f'y + c' = 0 \tag{4.6}$$

Then, solving the involved parameters and dividing by the constant c', it yields:

$$ax^2 + 2hxy + by^2 + 2gx + 2fy + 1 = 0 \tag{4.7}$$

Considering the configuration of the edge points shown by Fig. 4.2, the ellipse center (x_0, y_0), the radius maximum $(r_{\max})$, the radius minimum $(r_{\min})$ and the ellipse orientation (θ) can be calculated as follows:

$$x_0 = \frac{hf - bg}{C}, \tag{4.8}$$

$$y_0 = \frac{gh - af}{C}, \tag{4.9}$$

$$r_{\max} = \sqrt{\frac{-2\Delta}{C(a + b - R)}}, \tag{4.10}$$

$$r_{\min} = \sqrt{\frac{-2\Delta}{C(a + b + R)}}, \tag{4.11}$$

$$\theta = \frac{1}{2}\arctan\left(\frac{2h}{a - b}\right) \tag{4.12}$$

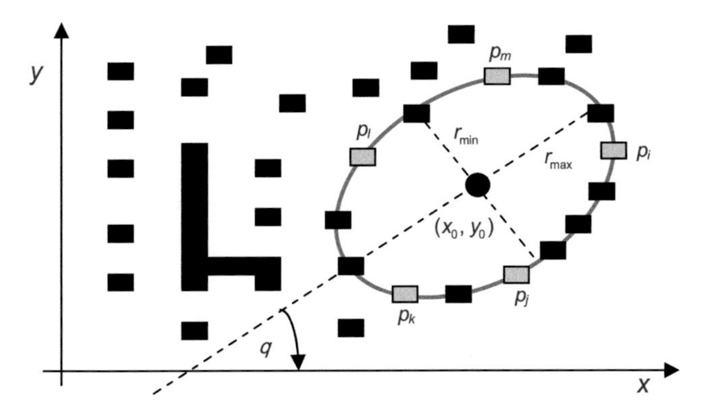

Fig. 4.2 Ellipse candidate (individual) built from the combination of points p_i, p_j, p_k, p_l, and p_m

where

$$R^2 = (a - b)^2 + 4h^2 \quad \text{and} \quad C = ab - h^2 \tag{4.13}$$

4.3.3 Objective Function

Optimization refers to choosing the best element from one set of available alternatives. In the simplest case, it means to minimize an objective function or error by systematically choosing the values of variables from their valid ranges. In order to calculate the error produced by a candidate solution E, the ellipse coordinates are calculated as a virtual shape which, in turn, must also be validated, i.e. if it really exists in the edge image. The test set is represented by $S = \{s_1, s_2, \ldots, s_{N_s}\}$, where N_s are the number of points over which the existence of an edge point, corresponding to E, should be tested. The set S is generated by the Midpoint Ellipse Algorithm (MEA) [33] which is a searching method that seeks required points for drawing an ellipse. Any point (x, y) on the boundary of the ellipse with a, h, b, g and f satisfies the equation $f_{ellipse}(x, y) \cong r_x x^2 + r_y y^2 - r_x^2 r_y^2$. However, MEA avoids computing square-root calculations by comparing the pixel separation distances. A method for direct distance comparison is to test the halfway position between two pixels (sub-pixel distance) to determine if this midpoint is inside or outside the ellipse boundary. If the point is in the interior of the ellipse, the ellipse function is negative. Thus, if the point is outside the ellipse, the ellipse function is positive. Therefore, the error involved in locating pixel positions using the midpoint test is limited to one-half the pixel separation (sub-pixel precision). To summarize, the relative position of any point (x, y) can be determined by checking the sign of the ellipse function:

$$f_{Circle}(x, y) \begin{cases} <0 & \text{if } (x, y) \text{ is inside the ellipse boundary} \\ =0 & \text{if } (x, y) \text{ is on the ellipse boundary} \\ >0 & \text{if } (x, y) \text{ is outside the ellipse boundary} \end{cases} \tag{4.14}$$

The ellipse-function test in Eq. 4.14 is applied to mid-positions between pixels nearby the ellipse path at each sampling step. Figures 4.3a and 4.4a are shows the midpoint between the two candidate pixels at sampling position. The ellipse is used to divide the quadrants into two regions the limit of the two regions is the point at which the curve has a slope of -1 as shown in Fig. 4.4.

In MEA the computation time is reduced by considering the symmetry of ellipses. Ellipses sections in adjacent octants within one quadrant are symmetric with respect to the dy/dy = −1 line dividing the two octants. These symmetry conditions are illustrated in Fig. 4.4. The algorithm can be considered as the quickest providing a sub-pixel precision [34]. However, in order to protect the

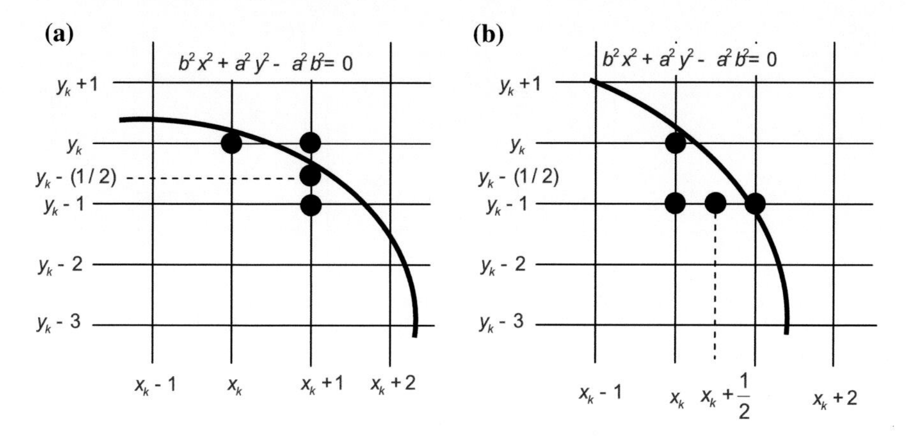

Fig. 4.3 **a** symmetry of the ellipse: an estimated one octant which belong to the first region where the slope is greater than −1, **b** in this region the slope will be less than −1 to complete the octant and continue to calculate the same so the remaining octants

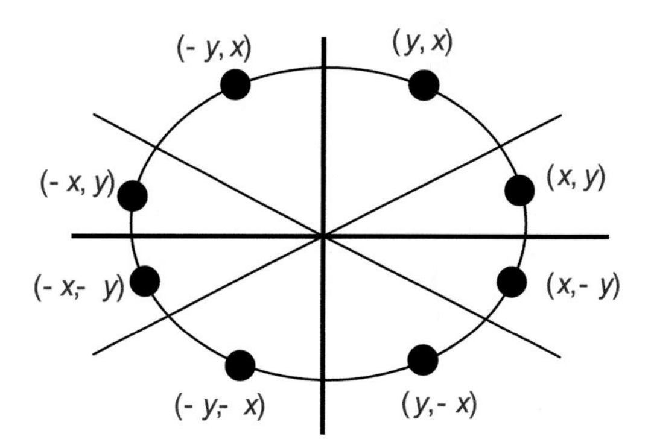

Fig. 4.4 Midpoint between candidate pixels at sampling position x_k along an elliptical path

MEA operation, it is important to assure that points lying outside the image plane must not be considered in S. The objective function *J*(*E*) represents the matching error produced between the pixels *S* of the ellipse candidate *E* and the pixels that actually exist in the edge image, yielding:

$$J(E) = 1 - \frac{\sum_{v=1}^{Ns} G(x_v, y_v)}{Ns}, \tag{4.15}$$

where $G(x_i, y_i)$ is a function that verifies the pixel existence in (x_v, y_v), with $(x_v, y_v) \in S$ and N_s being the number of pixels lying on the perimeter corresponding to *E* currently under testing. Hence, function $G(x_v, y_v)$ is defined as:

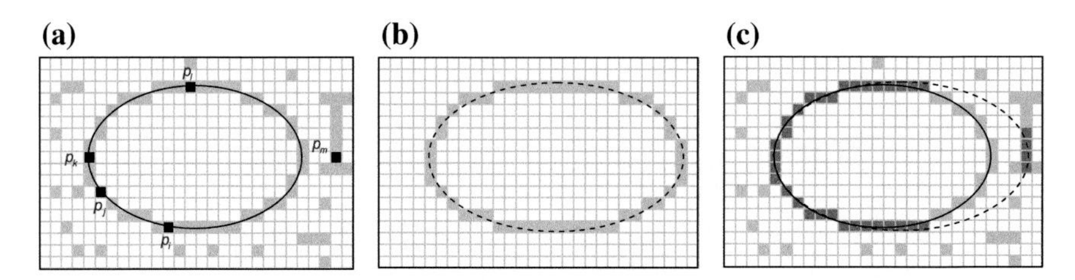

Fig. 4.5 Evaluation of a candidate solution *E*: the image in **a** shows the original image while **b** presents the generated virtual shape drawn from points p_i, p_j, p_k, p_l and p_m. The image in **c** shows coincidences between both images which have been marked by *darker pixels* while the virtual shape is also depicted through a *dashed line*

$$G(x_v, y_v) = \begin{cases} 1 & \text{if the pixel } (x_v, y_v) \text{ is an edge point} \\ 0 & \text{otherwise} \end{cases} \tag{4.16}$$

A value of *J*(*E*) near to zero implies a better response from the "ellipsoid" operator. Figure 4.5 shows the procedure to evaluate a candidate action *E* with its representation as a virtual shape *S*. Figure 4.5a shows the original edge map, while Fig. 4.5b presents the virtual shape *S* representing the individual $E = \{p_i, p_j, p_k, p_l, p_m\}$. In Fig. 4.5c, the virtual shape *S* is compared to the original image, point by point, in order to find coincidences between virtual and edge points. The individual has been built from points p_i, p_j, p_k, p_l and p_m which are shown by Fig. 4.5a. The virtual shape *S*, obtained by MEA, gathers 52 points ($N_s = 52$) with only 35 of them existing in both images (shown as darker points in Fig. 4.5c) and yielding: $\sum_{v=1}^{Ns} G(x_v, y_v) = 35$, therefore $J(E) = 0.327$.

4.3.4 Implementation of CAB for Ellipse Detection

The implementation of the presented algorithm can be summarized in the following steps:

Step 1	Adjust the algorithm parameters N_p, *B*, *H*, *P* and *NI*.
Step 2	Randomly generate a set of candidate ellipses (position of each animal) $\mathbf{E} = \{E_1, E_2, \ldots, E_{N_p}\}$ set using Eq. 4.1
Step 3	Sort **E** according to the objective function (dominance) to build $\mathbf{X} = \{\mathbf{x}_1, \mathbf{x}_2, \ldots, \mathbf{x}_{N_p}\}$.
Step 4	Choose the first *B* positions of **X** and store them into the memory $\mathbf{M}_g$
Step 5	Update $\mathbf{M}_h$ according to Sect. 2.1.5 (during the first iteration: $\mathbf{M}_h = \mathbf{M}_g$)
Step 6	Generate the first *B* positions of the new solution set **E** $\{E_1, E_2, \ldots, E_B\}$. Such positions correspond to the elements of $\mathbf{M}_h$ making a slight random perturbation around them $E_l = \mathbf{m}_h^l + \mathbf{v}$; being **v** a random vector of a small enough length

(continued)

(continued)

Step 7	Generate the rest of the **E** elements using the attraction, repulsion and random movements
	for i=B+1: N_p
	if (r_1<P) then
	attraction and repulsion movement
	{ if (r_2<H) then
	$E_i = \mathbf{x}_i \pm r \cdot (\mathbf{m}_h^{nearest} - \mathbf{x}_i)$
	else if
	$E_i = \mathbf{x}_i \pm r \cdot (\mathbf{m}_g^{nearest} - \mathbf{x}_i)$
	}
	else if
	random movement
	{
	$E_i = \mathbf{r}$
	}
	end for where $r_1, r_2, r \in \text{rand}(0,1)$

Step 8	If NI is completed, the process is finished; otherwise go back to step 3. The best values in $\mathbf{M}_h$ represents the best solutions (the best found ellipses)

4.4 The Multiple Ellipse Detection Procedure

Most detectors simply apply a one-minimum optimization algorithm to detect multiple ellipses yielding only one ellipse at a time and repeating the same process several times as previously detected primitives are removed from the image. The algorithm iterates until no more candidates are left in the image.

On the other hand, the method at this chapter is able to detect single or multiples ellipses through only one optimization step. The multi-detection procedure can be summarized as follows: guided by the values of a matching function, the whole group of encoded candidate ellipses is evolved through the set of evolutionary operators. The best ellipse candidate (global optimum) is considered to be the first detected ellipse over the edge-only image. An analysis of the incorporated historical memory $\mathbf{M}_h$ is thus executed in order to identify other local optima (other ellipses).

In order to find other possible ellipses contained in the image, the historical memory $\mathbf{M}_h$ is carefully examined. The approach aims to explore all elements, one at a time, assessing which of them represents an actual ellipse in the image. Since several elements can represent the same ellipse (i.e. ellipses slightly shifted or holding small deviations), a distinctiveness factor $D_{A,B}$ is required to measure the mismatch between two given ellipses (A and B). Such distinctiveness factor is defined as follows:

$$D_{A,B} = \left|x_0^A - x_0^B\right| + \left|y_0^A - y_0^B\right| + \left|r_{\min}^A - r_{\min}^B\right| + \left|r_{\max}^A - r_{\max}^B\right| + \left|\theta_A - \theta_B\right|, \quad (4.17)$$

being $\left(x_0^A, y_0^A\right), r_{\min}^A, r_{\max}^A$ and θ_A, the central coordinates, the lengths of the semi-axes and the orientation of the ellipse E_A, respectively. Likewise, $\left(x_0^B, y_0^B\right), r_{\min}^B, r_{\max}^B$ and θ_B represent the corresponding parameters of the ellipse E_B. One threshold value *Th* is also calculated to decide whether two ellipses must be considered different or not. *Th* is computed as:

$$Th = \frac{\left(\left|r_{\max}^h - r_{\max}^l\right| + \left|r_{\min}^h - r_{\min}^l\right|\right)/2}{s} \quad (4.18)$$

where $\left[r_{\max}^l, r_{\max}^h\right]$ and $\left[r_{\min}^l, r_{\min}^h\right]$ are the feasible semi-axes ranges and s is a sensitivity parameter. By using a high value for s, two very similar ellipses would be considered different while a smaller value for s would consider them as similar. In this work, after several experiments, the s value has been set to 2.

Thus, since the historical memory $\mathbf{M}_h\left\{E_1^{\mathbf{M}}, E_2^{\mathbf{M}}, \ldots, E_B^{\mathbf{M}}\right\}$ groups the elements in descending order according to their fitness values, the first element $E_1^{\mathbf{M}}$, whose fitness value represents the best value $F_1^{\mathbf{M}}$, is assigned to the first ellipse. Then, the distinctiveness factor $(D_{E_1^{\mathbf{M}}, E_2^{\mathbf{M}}})$ over the next element $E_2^{\mathbf{M}}$ is evaluated with respect to the prior $E_1^{\mathbf{M}}$. If $D_{E_1^{\mathbf{M}}, E_2^{\mathbf{M}}} > Th$, then $E_2^{\mathbf{M}}$ is considered as a new ellipse otherwise the next element $E_3^{\mathbf{M}}$ is selected. This process is repeated until the fitness value $F_i^{\mathbf{M}}$ reaches a minimum threshold F_{TH}. According to such threshold, other values above F_{TH} represent individuals (ellipses) that are considered as significant while other values lying below such boundary are considered as false ellipses and hence they are not contained in the image. After several experiments the value of F_{TH} is set to $(F_1^{\mathbf{M}}/10)$.

4.5 Experimental Results

Experimental tests have been developed in order to evaluate the performance of the ellipse detector. The experiments address the following tasks:

(1) Ellipse localization,
(2) Shape discrimination,
(3) Ellipse approximation: occluded ellipse and ellipsoidal detection.

Table 4.1 presents the parameters for the CAB algorithm at this work. They have been kept for all test images after being experimentally defined.

Table 4.1 CAB detector parameters

N_p	H	P	B	NI
30	0.5	0.1	12	200

4.5.1 *Ellipse Localization*

4.5.1.1 Synthetic Images

The experimental setup includes the use of several synthetic images of 400 × 300 pixels. All images contain a different amount of ellipsoidal shapes and some have also been contaminated by added noise as to increase the complexity of the localization task. The algorithm is executed over 50 times for each test image, successfully identifying and marking all required ellipse in the image. The detection has proved to be robust to translation and scaling still offering a reasonably low execution time. Figure 4.6 shows the outcome after applying the algorithm to two images from the experimental set.

4.5.1.2 Natural Images

This experiment tests the ellipse detection on several real images which contain a different number of ellipse shapes. The image set has been captured by a digital

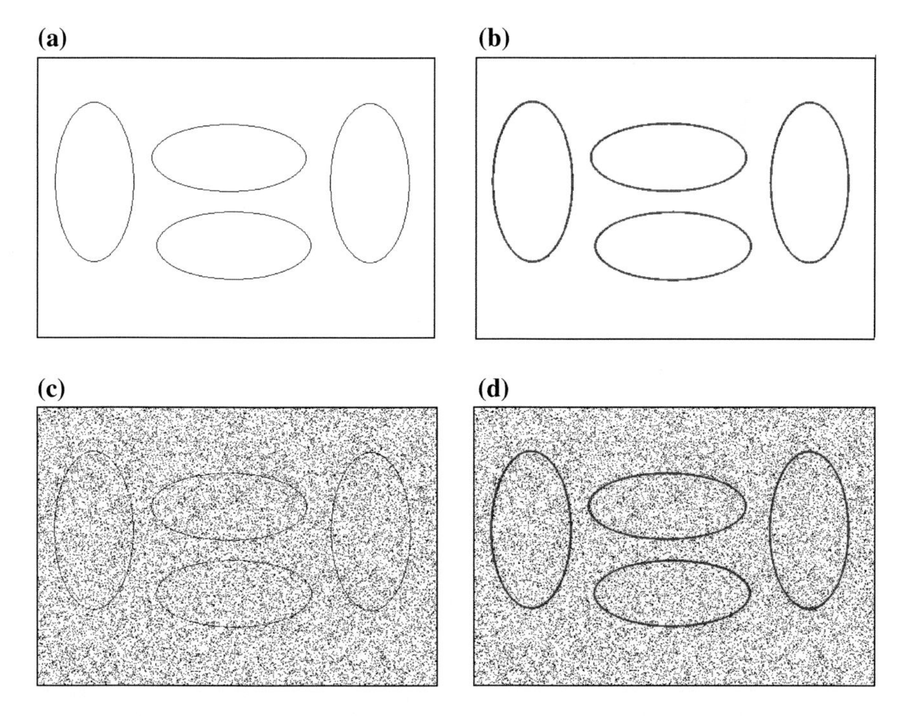

Fig. 4.6 Ellipse localization over synthetic images. The image **a** shows the original image while **b** presents the detected ellipses. The image in **c** shows a second image with salt and pepper noise, and **d** shows the detected ellipses in *red* overlay

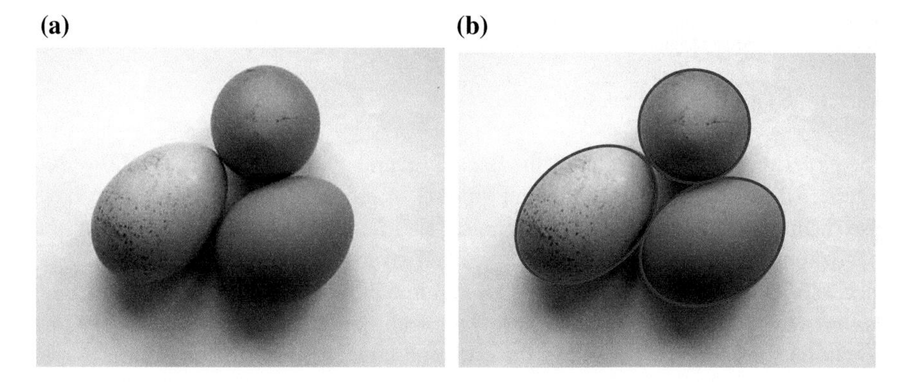

Fig. 4.7 Ellipse detection algorithm over natural images: the image in **a** shows the original image while **b** presents the detected ellipses in a *red* overlay

camera in an 8-bit format. Each image is pre-processed by the Canny edge detection algorithm. Figure 4.7 shows the results after the algorithm CAB has been applied to one image from the experimental set.

4.5.2 Shape Discrimination Tests

This section discusses on the algorithm's ability to detect elliptical patterns despite some other different shapes being present in the image. Figure 4.8 shows four shapes in the image of 400 × 300 pixels. The image has been contaminated by local noise in order to increase the complexity on the localization task. Figure 4.9 repeats the experiment over a real-life image.

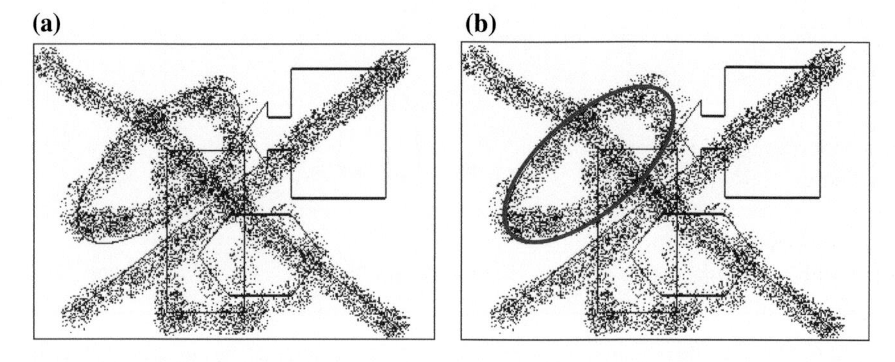

Fig. 4.8 Shape discrimination over synthetic images: **a** shows the original image **b** presents the detected ellipse as a *red* overlay

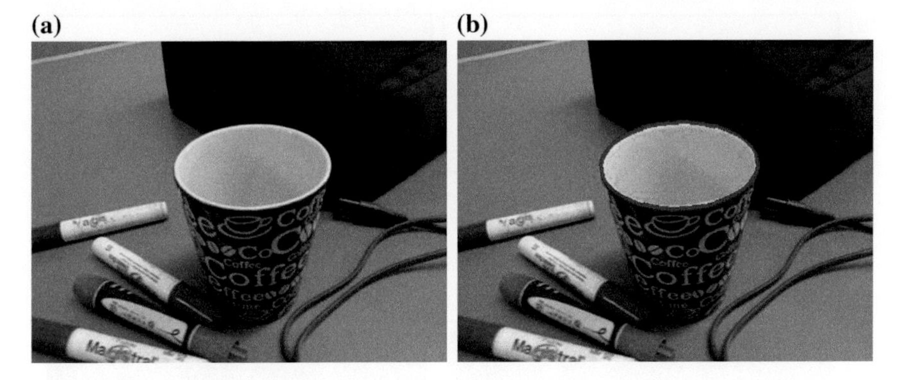

Fig. 4.9 Shape discrimination in real-life images: **a** shows the original image and **b** presents the detected ellipse as an overlay

4.5.3 Ellipse Approximation: Occluded Ellipse and Ellipsoidal Detection

The CAB detector algorithm is able to detect occluded or imperfect ellipses as well as partially defined shapes such as arc segments. The relevance of such functionality comes from the fact that imperfect ellipses are commonly found in typical computer vision applications. Since ellipse detection has been considered as an optimization problem, the CAB algorithm allows finding ellipses that may approach a given shape according to fitness values for each candidate. Figure 4.10a shows some examples of ellipse approximation. Likewise, the presented algorithm is able to find ellipse parameters that better approach an arc or an occluded ellipse. Figure 4.10b, c show some examples of this functionality. A small value for $J(C)$, i.e., near zero, refers to an ellipse while a slightly bigger value accounts for an arc or an occluded circular shape. Such a fact does not represent any problem as ellipses can be shown following the obtained $J(C)$ values.

4.5.4 Performance Comparison

In order to enhance the performance analysis, the presented approach is compared to the GA-based algorithm [14] and the RHT [10] method over a common image set.

The GA-based algorithm follows the proposal of Yao et al. [14], which considers two different subpopulations (each one of 50 individuals), a crossover probability of 0.55, the mutation probability being 0.10 and the number of elite individuals chosen as 5. The roulette wheel selection and the 1-point crossover operator are also applied. The parameter setup and the fitness function follow the configuration

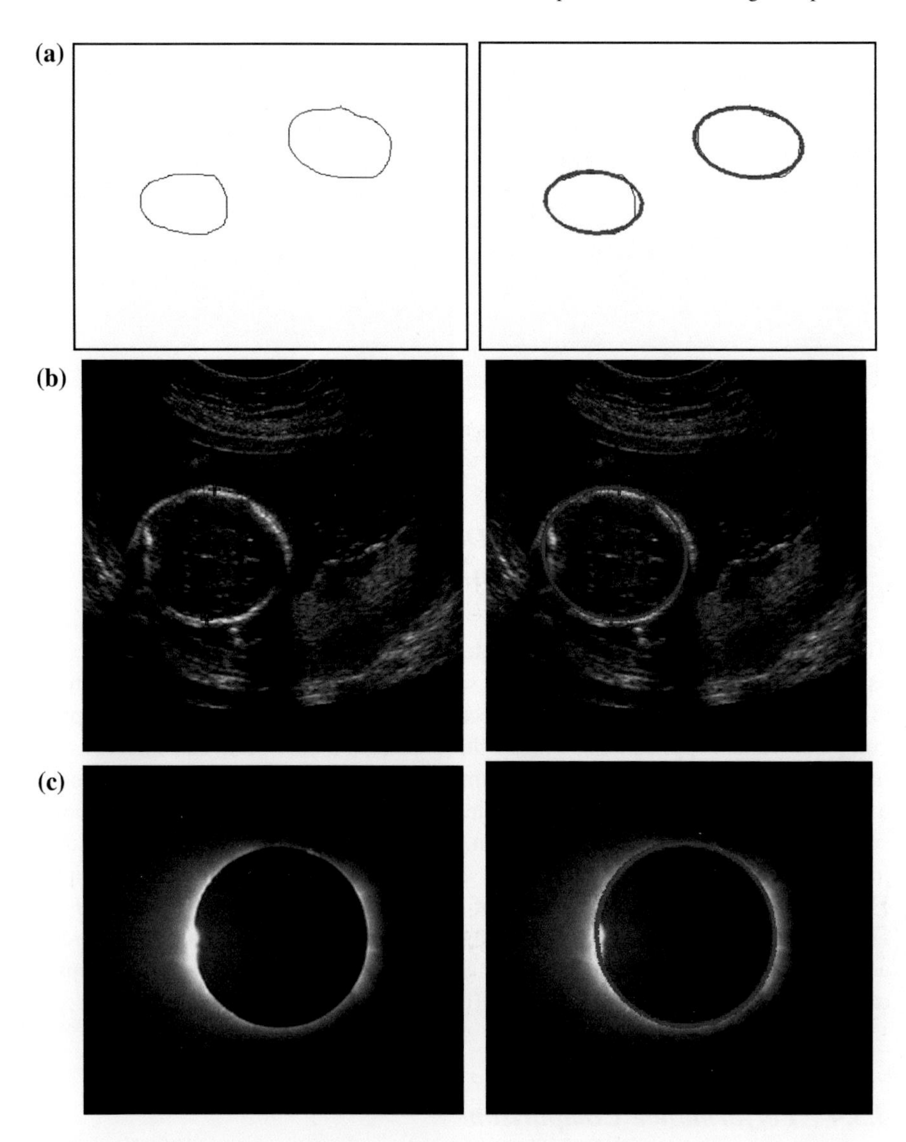

Fig. 4.10 CAB approximating ellipsoidal shapes and arc detections

suggested in [14]. Likewise, the RHT method has been implemented as it is described in [10].

Images rarely contain perfectly-shaped ellipses. In order to test the accuracy for a single-ellipse, the detection is challenged by a ground-truth ellipse, which is determined manually from the original edge-map. The parameters $(x_0^{true}, y_0^{true}, r_{\max}^{true}, r_{\min}^{true}, \theta_{true})$ of the ground-truth ellipse are computed considering the best fitted ellipse that a human observer can identify through a drawing software

(MSPaint©, CorelDraw©, etc.). If the parameters of the detected ellipse are defined as $x_0^D, y_0^D, r_{\max}^D, r_{\min}^D$ and θ_D, then an error score (Es) is defined as follows:

$$\mathrm{Es} = P_1\left(\left|x_0^{true} - x_0^D\right| + \left|y_0^{true} - y_0^D\right|\right) + P_2\left(\frac{\left|r_{\max}^{true} - r_{\max}^D\right| + \left|r_{\min}^{true} - r_{\min}^D\right|}{2}\right) + P_3\left|\theta_{true} - \theta_D\right| \tag{4.19}$$

The central point difference $\left(\left|x_0^{true} - x_0^D\right| + \left|y_0^{true} - y_0^D\right|\right)$ represents the center shift for the detected ellipse as it is compared to the ground-truth ellipse. The averaged radio mismatch $\left(\left(\left|r_{\max}^{true} - r_{\max}^D\right| + \left|r_{\min}^{true} - r_{\min}^D\right|\right)/2\right)$ accounts for the difference between their radii. Finally, the angle error $\left(\left|\theta_{true} - \theta_D\right|\right)$ corresponds to the orientation difference between the ground-truth and detected ellipses. P_1, P_2 and P_3 represent three weighting parameters, which are applied to the central point difference, to the radius mismatch and to the angle error in order to impose a relative importance in the final error Es. In this study, they are chosen as $P_1 = 0.05$, $P2 = 0.1$ and $P_3 = 0.2$. This particular choice ensures that the angle error would be strongly weighted in comparison to the previous terms which account for the difference in the central ellipse positions and the averaged radii differences of manually detected and machine-detected ellipses, respectively. In order to use an error metric for multiple-ellipse detection, the averaged Es resulting from each ellipse in the image, is considered. The multiple error (ME) is thus calculated as follows:

$$\mathrm{ME} = \left(\frac{1}{NC}\right) \cdot \sum_{R=1}^{NC} \mathrm{Es}_R, \tag{4.20}$$

where NC represents the number of ellipses actually present the image. In case of ME being less than 1, the algorithm is considered successful; otherwise it is said to have failed in the detection of the ellipse set. Notice that for $P_1 = 0.05$, $P_2 = 0.1$ and $P_3 = 0.2$, an ME < 1 is generated accounting for a maximal tolerated average difference for radius (10 pixels long) and a mismatch limit of 5° whereas the maximum average mismatch for the center location can be up to 20 pixels. In general, the Success Rate (SR) can thus be defined as the percentage of achieving success after a certain number of trials.

Figures 4.11 and 4.12 show six images (three synthetic and three natural) that have been used to compare the performance of the GA-based algorithm [14], the RHT method [10] and the presented approach. The performance is analyzed by considering 35 different executions for each algorithm over six images. Table 4.2 presents the averaged execution time, the success rate (SR) in percentage and the averaged multiple error (ME). The best entries are bold-cased in Table 4.2. Closer inspection reveals that the presented method is able to achieve the highest success rate with the smallest error and still requires less computational time for most cases.

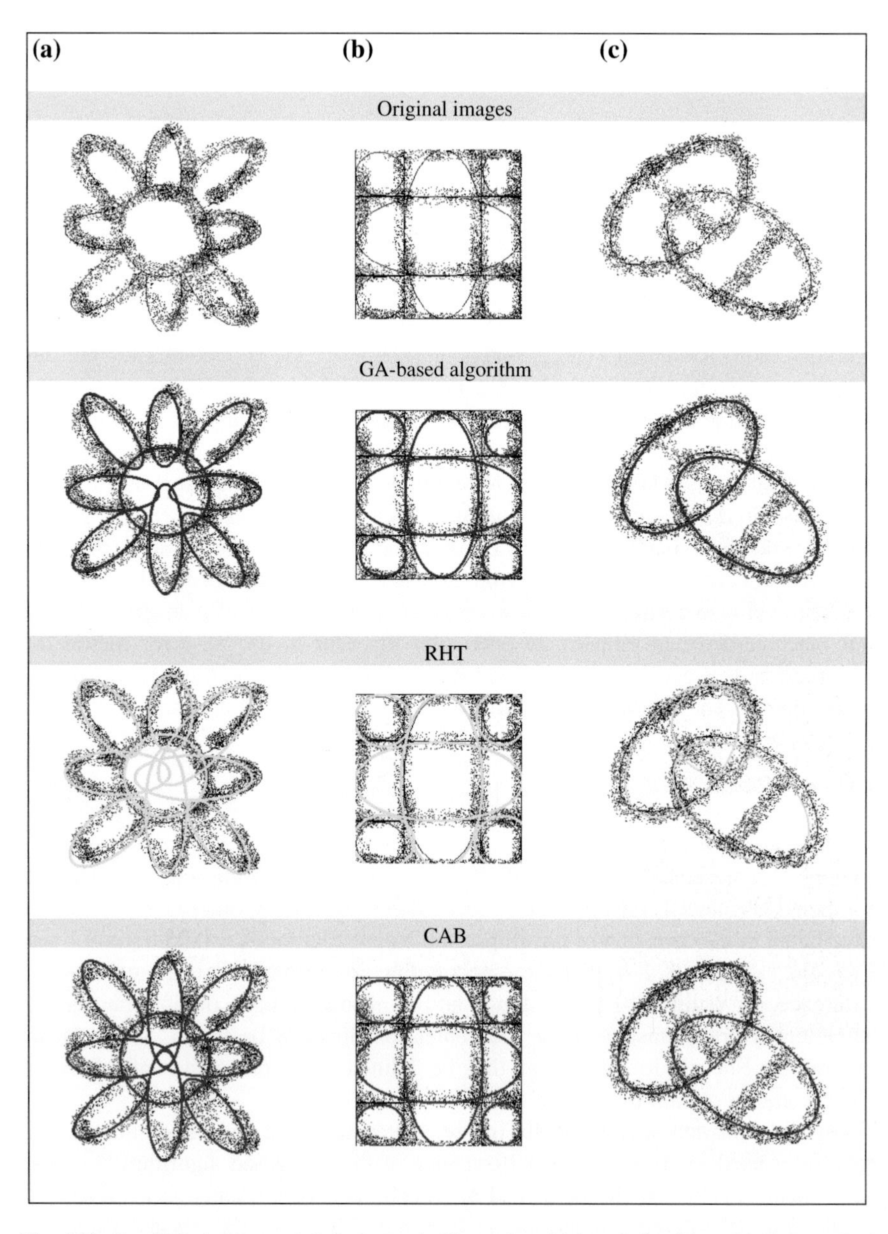

Fig. 4.11 Synthetic images and their detected ellipses for: *GA*-based algorithm, the *RHT* method and the presented *CAB* algorithm

Figures 4.11 and 4.12 also exhibit the resulting images after applying the correspondent detector over each image. Such results present the median cases obtained throughout 35 runs.

Fig. 4.12 Real-life images and their detected ellipses for: *GA*-based algorithm, the *RHT* method and the presented *CAB* algorithm

In order to statistically analyze the results in Table 4.2, a non-parametric significance proof known as the Wilcoxon's rank test [35–37] has been conducted. Such proof allows assessing result differences among two related methods. The analysis is performed considering a 5 % significance level over multiple error (ME) data. Table 4.3 reports the p-values produced by Wilcoxon's test for a pair-wise comparison of the multiple error (ME), considering two groups gathered as CAB versus GA and CAB versus RHT. As a null hypothesis, it is assumed that

Table 4.2 The averaged execution-time, success rate and the averaged multiple error for the GA-based algorithm, the RHT method and the presented CAB algorithm, considering the six test images shown in Fig. 4.11

Image	Averaged execution time ± Standard deviation (s)			Success rate (SR) (%)			Averaged ME ± Standard deviation		
	GA	RHT	CAB	GA	RHT	CAB	GA	RHT	CAB
(a)	4.32 ± (0.81)	2.82 ± (0.77)	**0.51 ± (0.31)**	94	**90**	**100**	0.78 ± (0.023)	0.98 ± (0.076)	**0.31 ± (0.052)**
(b)	6.83 ± (1.23)	3.75 ± (0.92)	**0.62 ± (0.63)**	97	88	**100**	0.85 ± (0.053)	0.93 ± (0.087)	**0.19 ± (0.027)**
(c)	5.28 ± (0.11)	3.18 ± (0.89)	**0.57 ± (0.68)**	**100**	91	**100**	0.89 ± (0.096)	0.92 ± (0.051)	**0.26 ± (0.018)**
(d)	7.08 ± (1.27)	3.82 ± (0.67)	**0.60 ± (0.42)**	79	**70**	**100**	0.84 ± (0.039)	0.89 ± (0.092)	**0.19 ± (0.061)**
(e)	6.83 ± (1.77)	3.55 ± (0.54)	**0.57 ± (0.36)**	**100**	89	**100**	0.92 ± (0.058)	0.98 ± (0.023)	**0.29 ± (0.061)**
(f)	6.77 ± (1.23)	3.88 ± (0.78)	**0.54 ± (0.51)**	93	90	**100**	0.80 ± (0.092)	0.88 ± (0.062)	**0.28 ± (0.071)**

Table 4.3 p-values produced by Wilcoxon's test comparing CAB to GA and RHT over the averaged ME from Table 4.2

Image	p-Value	
	CAB versus GA	CAB versus RHT
Synthetic images		
(a)	1.7401e-004	1.7245e-004
(b)	1.4048e-004	1.6287e-004
(c)	1.8261e-004	1.4287e-004
Natural images		
(a)	1.4397e-004	1.0341e-004
(b)	1.5481e-004	1.6497e-004
(c)	1.7201e-004	1.4201e-004

there is no difference between the values of the two algorithms. The alternative hypothesis considers an existent difference between the values of both approaches. All p-values reported in the Table 4.3 are less than 0.05 (5 % significance level) which is a strong evidence against the null hypothesis, indicating that the best CAB mean values for the performance are statistically significant which has not occurred by chance.

4.6 Conclusions

This chapter discusses an effective technique for extracting multiple ellipses from an image that considers the overall detection process as a multi-modal optimization problem. In the detection, the approach employs an evolutionary algorithm based on the way in which animals behave collectively. In such algorithm, searcher agents emulate a group of animals which interact to each other using simple biological rules that are modeled as evolutionary operators. Such operators are applied to each agent considering that the complete group has a memory to store the optimal solutions (ellipses) seen so-far by applying a competition principle. The detector uses a combination of five edge points as parameters to determine ellipse candidates (possible solutions). A matching function determines if such ellipse candidates are actually present in the image. Guided by the values of such matching functions, the set of encoded candidate ellipses are evolved through the evolutionary algorithm so that the best candidates can be fitted into actual ellipses within the image. Just after the optimization process ends, an analysis over the embedded memory is executed in order to find the best obtained solution (the best ellipse) and significant local minima (remaining ellipses). The overall approach generates a fast sub-pixel detector which can effectively identify multiple ellipses in real images despite ellipsoidal objects exhibiting a significant occluded or distorted portion.

Classical Hough transform methods for ellipse detection use five edge points to cast a vote for the potential elliptical shape in the parameter space. However, they would require a huge amount of memory and longer computational time to obtain a sub-pixel resolution. Moreover, HT-based methods rarely find a precise parameter set for a given ellipse in the image [38]. In our approach, the detected ellipses hold a sub-pixel accuracy inherited directly from the ellipse equation and the MEA method.

In order to test the ellipse detection performance, both its speed and accuracy have been compared. Score functions are defined by Eqs. 4.19 and 4.20 in order to measure accuracy and effectively evaluate the mismatch between manually detected and machine-detected ellipses. We have demonstrated that the CAB method outperforms both the GA (as described in [14]) and the RHT (as described in [10]) within a statistically significant framework (Wilcoxon test).

References

1. Pietrocew, A.: Face detection in colour images using fuzzy Hough transform. Opto-Electron. Rev. **11**(3), 247–251 (2003)
2. Toennies, K., Behrens, F., Aurnhammer, M.: Feasibility of Hough-transform based iris localisation for real-time-application. In: Proceedings of the 16th International Conference on Pattern Recognition (2002)
3. Hardzeyeu, V., Klefenz, F.: On using the Hough transform for driving assistance applications. In: Proceedings of the 4th International Conference on Intelligent Computer Communication and Processing (2008)
4. Teutsch, C., Berndt, D., Trostmann, E., Weber, M.: Real-time detection of elliptic shapes for automated object recognition and object tracking. In: Proceedings of Machine Vision Applications in Industrial Inspection XIV, pp. 171–179, San Jose, CA, USA (2006)
5. Soetedjo, A., Yamada, K.: Fast and robust traffic sign detection. In: Proceedings of IEEE International Conference on Systems, Man and Cybernetics, pp. 1341–1346, Waikoloa, Hawaii (2005)
6. Ho, C.-T., Chen, L.-H.: A fast ellipse/circle detector using geometric symmetry. Pattern Recognit. **28**(1), 117–124 (1995)
7. Chen, K.C., Bouguila, N., Ziou, D.: Quantization-free parameter space reduction in ellipse detection. Expert Syst. Appl. **38**, 7622–7632 (2011)
8. Hough, P.V.C.: Method and means for recognizing complex patterns. US Patent 3069654, December 18 (1962)
9. Wang, C.M., Hwang, N.C., Tsai, Y.Y., Chang, C.H.: Ellipse sampling for montecarlo applications. Electron. Lett. **40**(1), 21–22 (2004)
10. Lu, W., Tan, J.: Detection of incomplete ellipse in images with strong noise by iterative randomized Hough transform (IRHT). Pattern Recognit. **41**, 1268–1279 (2008)
11. McLaughlin, R.A.: Randomized hough transform: improved ellipse detection with comparison. Pattern Recognit. Lett. **19**(3–4), 299–305 (1998)
12. Ayala-Ramirez, V., Garcia-Capulin, C.H., Perez-Garcia, A., Sanchez-Yanez, R.E.: Circle detection on images using genetic algorithms. Pattern Recogn. Lett. **27**, 652–657 (2006)
13. Lutton, E., Martinez, P.: A genetic algorithm for the detection of 2D geometric primitives in images. In: Proceedings of the 12th International Conference on Pattern Recognition, Jerusalem, Israel, 9–13 October 1994, vol. 1, pp. 526–528 (1994)
14. Yao, J., Kharma, N., Grogono, P.: A multi-population genetic algorithm for robust and fast ellipse detection. Pattern Anal. Appl. **8**, 149–162 (2005)
15. Cheng, H.D., Guo, Y., Zhang, Y.: A novel Hough transform based on eliminating particle swarm optimization and its applications. Pattern Recognit, **42**(9), 1959–1969 (2009)

16. Petalas, Y.G., Antonopoulos, C.G., Bountis, T.C., Vrahatis, M.N.: Detecting resonances in conservative maps using evolutionary algorithms. Phys. Lett. A **373**, 334–341 (2009)
17. Goldberg, D.E., Richardson, J.: Genetic algorithms with sharing for multimodal function optimization. In: Proceedings of the Second International Conference on Genetic Algorithms, ICGA, New Jersey (1987)
18. Kennedy, J., Eberhart, R.C.: Particle swarm optimization. In: Proceedings of the 1995 IEEE International Conference on Neural Networks, vol. 4, pp. 1942–1948 (1995)
19. Liang, J., Qin, A.K., Suganthan, P.N.: Comprehensive learning particle swarm optimizer for global optimization of multimodal functions. IEEE Trans. Evol. Comput. **10**(3), 281–295 (2006)
20. Chen, D.B., Zhao, C.X.: Particle swarm optimization with adaptive population size and its application. Appl. Soft Comput. **9**(1), 39–48 (2009)
21. Xu, Q., Lei, W., Si, J.: Predication based immune network for multimodal function optimization. Eng. Appl. Artif. Intell. **23**, 495–504 (2010)
22. Sumper, D.: The principles of collective animal behaviour. Philos. Trans. R Soc. Lond. B Biol. Sci. **361**(1465), 5–22 (2006)
23. Petit, O., Bon, R.: Decision-making processes: the case of collective movements. Behav. Process. **84**, 635–647 (2010)
24. Kolpas, A., Moehlis, J., Frewen, T., Kevrekidis, I.: Coarse analysis of collective motion with different communication mechanisms. Math. Biosci. **214**, 49–57 (2008)
25. Couzin, I.: Collective cognition in animal groups. Trends Cognit. Sci. **13**(1), 36–43 (2008)
26. Couzin, I.D., Krause, J.: Self-organization and collective behavior in vertebrates. Adv. Stud. Behav. **32**, 1–75 (2003)
27. Bode, N., Franks, D., Wood, A.: Making noise: emergent stochasticity in collective motion. J. Theor. Biol. **267**, 292–299 (2010)
28. Couzi, I., Krause, I., James, R., Ruxton, G., Franks, N.: Collective memory and spatial sorting in animal groups. J. Theor. Biol. **218**, 1–11 (2002)
29. Couzin, I.D.: Collective minds. Nature **445**, 715–728 (2007)
30. Bazazi, S., Buhl, J., Hale, J.J., Anstey, M.L., Sword, G.A., Simpson, S.J., Couzin, I.D.: Collective motion and cannibalism in locust migratory bands. Curr. Biol. **18**, 735–739 (2008)
31. Hsu, Y., Earley, R., Wolf, L.: Modulation of aggressive behaviour by fighting experience: mechanisms and contest outcomes. Biol. Rev. **81**(1), 33–74 (2006)
32. Ballerini, M.: Interaction ruling collective animal behavior depends on topological rather than metric distance: evidence from a field study. Proc. Natl. Acad. Sci. U.S.A. **105**, 1232–1237 (2008)
33. Bresenham, J.E.: A linear algorithm for incremental digital display of circular arcs. Commun. ACM **20**, 100–106 (1987)
34. Van Aken, J.R.: Efficient ellipse-drawing algorithm. IEEE Comput. Graph. Appl. **4**(9), 24–35 (2005)
35. Wilcoxon, F.: Individual comparisons by ranking methods. Biometrics **1**, 80–83 (1945)
36. Garcia, S., Molina, D., Lozano, M., Herrera, F.: A study on the use of non-parametric tests for analyzing the evolutionary algorithms' behaviour: a case study on the CEC'2005 Special session on real parameter optimization. J Heurist. (2008). doi:10.1007/s10732-008-9080-4
37. Santamaría, J., Cordón, O., Damas, S., García-Torres, J.M., Quirin, A.: Performance evaluation of memetic approaches in 3D reconstruction of forensic objects. Soft Comput. doi:10.1007/s00500-008-0351-7, in press (2008)
38. Chen, T.-C., Chung, K.-L.: An efficient randomized algorithm for detecting circles. Comput. Vis. Image Underst. **83**, 172–191 (2001)

Chapter 5
Template Matching by Using the States of Matter Algorithm

Abstract Template matching (TM) plays an important role in several image processing applications such as feature tracking, object recognition, stereo matching and remote sensing. The TM approach seeks the best possible resemblance between a sub-image, known as template, and its coincident region within a source image. TM has two critical aspects: similarity measurement and search strategy. The simplest available TM method finds the best possible coincidence between the images through an exhaustive computation of the Normalized Cross-Correlation (NCC) value (similarity measurement) for all elements in the source image (search strategy). Unfortunately, the use of such approach is strongly restricted since the NCC evaluation is a computationally expensive operation. Recently, several TM algorithms that are based on evolutionary approaches have been proposed to reduce the number of NCC operations by calculating only a subset of search locations. In this chapter, an algorithm based on the States of Matter phenomenon is presented to reduce the number of search locations in the TM process. In the approach, individuals emulate molecules that experiment state transitions which represent different exploration–exploitation levels. In the algorithm, the computation of search locations is drastically reduced by incorporating a fitness calculation strategy which indicates when it is feasible to calculate or to only estimate the NCC value for new search locations. Conducted simulations show that the resultant method achieves the best balance in comparison to other TM algorithms considering the estimation accuracy and the computational cost.

5.1 Introduction

Template matching (TM) is employed to measure the degree of similarity between two image sets that are superimposed, one over the other. TM is one of the most important and challenging subjects in digital photogrammetry, object recognition, stereo matching, feature tracking, remote sensing and computer vision [1]. It relies on calculating the degree of similarity between the image under examination and the template image for each position. Thus, the best matching is obtained when the

E. Cuevas et al., *Applications of Evolutionary Computation in Image Processing and Pattern Recognition*, Intelligent Systems Reference Library 100,
DOI 10.1007/978-3-319-26462-2_5

similarity value is maximized. Generally, template matching involves two critical aspects: similarity measurement and search strategy [2]. The most used matching criterion is the Normalized cross-correlation (NCC) which is computationally expensive and represents the most consuming operation in the TM process [3].

The full search algorithm [4–6] is the simplest TM algorithm that can deliver the optimal detection with respect to a maximal NCC coefficient as it checks all pixel-candidates one at a time. However, such exhaustive search and the NCC calculation at each checking point, yields an extremely computational expensive TM method that seriously constraints its use for several image processing applications.

Recently, several TM algorithms, based on evolutionary approaches, have been proposed to reduce the number of NCC operations by calculating only a subset of search locations. Such approaches have produced several robust detectors using different optimization methods such as Genetic algorithms (GA) [7], Particle Swarm Optimization (PSO) [8, 9] and Imperialist competitive algorithm (ICA) [10]. Although these algorithms allow reducing the number of search locations, they do not explore the whole region effectively and often suffers premature convergence which conducts to sub-optimal detections. The reason of these problems is the operators used for modifying the particles. In such algorithms, during their evolution, the position of each agent in the next iteration is updated yielding an attraction towards the position of the best particle seen so-far [11, 12]. This behavior produces that the entire population, as the algorithm evolves, concentrates around the best particle, favoring the premature convergence and damaging the particle diversity.

Every evolutionary algorithm (EA) needs to address the issue of exploration-exploitation of the search space. Exploration is the process of visiting entirely new points of a search space whilst exploitation is the process of refining those points within the neighborhood of previously visited locations, in order to improve their solution quality. Pure exploration degrades the precision of the evolutionary process but increases its capacity to find new potential solutions. On the other hand, pure exploitation allows refining existent solutions but adversely driving the process to local optimal solutions. Therefore, the ability of an EA to find a global optimal solution depends on its capacity to find a good balance between the exploitation of found-so-far elements and the exploration of the search space [13]. So far, the exploration–exploitation dilemma has been an unsolved issue within the framework of EA.

In this chapter, a novel nature-inspired algorithm, called the States of Matter Search (SMS) is presented for solving the TM problem. The SMS algorithm is based on the simulation of the states of matter phenomenon. In SMS, individuals emulate molecules which interact to each other by using evolutionary operations based on the physical principles of the thermal-energy motion mechanism. Such operations allow the increase of the population diversity and avoid the concentration of particles within a local minimum. The presented approach combines the use of the defined operators with a control strategy that modifies the parameter setting of each operation during the evolution process. The algorithm is devised by considering each state of matter at one different exploration–exploitation rate. Thus, the evolutionary process is divided into three stages which emulate the three states of matter: gas, liquid and solid. At each state, molecules (individuals) exhibit different behaviors. Beginning from the gas state (pure exploration), the algorithm modifies the intensities of exploration and exploitation until the solid state (pure exploitation) is reached. As a result, the approach can substantially improve the balance between exploration–exploitation, yet preserving the good search capabilities of an evolutionary approach.

However, one particular difficulty in applying any EA to real-world problems is about its demand for a large number of fitness evaluations before delivering a satisfying result. Fitness evaluations are not always straightforward in many applications as either an explicit fitness function does not exist or the fitness evaluation is computationally expensive. Furthermore, since random numbers are involved in the calculation of new individuals, they may encounter same positions (repetition) that have been visited by other individuals at previous iterations, particularly when individuals are confined to a finite area.

The problem of considering expensive fitness evaluations has already been faced in the field of evolutionary algorithms (EA) and is better known as fitness approximation [14]. In such approach, the idea is to estimate the fitness value of so many individuals as it is possible instead of evaluating the complete set. Such estimations are based on an approximate model of the fitness landscape. Thus, the individuals to be evaluated and those to be estimated are determined following some fixed criteria which depend on the specific properties of the approximate model [15]. The models involved at the estimation can be built during the actual EA run, since EA repeatedly samples the search space at different points [16]. There are many possible approximation models which have been used in combination with EA (e.g. polynomials [17], the kriging model [18], the feed-forward neural networks that includes multi-layer Perceptrons [19] and radial basis-function networks [20]).

In this chapter, an algorithm based on SMS is presented to reduce the number of search locations in the TM process. The algorithm uses a simple fitness calculation approach which is based on the Nearest Neighbor Interpolation (NNI) algorithm in order to estimate the fitness value (NCC operation) for several candidate solutions (search locations). As a result, the approach can not only substantially reduce the number search positions (by using the SMS approach), but also to avoid the NCC evaluation for many of them (by incorporating the NNI strategy). The resultant method achieves the best balance over other TM algorithms, in terms of both estimation accuracy and computational cost.

5.2 States of Matter

The matter can take different phases which are commonly known as states. Traditionally, three states of matter are known: solid, liquid, and gas. The differences among such states are based on forces which are exerted among particles composing a material [21].

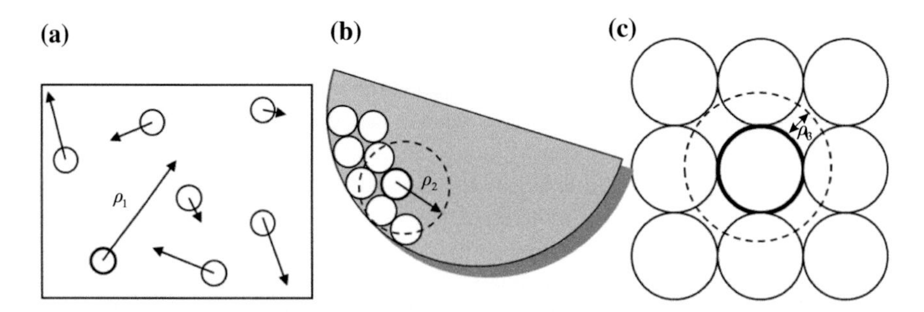

Fig. 5.1 Different states of matter: **a** gas, **b** liquid, and **c** solid

In the gas phase, molecules present enough kinetic energy so that the effect of intermolecular forces is small (or zero for an ideal gas), while the typical distance between neighboring molecules is greater than the molecular size. A gas has no definite shape or volume, but occupies the entire container in which it is confined. Figure 5.1a shows the movements exerted by particles in a gas state. The movement experimented by the molecules represent the maximum permissible displacement ρ_1 among particles [22]. In a liquid state, intermolecular forces are more restrictive than those in the gas state. The molecules have enough energy to move relatively to each other still keeping a mobile structure. Therefore, the shape of a liquid is not definite but is determined by its container. Figure 5.1b presents a particle movement ρ_2 within a liquid state. Such movement is smaller than those considered by the gas state but larger than the solid state. In the solid state, particles (or molecules) are packed together closely with forces among particles being strong enough so that the particles cannot move freely but only vibrate. As a result, a solid has a stable, definite shape and a definite volume. Solids can only change their shape by force, as when they are broken or cut. Figure 5.1c shows a molecule configuration in a solid state. Under such conditions, particles are able to vibrate (being perturbed) considering a minimal ρ_3 distance [22].

In this chapter, a novel nature-inspired algorithm known as the States of Matter Search (SMS) is presented for solving global optimization problems. The SMS algorithm is based on the simulation of the states of matter phenomenon that considers individuals as molecules which interact to each other by using evolutionary operations based on the physical principles of the thermal-energy motion mechanism. The algorithm is devised by considering each state of matter at one different exploration–exploitation ratio. Thus, the evolutionary process is divided into three stages which emulate the three states of matter: gas, liquid and solid. In each state, individuals exhibit different behaviors.

5.3 States of Matter Search (SMS)

5.3.1 *Definition of Operators*

In the approach, individuals are considered as molecules whose positions on a multidimensional space are modified as the algorithm evolves. The movement of such molecules is motivated by the analogy to the motion of thermal-energy.

The velocity and direction of each molecule's movement are determined by considering the collision, the attraction forces and the random phenomena experimented by the molecule set [23]. In our approach, such behaviors have been implemented by defining several operators such as the direction vector, the collision and the random positions operators, all of which emulate the behavior of actual physics laws.

The direction vector operator assigns a direction to each molecule in order to lead the particle movement as the evolution process takes place. On the other side, the collision operator mimics those collisions that are experimented by molecules as they interact to each other. A collision is considered when the distance between two molecules is shorter than a determined proximity distance. The collision operator is thus implemented by interchanging directions of the involved molecules. In order to simulate the random behavior of molecules, the presented algorithm generates random positions following a probabilistic criterion that considers random locations within a feasible search space. The next section presents all operators that are used in the algorithm. Although such operators are the same for all the states of matter, they are employed over a different configuration set depending on the particular state under consideration.

5.3.1.1 Direction Vector

The direction vector operator mimics the way in which molecules change their positions as the evolution process develops. For each n-dimensional molecule $\mathbf{p}_i$ from the population $\mathbf{P}$, it is assigned an n-dimensional direction vector $\mathbf{d}_i$ which stores the vector that controls the particle movement. Initially, all the direction vectors ($\mathbf{D} = \{\mathbf{d}_1, \mathbf{d}_2, \ldots, \mathbf{d}_{N_p}\}$) are randomly chosen within the range of $[-1, 1]$.

As the system evolves, molecules experiment several attraction forces. In order to simulate such forces, the presented algorithm implements the attraction phenomenon by moving each molecule towards the best so-far particle. Therefore, the new direction vector for each molecule is iteratively computed considering the following model:

$$\mathbf{d}_i^{k+1} = \mathbf{d}_i^k \cdot \left(1 - \frac{k}{gen}\right) \cdot 0.5 + \mathbf{a}_i, \qquad (5.1)$$

where $\mathbf{a}_i$ represents the attraction unitary vector calculated as $\mathbf{a}_i = (\mathbf{p}^{best} - \mathbf{p}_i)/\left\|\mathbf{p}^{best} - \mathbf{p}_i\right\|$, being $\mathbf{p}^{best}$ the best individual seen so-far, while $\mathbf{p}_i$ is the molecule i of population $\mathbf{P}$. k represents the iteration number whereas gen involves the total iteration number that constitutes the complete evolution process.

Under this operation, each particle is moved towards a new direction which combines the past direction, which was initially computed, with the attraction vector over the best individual seen so-far. It is important to point out that the relative importance of the past direction decreases as the evolving process advances. This particular type of interaction avoids the quick concentration of information among particles and encourages each particle to search around a local candidate region in its neighborhood, rather than interacting to a particle lying at distant region of the domain. The use of this scheme has two advantages: first, it prevents the particles from moving toward the global best position in early stages of algorithm and thus makes the algorithm less susceptible to premature convergence; second, it encourages particles to explore their own neighborhood thoroughly, just before they converge towards a global best position. Therefore, it provides the algorithm with local search ability enhancing the exploitative behavior.

In order to calculate the new molecule position, it is necessary to compute the velocity $\mathbf{v}_i$ of each molecule by using:

$$\mathbf{v}_i = \mathbf{d}_i \cdot v_{init} \tag{5.2}$$

being v_{init} the initial velocity magnitude which is calculated as follows:

$$v_{init} = \frac{\sum_{j=1}^{n} (b_j^{high} - b_j^{low})}{n} \cdot \beta \tag{5.3}$$

where b_j^{low} and b_j^{high} are the low j parameter bound and the upper j parameter bound respectively, whereas $\beta \in [0, 1]$.

Then, the new position for each molecule is updated by:

$$p_{i,j}^{k+1} = p_{i,j}^{k} + v_{i,j} \cdot \mathrm{rand}(0, 1) \cdot \rho \cdot (b_j^{high} - b_j^{low}) \tag{5.4}$$

where $0.5 \leq \rho \leq 1$.

5.3.1.2 Collision

The collision operator mimics the collisions experimented by molecules while they interact to each other. Collisions are calculated if the distance between two molecules is shorter than a determined proximity value. Therefore, if $\left\|\mathbf{p}_i - \mathbf{p}_q\right\| < r$, a collision between molecules i and q is assumed; otherwise, there is no collision, considering $i, q \in \{1, \ldots, N_p\}$ such that $i \neq q$. If a collision occurs, the direction vector for each particle is modified by interchanging their respective direction vectors as follows:

$$\mathbf{d}_i = \mathbf{d}_q \quad \text{and} \quad \mathbf{d}_q = \mathbf{d}_i \tag{5.5}$$

The collision radius is calculated by:

$$r = \frac{\sum_{j=1}^{n} (b_j^{high} - b_j^{low})}{n} \cdot \alpha \tag{5.6}$$

where $\alpha \in [0, 1]$.

Under this operator, a spatial region enclosed within the radius r is assigned to each particle. In case the particle regions collide to each other, the collision operator acts upon particles by forcing them out of the region. The radio r and the collision operator provide the ability to control diversity throughout the search process. In other words, the rate of increase or decrease of diversity is predetermined for each stage. Unlike other diversity-guided algorithms, it is not necessary to inject diversity into the population when particles gather around a local optimum because the diversity will be preserved during the overall search process. The collision incorporation therefore enhances the exploratory behavior in the presented approach.

5.3.1.3 Random Positions

In order to simulate the random behavior of molecules, the presented algorithm generates random positions following a probabilistic criterion within a feasible search space. For this operation, a uniform random number r_m is generated within the range [0, 1]. If r_m is smaller than a threshold H, a random molecule's position is generated; otherwise, the element remains with no change. Therefore such operation can be modeled as follows:

$$p_{i,j}^{k+1} = \begin{cases} b_j^{low} + \text{rand}(0,1) \cdot (b_j^{high} - b_j^{low}) & \text{with probability } H \\ p_{i,j}^{k+1} & \text{with probability } (1 - H) \end{cases} \tag{5.7}$$

where $i \in \{1, \ldots, N_p\}$ and $j \in \{1, \ldots, n\}$.

5.3.1.4 Best Element Updating

Despite this updating operator does not belong to State of Matter metaphor, it is used to simply store the best so-far solution. In order to update the best molecule $\mathbf{p}^{best}$ seen so-far, the best found individual from the current k population $\mathbf{p}^{best,k}$ is compared to the best individual $\mathbf{p}^{best,k-1}$ of the last generation. If $\mathbf{p}^{best,k}$ is better than $\mathbf{p}^{best,k-1}$ according to its fitness value, $\mathbf{p}^{best}$ is updated with $\mathbf{p}^{best,k}$, otherwise $\mathbf{p}^{best}$ remains with no change. Therefore, $\mathbf{p}^{best}$ stores the best historical individual found so-far.

5.3.2 *SMS Algorithm*

The overall SMS algorithm is composed of three stages corresponding to the three States of Matter: the gas, the liquid and the solid state. Each stage has its own behavior. In the first stage (gas state), exploration is intensified whereas in the second one (liquid state) a mild transition between exploration and exploitation is executed. Finally, in the third phase (solid state), solutions are refined by emphasizing the exploitation process.

5.3.2.1 General Procedure

At each stage, the same operations are implemented. However, depending on which state is referred, they are employed considering a different parameter configuration. The general procedure in each state is shown as pseudo-code in Algorithm 1. Such procedure is composed by five steps and maps the current population $\mathbf{P}^k$ to a new population $\mathbf{P}^{k+1}$. The algorithm receives as input the current population $\mathbf{P}^k$ and the configuration parameters ρ, β, α, and H, whereas it yields the new population $\mathbf{P}^{k+1}$.

5.3.2.2 The Complete Algorithm

The complete algorithm is divided into four different parts. The first corresponds to the initialization stage, whereas the last three represent the States of Matter. All the optimization process, which consists of a *gen* number of iterations, is organized into three different asymmetric phases, employing 50 % of all iterations for the gas state (exploration), 40 % for the liquid state (exploration-exploitation) and 10 % for the solid state (exploitation). The overall process is graphically described by Fig. 5.2. At each state, the same general procedure (see Algorithm 5.1) is iteratively used considering the particular configuration predefined for each State of Matter. Figure 5.3 shows the data flow for the complete SMS algorithm.

Algorithm 5.1 General procedure executed by all the states of matter.

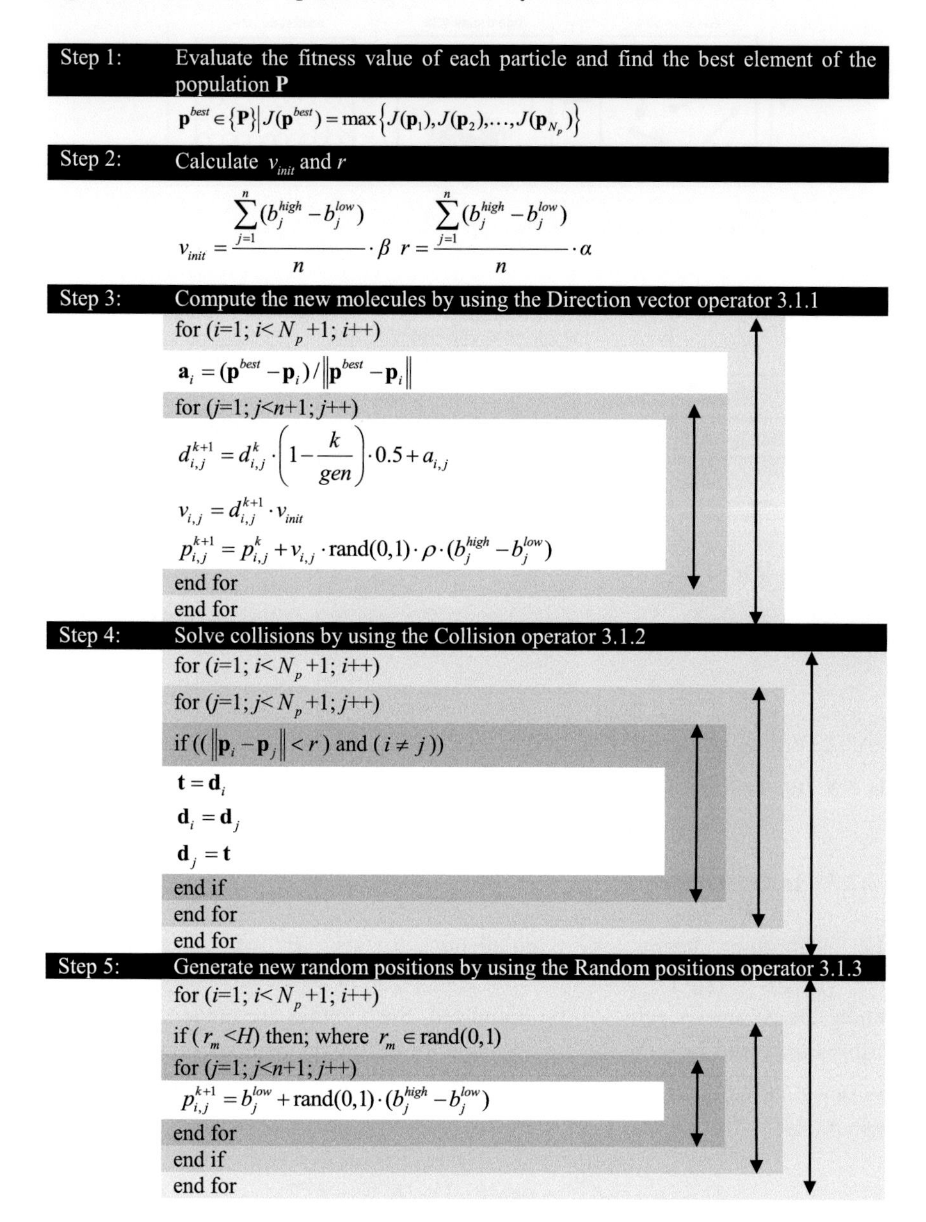

Step 1: Evaluate the fitness value of each particle and find the best element of the population **P**

$$\mathbf{p}^{best} \in \{\mathbf{P}\} \,|\, J(\mathbf{p}^{best}) = \max\{J(\mathbf{p}_1), J(\mathbf{p}_2), \ldots, J(\mathbf{p}_{N_p})\}$$

Step 2: Calculate v_{init} and r

$$v_{init} = \frac{\sum_{j=1}^{n}(b_j^{high} - b_j^{low})}{n} \cdot \beta \quad r = \frac{\sum_{j=1}^{n}(b_j^{high} - b_j^{low})}{n} \cdot \alpha$$

Step 3: Compute the new molecules by using the Direction vector operator 3.1.1

for (i=1; i< N_p+1; i++)

$\mathbf{a}_i = (\mathbf{p}^{best} - \mathbf{p}_i)/\|\mathbf{p}^{best} - \mathbf{p}_i\|$

for (j=1; j<n+1; j++)

$d_{i,j}^{k+1} = d_{i,j}^{k} \cdot \left(1 - \frac{k}{gen}\right) \cdot 0.5 + a_{i,j}$

$v_{i,j} = d_{i,j}^{k+1} \cdot v_{init}$

$p_{i,j}^{k+1} = p_{i,j}^{k} + v_{i,j} \cdot \text{rand}(0,1) \cdot \rho \cdot (b_j^{high} - b_j^{low})$

end for

end for

Step 4: Solve collisions by using the Collision operator 3.1.2

for (i=1; i< N_p+1; i++)

for (j=1; j< N_p+1; j++)

if (($\|\mathbf{p}_i - \mathbf{p}_j\| < r$) and ($i \neq j$))

$\mathbf{t} = \mathbf{d}_i$

$\mathbf{d}_i = \mathbf{d}_j$

$\mathbf{d}_j = \mathbf{t}$

end if

end for

end for

Step 5: Generate new random positions by using the Random positions operator 3.1.3

for (i=1; i< N_p+1; i++)

if (r_m <H) then; where $r_m \in \text{rand}(0,1)$

for (j=1; j<n+1; j++)

$p_{i,j}^{k+1} = b_j^{low} + \text{rand}(0,1) \cdot (b_j^{high} - b_j^{low})$

end for

end if

end for

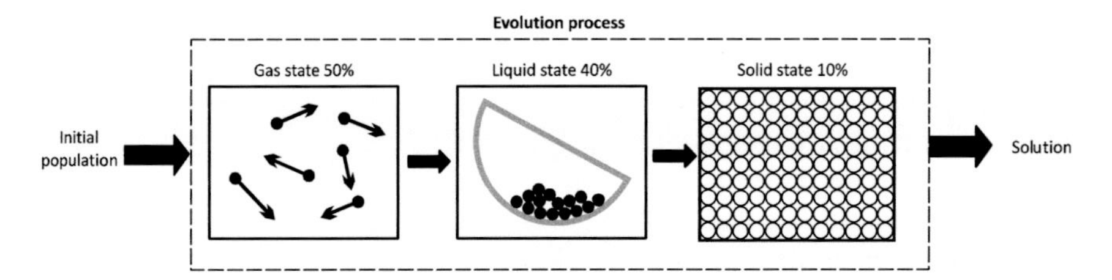

Fig. 5.2 Evolution processing the presented approach

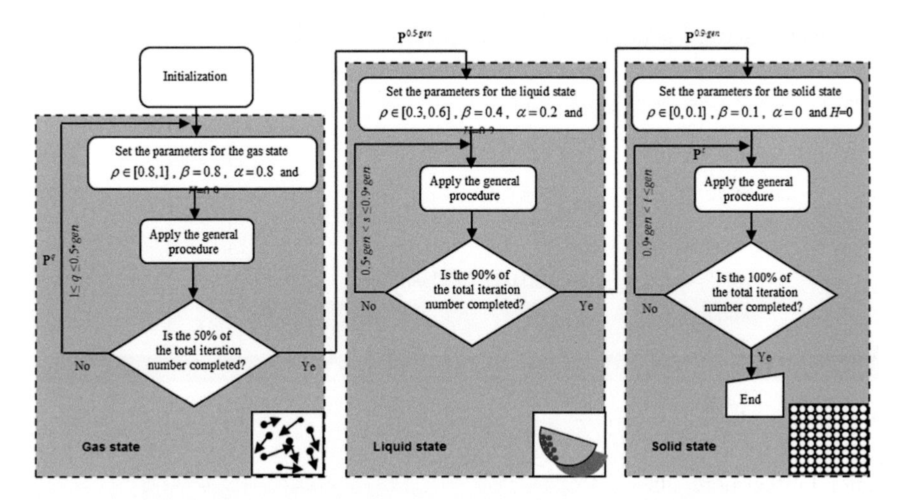

Fig. 5.3 Data flow in the complete SMS algorithm

5.3.2.3 Initialization

The algorithm begins by initializing a set $\mathbf{P}$ of N_p molecules ($\mathbf{P} = \{\mathbf{p}_1, \mathbf{p}_2, \ldots, \mathbf{p}_{N_p}\}$). Each molecule position $\mathbf{p}_i$ is an-dimensional vector containing the parameter values to be optimized. Such values are randomly and uniformly distributed between the pre-specified lower initial parameter bound b_j^{low} and the upper initial parameter bound b_j^{high}, just as it is described by the following expressions:

$$p_{i,j}^0 = b_j^{low} + \text{rand}(0,1) \cdot (b_j^{high} - b_j^{low}) \qquad (5.8)$$
$$j = 1, 2, \ldots, n; \quad i = 1, 2, \ldots, N_p,$$

where j and i, are the parameter and molecule index respectively whereas zero indicates the initial population. Hence, p_i^j is the j-th parameter of the i-th molecule.

5.3.2.4 Gas State

In the gas state, molecules experiment severe displacements and collisions. Such state is characterized by random movements produced by non-modeled molecule phenomena [23]. Therefore, the ρ value from the direction vector operator is set to a value near to one so that the molecules can travel longer distances. Similarly, the H value representing the random positions operator is also configured to a value around one, in order to allow the random generation for other molecule positions. The gas state is the first phase and lasts for the 50 % of all iterations which compose the complete optimization process. The computational procedure for the gas state can be summarized as follows:

Step 1	Set the parameters $\rho \in [0.8, 1]$, $\beta = 0.8$, $\alpha = 0.8$ and $H = 0.9$ being consistent with the gas state
Step 2	Apply the general procedure which is illustrated in Algorithm 5.1
Step 3	If the 50 % of the total iteration number is completed $(1 \leq k \leq 0.5 \cdot gen)$, then the process continues to the liquid state procedure; otherwise go back to step 2

5.3.2.5 Liquid State

Although molecules currently at the liquid state exhibit restricted motion in comparison to the gas state, they still show a higher flexibility with respect to the solid state. Furthermore, the generation of random positions which are produced by non-modeled molecule phenomena is scarce [24]. For this reason, the ρ value from the direction vector operator is bounded to a value between 0.3 and 0.6. Similarly, the random position operator H is configured to a value near to cero in order to allow the random generation of fewer molecule positions. In the liquid state, collisions are also less common than in gas state, so the collision radius, that is controlled by α, is set to a smaller value in comparison to the gas state. The liquid state is the second phase and lasts the 40 % of all iterations which compose the complete optimization process. The computational procedure for the liquid state can be summarized as follows:

Step 4	Set the parameters $\rho \in [0.3, 0.6]$, $\beta = 0.4$, $\alpha = 0.2$ and $H = 0.2$ being consistent with the liquid state
Step 5	Apply the general procedure that is defined in Algorithm 5.1
Step 6	If the 90 % (50 % from the gas state and 40 % from the liquid state) of the total iteration number is completed $(0.5 \cdot gen < k \leq 0.9 \cdot gen)$, then the process continues to the solid state procedure; otherwise go back to step 5

5.3.2.6 Solid State

In the solid state, forces among particles are stronger so that particles cannot move freely but only vibrate. As a result, effects such as collision and generation of random positions are not considered [25]. Therefore, the ρ value of the direction vector operator is set to a value near to zero indicating that the molecules can only vibrate around their original positions. The solid state is the third phase and lasts for the 10 % of all iterations which compose the complete optimization process. The computational procedure for the solid state can be summarized as follows:

Step 7	Set the parameters $\rho \in [0.0, 0.1]$ and $\beta = 0.1$, $\alpha = 0$ and $H = 0$ being consistent with the solid state
Step 8	Apply the general procedure that is defined in Algorithm 5.1
Step 9	If the 100 % of the total iteration number is completed $(0.9 \cdot gen < k \leq gen)$, the process is finished; otherwise go back to step 8

It is important to clarify that the use of this particular configuration ($\alpha = 0$ and $H = 0$) disables the collision and generation of random positions operators which have been illustrated in the general procedure.

5.4 Fitness Approximation Method

Evolutionary methods based on fitness approximation aim to find the global optimum of a given function considering only a very few number of function evaluations. In order to apply such approach, it is necessary that the objective function portrait the following conditions: [18]: (1) it must be very costly to evaluate and (2) must have few dimensions (up to five). Recently, several fitness estimators have been reported in the literature [14–17], where the function evaluation number is considerably reduced (to hundreds, dozens, or even less). However, most of these methods produce complex algorithms whose performance is conditioned to the quality of the training phase and the learning algorithm in the construction of the approximation model.

In this chapter, we explore the use of a local approximation scheme, based on the nearest-neighbor-interpolation (NNI), in order to reduce the function evaluation number. The model estimates the fitness values based on previously evaluated neighboring individuals, stored during the evolution process. In each generation, some individuals of the population are evaluated with the accurate (real) objective function, while the remaining individuals' fitnesses are estimated. The individuals to be evaluated accurately are determined based on their proximity to the best fitness value or uncertainty.

5.4.1 Updating Individual Database

In our fitness calculation approach, during de evolution process, every evaluation or estimation of an individual produces a data point (individual position and fitness value) that is potentially taken into account for building the approximation model. Therefore, we keep all seen so far evaluations in a history array **T**, and then just select the closest neighbor to estimate the fitness value of a new individual. Thus, all data are preserved and potentially available for use, while the construction of the model is still fast since only the most relevant data points are actually used to construct the model.

5.4.2 Fitness Calculation Strategy

In the presented fitness calculation scheme, most of the fitness values are estimated to reduce the calculation time in each generation. In the model, it is evaluated (using the real fitness function) those individuals that are near the individual with the best fitness value contained in **T** (rule 1). Such individuals are important, since they will have a stronger influence on the evolution process than other individuals. Moreover, it is also evaluated those individuals in regions of the search space with few previous evaluations (rule 2). The fitness values of these individuals are uncertain; since there is no close reference (close points contained in **T**) in order to calculate their estimates.

The rest of the individuals are estimated using NNI (rule 3). Thus, the fitness value of an individual is estimated assigning it the same fitness value that the nearest individual stored in **T**. For the sake of clarity, it is considered that the fitness value of i is evaluated by the true fitness function using the representation $J(i)$ whereas $\tilde{J}(i)$ indicates that the fitness value of the individual i has been estimated using an alternative model.

Therefore, the estimation model follows 3 different rules in order to evaluate or estimate the fitness values:

1. If the new individual (search position) P is located closer than a distance d with respect to the nearest individual location L_q whose fitness value F_{L_q} corresponds to the best fitness value stored in **T**, then the fitness value of P is evaluated using the true fitness function ($J(P)$). Figure 5.4a draws the rule procedure.
2. If the new individual P is located longer than a distance d with respect to the nearest individual location L_q whose fitness value F_{L_q} has been already stored in **T**, then its fitness value is evaluated using the true fitness function ($J(P)$). Figure 5.4b outlines the rule procedure.
3. If the new individual P is located closer than a distance d with respect to the nearest individual location L_q whose fitness value F_{L_q} has been already stored in **T**, then its fitness value is estimated $(\tilde{J}(P))$ assigning it the same fitness that $L_q(F_P = F_{L_q})$. Figure 5.4c sketches the rule procedure.

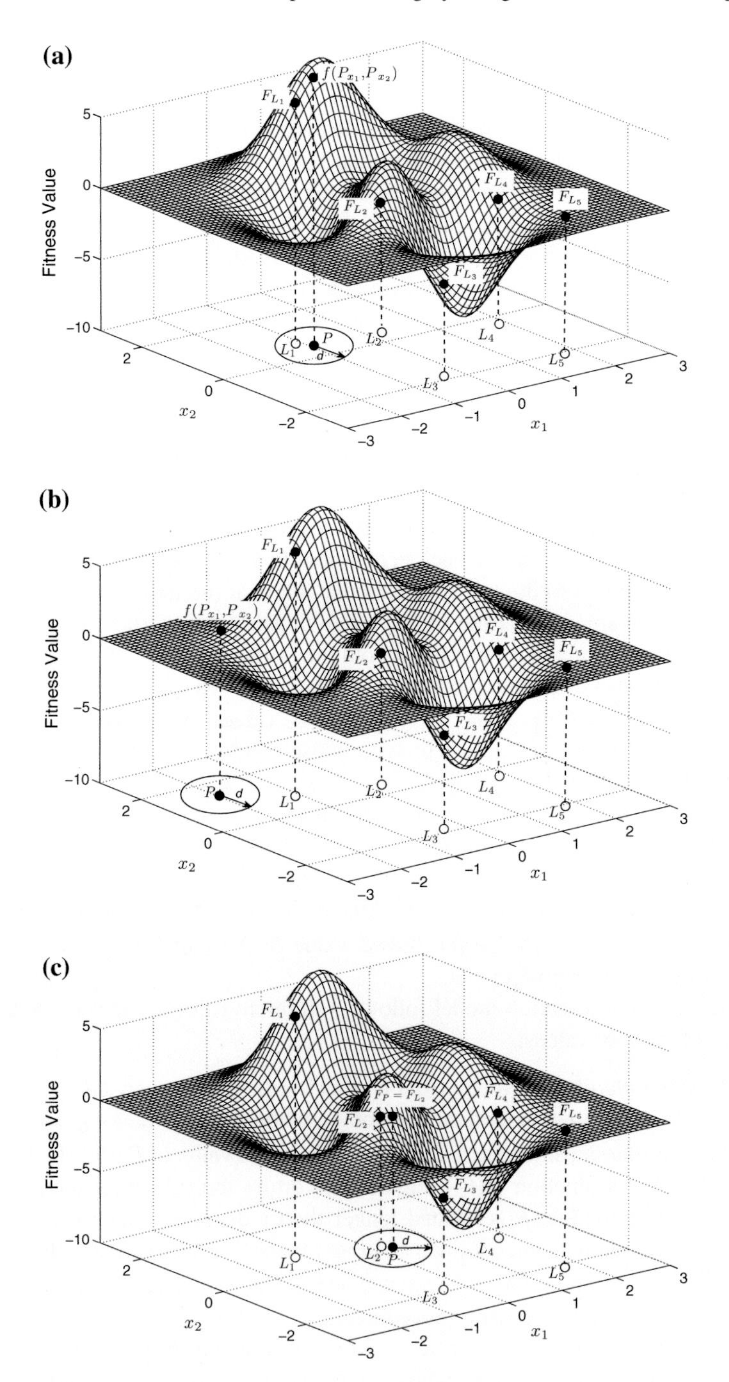

Fig. 5.4 The fitness calculation strategy. **a** According to the rule 1, the individual (search position) P is evaluated ($J(P)$), since it is located closer than a distance d with respect to the nearest individual location L_1 whose fitness value F_{L_1} corresponds to the best fitness value (maximum so-far). **b** According to the rule 2, the search point P is evaluated ($J(P)$), as there is no close reference in its neighborhood. **c** According to the rule 3, the fitness value of P is estimated $(\tilde{J}(P))$ by means of the NNI-estimator, assigning $F_P = F_{L_2}$

The d value controls the tradeoff between the evaluation and estimation of search locations. Typical values of d range from 5 to 10; however, in this chapter, the value of 7 has been selected. Thus, the presented approach favors the exploitation and exploration in the search process. For the exploration, the estimator evaluates the true fitness function of new search locations that have been located far from the positions already calculated. Meanwhile, it also estimates those that are closer. For the exploitation, the presented method evaluates the effective fitness function of those new searching locations that are placed near to the position with the minimum fitness value seen so far, aiming to improve its minimum.

Algorithm 5.2 Enhanced general procedure executed by all the states of matter. The procedure incorporates the fitness calculation strategy in order to reduce the number of function evaluations.

Step 1: Evaluate or estimate the fitness value of each particle and find the best element of the population $\mathbf{P}$

for (i=1; i< N_p+1; i++)

If ($\mathbf{p}_i$ fulfils rule 1 or rule 2) then $J(\mathbf{p}_i)$

If ($\mathbf{p}_i$ fulfils rule 3) then $\tilde{J}(\mathbf{p}_i)$

update $\mathbf{T}$

end for

$$\mathbf{p}^{best} \in \{\mathbf{T}\} \big| J(\mathbf{p}^{best}) = \max\left\{J(\mathbf{p}_1), \tilde{J}(\mathbf{p}_2), \ldots, J(\mathbf{p}_{N_p})\right\}$$

Step 2: Calculate v_{init} and r

$$v_{init} = \frac{\sum_{j=1}^{n}(b_j^{high} - b_j^{low})}{n} \cdot \beta \quad r = \frac{\sum_{j=1}^{n}(b_j^{high} - b_j^{low})}{n} \cdot \alpha$$

The other steps are similar to those presented in Algorithm 1.

The three rules show that the fitness calculation strategy is simple and straightforward. Figure 5.4 illustrates the procedure of fitness computation for a new solution (point P) considering the three different rules. In the problem the objective function J is maximized with respect to two parameters (x_1, x_2). In all figures (Fig. 5.4a, b, c) the individual database array $\mathbf{T}$ contains five different elements $(L_1, L_2, L_3, L_4, L_5)$ with their corresponding fitness values $(F_{L_1}, F_{L_2}, F_{L_3}, F_{L_4}, F_{L_5})$. Figure 5.4a, b show the fitness evaluation $(J(P_{x_1}, P_{x_2}))$ of the new solution P following the rule 1 and 2 respectively, whereas Fig. 5.4c present the fitness estimation of P $(\tilde{J}(P))$ using the NNI approach considered by rule 3.

5.4.3 Presented Optimization SMS Method

In this section, it has been presented a fitness calculation approach in order to accelerate the SMS algorithm. Only the fitness calculation scheme shows difference between the presented SMS and the enhanced one. In the modified SMS, only some individuals are actually evaluated (rules 1 and 2) in each generation. The fitness values of the rest are estimated using the NNI-approach (rule 3). The estimation is executed using the individual database (array **T**).

Figure 5.5 shows the difference between the original SMS and the modified one. In the Figure, it is clear that two new blocks have been added, the fitness estimation and the updating individual database. Both elements, together with the actual evaluation block, represent the fitness calculation strategy presented in this sub-section. The incorporation of the fitness calculation strategy modifies only the step 1 of the general procedure shown in Algorithm 5.1. Such step is extended by incorporating the decision rules (whether the individual i is $J(i)$ or $\tilde{J}(i)$) and the sub-system that updates the **T** array. Algorithm 5.2 illustrates the enhanced procedure. As a result, the SMS approach can substantially reduce the number of function evaluations preserving its good search capabilities.

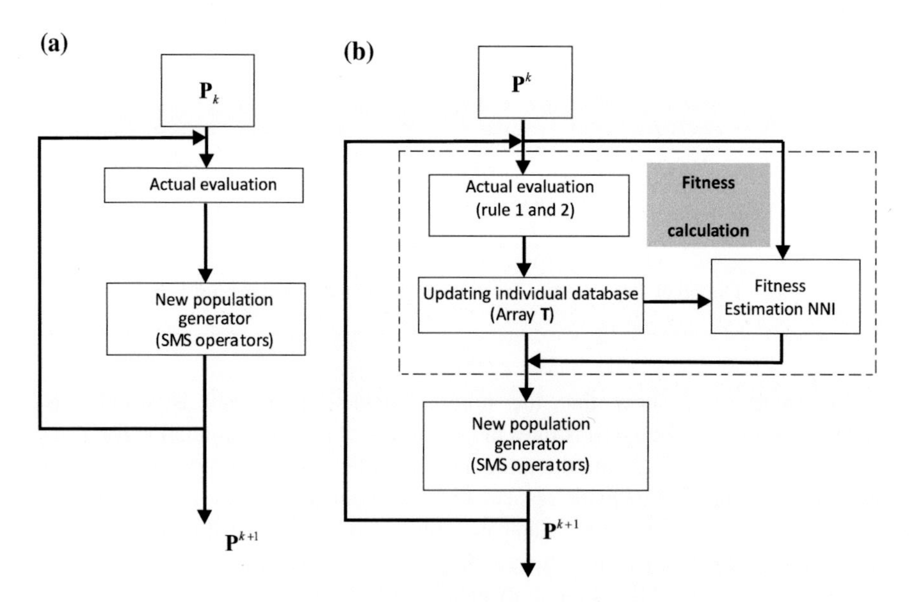

Fig. 5.5 Differences between the original SMS and the modified SMS. **a** Conventional SMS and **b** SMS algorithm included the fitness calculation strategy

5.5 Template Matching Process

Beginning from the problem of locating a given reference image (template) R over a larger intensity image I, the task is to find those positions at image I whose coincident region matches with R or at least is most similar.

If it is denoted by $R_{u,v}(x, y) = R(x - u, y - v)$, the reference image R is shifted by the distance (u, v) towards the horizontal and vertical directions, respectively. Then the matching problem that is illustrated by Fig. 5.6 can be summarized as follows: considering the source image I and the reference image R, find the offset (u, v) within the search region S such that the similarity between the shifted reference image $R_{u,v}$ and the corresponding sub-image of I is the maximum.

In order to successfully solve this task, several issues need to be addressed such as determining a minimum similarity value to validate that a match has occurred and developing a good search strategy to find, in a fast way, the optimal displacement. Several TM algorithms that use evolutionary approaches [7–10] have been proposed as a search strategy to reduce the number of search positions. In comparison to other similarity criteria, NCC is the most effective and robust method that supports the measurement of the resemblance between R and its coincident region at image I, at each displacement (u, v). The NCC value between a given image I of size $M \times N$ and a template image R of size $m \times n$, at the displacement (u, v), is given by:

$$NCC(u,v) = \frac{\sum_{i=1}^{m}\sum_{j=1}^{n}\left[I(u+i,v+j) - \bar{I}(u,v)\right] \cdot \left[R(i,j) - \bar{R}\right]}{\left[\sum_{i=1}^{m}\sum_{j=1}^{n} I(u+i,v+j) - \bar{I}(u,v)\right]^{\frac{1}{2}} \cdot \left[\sum_{i=1}^{m}\sum_{j=1}^{n} R(i,j) - \bar{R}\right]^{\frac{1}{2}}} \tag{5.9}$$

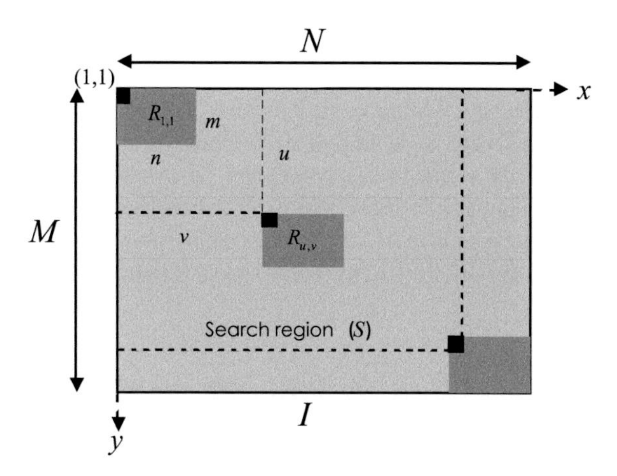

Fig. 5.6 Template matching geometry: the reference image R is shifted across the search image I by an offset (u, v) by using the origins of the two images as reference points. The dimensions of the source image (MxN) and the reference image (mxn) determine the maximal search region (S) for the comparison

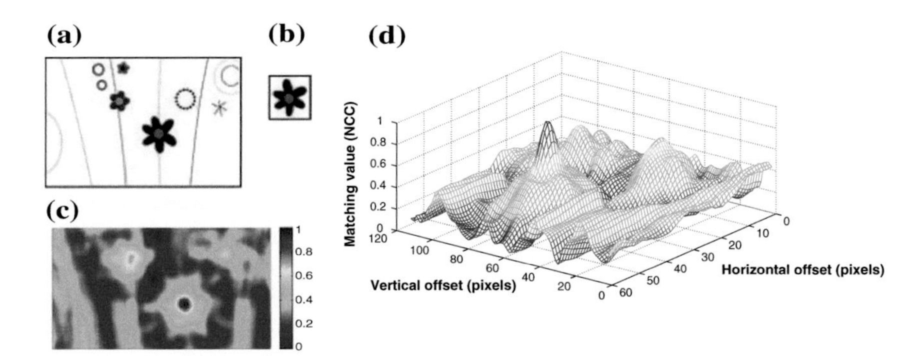

Fig. 5.7 Template matching process. **a** Example source image, **b** template image, **c** color-encoded NCC values and **d** NCC multi-modal surface

where $\bar{I}(u, v)$ is the grey-scale average intensity of the source-image for the coincident region of template image R whereas $\bar{R}$ is the grey-scale average intensity of the template image. These values are defined as follows:

$$\bar{I}(u,v) = \frac{1}{m \cdot n} \sum_{i=1}^{m} \sum_{j=1}^{n} I(u+i, v+j) \bar{R} = \frac{1}{m \cdot n} \sum_{i=1}^{m} \sum_{j=1}^{m} R(i,j) \tag{5.10}$$

The point (u, v) that presents the best possible resemblance between R and I is thus defined as follows:

$$(u, v) = \arg \max_{(\hat{u},\hat{v}) \in S} NCC(\hat{u}, \hat{v}) \tag{5.11}$$

where

$$S = \{(\hat{u}, \hat{v}) | \, 1 \leq \hat{u} \leq M - m, 1 \leq \hat{v} \leq N - n\}.$$

Figure 5.7 illustrates the TM process. Figure 5.7a, b show the source and the template image respectively. It is important to consider that the template image (5.7b) is similar but not equal to the coincident pattern that is contained in the source image (5.7a). Figure 5.7c shows the NCC values (color-encoded) calculated for all locations of the search region S. On the other hand, Fig. 5.7d presents the NCC surface which exhibits the highly multi-modal nature of the TM problem.

5.6 TM Algorithm Based on SMS with the Estimation Strategy

The simplest available TM method finds the global maximum (the accurate detection point (u, v)), considering all locations within the search space S. Nevertheless, the approach has a high computational cost for its practical use.

Several TM algorithms [7–10] have been proposed to accelerate the search process by calculating only a subset of search locations. Although these algorithms allow reducing the number of search locations, they do not explore the whole region effectively and often suffers premature convergence which conducts to sub-optimal detections. The cause of these problems is the operators used for modifying the particles. In such algorithms, during their evolution, the position of each agent in the next iteration is updated yielding an attraction towards the position of the best particle seen so-far [19, 20]. This behavior produces that the entire population, as the algorithm evolves, concentrates around the best coincidence seen so-far, favoring the premature convergence in a local minima of the multi-modal surface. Therefore, a better TM algorithm should spend less computational time on the search strategy and get the optimum match position.

In the SMS-based algorithm, individuals represent search positions (u, v) which move throughout the search space S. The NCC coefficient, used as a fitness value, evaluates the matching quality presented between the template image R and the source image I, for a determined search position (individual). The number of NCC evaluations is drastically reduced by considering a fitness calculation strategy which indicates when it is feasible to calculate or only estimate the NCC values for new search locations. Guided by the fitness values (NCC coefficients), the set of encoded candidate positions are evolved using the SMS operators until the best possible resemblance has been found.

In the algorithm, the search space S consists of a set of 2-D search positions $\hat{u}$ and $\hat{v}$ representing the x and y components of the detection locations, respectively. Each particle is thus defined as:

$$P_i = \{(\hat{u}_i, \hat{v}_i) | \, 1 \leq \hat{u}_i \leq M - m, 1 \leq \hat{v}_i \leq N - n\} \tag{5.12}$$

5.6.1 The SMS-TM Algorithm

The goal of our TM-approach is to reduce the number of evaluations of the NCC values (actual fitness function) avoiding any performance loss and achieving the optimal solution. The SMS-TM method is listed below:

Step 1	Set the SMS parameters
Step 2	Initialize the population of 5 random individuals $\mathbf{P} = \{P_1, \ldots, P_5\}$ inside of the search region S and the individual database array $\mathbf{T}$, as an empty array
Step 3	Compute the fitness values for each individual according to the fitness calculation strategy presented in Sect. 5.3
Step 4	Update new evaluations in the individual database array $\mathbf{T}$
Step 5	Set the parameters $\rho \in [0.8, 1]$, $\beta = 0.8$, $\alpha = 0.8$ and $H = 0.9$ being consistent with the gas state

(continued)

(continued)

Step 6	Apply the enhanced general procedure which is illustrated in Algorithm 5.2
Step 7	If the 50 % of the total iteration number is completed ($1 \leq k \leq 0.5 \cdot gen$), then the process continues to the liquid state procedure; otherwise go back to step 6
Step 8	Set the parameters $\rho \in [0.3, 0.6]$, $\beta = 0.4$, $\alpha = 0.2$ and $H = 0.2$ being consistent with the liquid state
Step 9	Apply the enhanced general procedure which is illustrated in Algorithm 5.2
Step 10	If the 90 % (50 % from the gas state and 40 % from the liquid state) of the total iteration number is completed ($0.5 \cdot gen < k \leq 0.9 \cdot gen$), then the process continues to the solid state procedure; otherwise go back to step 9
Step 11	Set the parameters $\rho \in [0.0, 0.1]$ and $\beta = 0.1$, $\alpha = 0$ and $H = 0$ being consistent with the solid state
Step 12	Apply the enhanced general procedure which is illustrated in Algorithm 5.2
Step 13	If the 100 % of the total iteration number is completed ($0.9 \cdot gen < k \leq gen$), the process is finished; otherwise go back to step 11
Step 14	If the number of target iterations has been reached, then determine the best individual (matching position) of the final population is $\hat{u}_{best}$, $\hat{v}_{best}$

The presented SMS-TM algorithm considers multiple search locations during the complete optimization process. However, only a few of them are evaluated using the true fitness function whereas all other remaining positions are just estimated. Figure 5.8 shows a search-pattern that has been generated by the SMS-TM approach considering the problem exposed in Fig. 5.7. Such pattern exhibits the evaluated search-locations (rule 1 and 2) in red-cells, whereas the minimum location is marked in green. Blue-cells represent those that have been estimated (rule 3) whereas gray-intensity-cells were not visited at all, during the optimization process.

Since most of fast TM methods employ optimization algorithms that face difficulties with multi-modal surfaces, they may get trapped into local minima and find

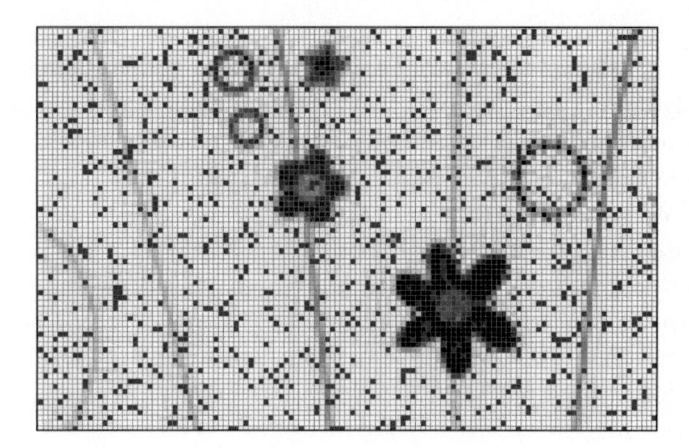

Fig. 5.8 Search-pattern generated by the SMS-TM algorithm. *Red* points represent the evaluated search positions whereas *blue* points indicate the estimated locations. The *Green* point exhibits the optimal match detection

sub-optimal detections. On the other hand, the presented approach allows finding out the optimal solution due to a better balance between the exploration and exploitation of the search space. Under the effect of the SMS operators, the search locations vary from generation to generation, avoiding to get trapped into a local minimum. Besides, since the presented algorithm uses a fitness calculation strategy for reducing the evaluation of the NCC values, it requires fewer search positions. As example Fig. 5.8 shows how the SMS-TM algorithm found the optimal detection, evaluating only the 11 % of the feasible search locations.

5.7 Experimental Results

In order to verify the feasibility and effectiveness of our algorithm, a series of comparative experiments with other TM algorithms are developed. Simulations have been performed over a set of images that is shown in Fig. 5.9. The presented approach has been applied to the experimental set and their results have been compared to those produced by the ICA-TM method [10] and the PSO-TM algorithm [8]. Both are considered as state-of-the-art algorithms whose results have been recently published. Notice that the maximum iteration number for the experimental work has been set to 300. Such stop criterion has been selected to maintain compatibility to similar works reported in the literature [7–10].

The parameter set for each algorithm is described as follows:

1. ICA-TM [10]: The parameters are set to: *NumOfCountries* = 100, *NumOfImper* = 10, *NumOf-Colony* = 90, $T_{\max} = 300$, $\xi = 0.1$, $\varepsilon_1 = 0.15$ and $\varepsilon_2 = 0.9$. Such values represent the best parameter set for this algorithm according to [10].
2. PSO-TM [8]: The parameters are set to: particle number = 100, $c_1 = 1.5$ and $c_2 = 1.5$; besides, the particle velocity is initialized between [−4, 4].
3. SMS-TM: The algorithm has been configured by using: Particle number = 100 and $d = 3$.

After all algorithms have been configured with its corresponding value set, they are used without modification for all experiments. The comparisons are analyzed considering three performance indexes: the average elapsed time (At), the success rate (Sr), the average number of checked locations (AsL) and the average number of function evaluations (AfE). The average elapsed time (At) indicates the time in seconds which is employed during the execution of each single experiment. The success rate (Sr) represents the number of executions in percentage for which the algorithm successfully locates the optimal detection point. The average number of checked locations (AsL) exhibits the number of search locations that have been visited during a single experiment. The average number of function evaluations (AfE) indicates the number of times the NCC coefficient is computed. In order to assure statistical consistency, all these performance indexes are averaged considering a determined number of executions.

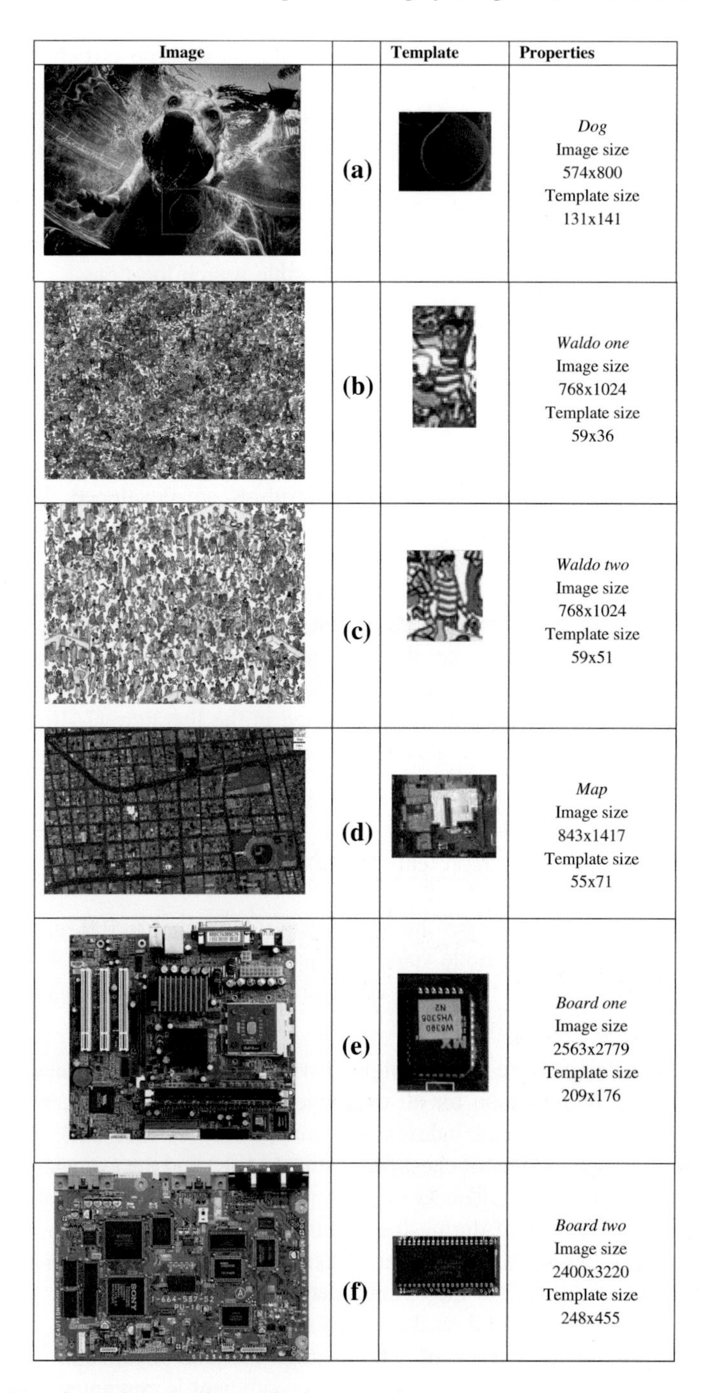

Fig. 5.9 Experimental set that is used in the comparisons

Table 5.1 Performance comparison of ICA-TM, PSO-TM and the presented approach for the experimental set shown in Fig. 5.8

Image	Algorithm	Average elapsed time (At)	Success rate (Sr)%	Average number of checked locations (AsL)	Average number of function evaluations (AfE)
(a)	ICA-TM	12.345	88.12	32,000	32,000
	PSO-TM	10.862	80.84	31,250	31,250
	SMS-TM	2.854	100	30,000	5640
(b)	ICA-TM	17.534	70.23	82,000	82,000
	PSO-TM	16.297	62.45	81,784	81,784
	SMS-TM	3.643	100	60,000	9213
(c)	ICA-TM	24.342	70.21	82,000	82,000
	PSO-TM	23.174	60.33	81,784	81,784
	SMS-TM	6.871	99	60,000	8807
(d)	ICA-TM	26.249	67.12	220,512	220,512
	PSO-TM	26.381	58.12	210,784	210,784
	SMS-TM	6.937	98	100,512	18,506
(e)	ICA-TM	37.231	90.54	578,400	578,400
	PSO-TM	35.925	85.27	578,400	578,400
	SMS-TM	10.214	100	220,512	54,341
(f)	ICA-TM	40.287	89.78	578,400	578,400
	PSO-TM	38.298	81.47	578,400	578,400
	SMS-TM	10.719	100	220,512	57,981

Results for 30 runs are reported in Table 5.1. According to this table, SMS-TM delivers better results than ICA and PSO for all images. In particular, the test remarks the largest difference in the success rate (Sr) and the average number of checked locations (AsL). Such facts are directly related to a better trade-off between exploration and exploitation, and the incorporation of the fitness calculation strategy respectively. Figure 5.10 presents the matching evolution curve for each image considering the averaged best NCC value seen so-far for all the algorithms that are employed in the comparison.

From Table 5.1, it turns out that the average cost of our algorithm is 6.873 s, while the average cost of the ICA-TM and the PSO-TM algorithms are 26.331 and 25.156 respectively. Such values demonstrate that SMS-TM spends less time on image matching than its counterparts. According to Table 5.2, the SMS-TM presents a better performance than the other two algorithms in terms of effectiveness, since it successfully detects the optimal detection point for all experiments. On the other hand, although the three algorithms visit the same number of search location approximately, the presented algorithm evaluates (NCC evaluation) a minimal number. It is important to recall that such evaluation represents the main computational cost which is commonly associated to the TM process. A non-parametric statistical significance proof known as the Wilcoxon's rank sum test for

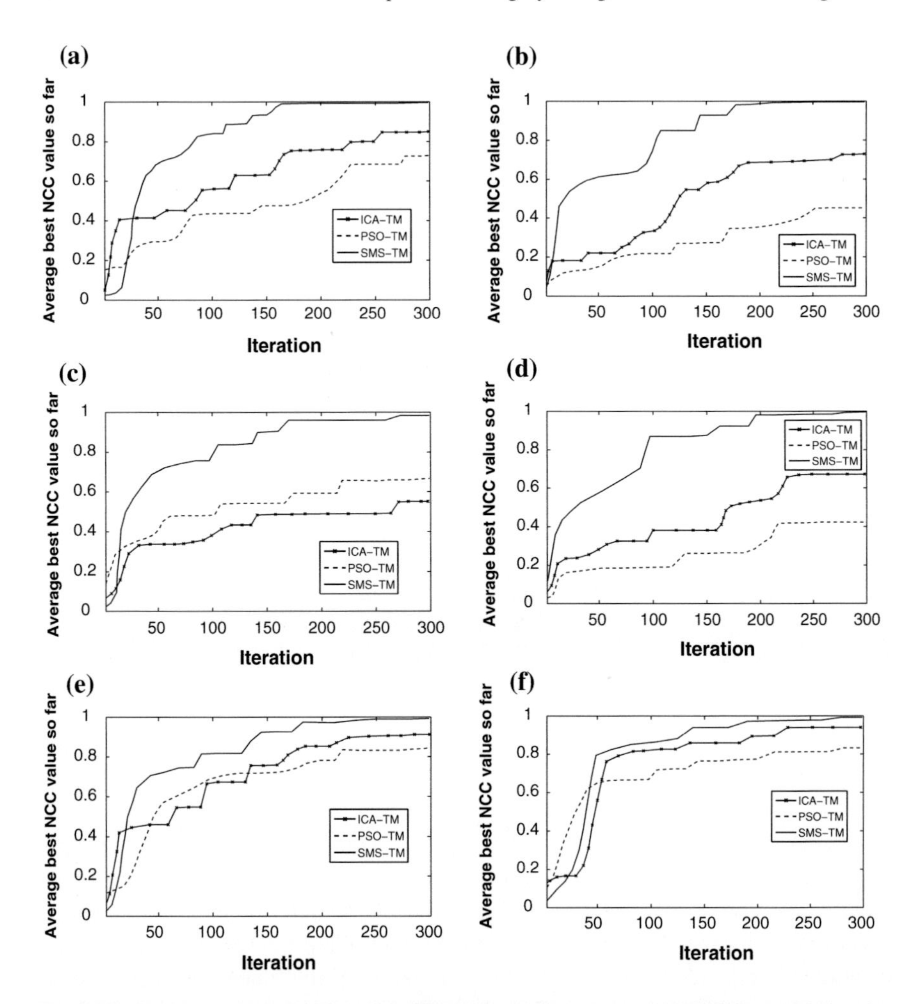

Fig. 5.10 Evolution curves for ICA-TM, PSO-TM and the presented SMS-TM considering the average best NCC value seen so-far, each curve corresponds to the image of the experimental set

Table 5.2 p-values produced by Wilcoxon's test comparing SMS-TM versus ICA-TM and SMS versus PSO-TM over the average number of function evaluations (AfE) values from Table 5.1

Image	SMS-TM versus ICA-TM	SMS-TM versus PSO-TM
(a)	1.52E-10	1.78E-10
(b)	3.23E-12	5.47E-12
(c)	1.56E-12	2.67E-12
(d)	3.21E-12	5.87E-12
(e)	4.87E-12	7.58E-12
(f)	2.11E-12	4.49E-12

independent samples [26, 27] has been conducted over the average number of function evaluations (AfE) data of Table 5.1, with an 5 % significance level. Table 5.2 reports the p-values produced by Wilcoxon's test for the pair-wise comparison of the average number of function evaluations (AfE) of four groups. Such groups are constituted by SMS-TM vs. ICA-TM and SMS vs. PSO-TM. As a null hypothesis, it is assumed that there is no significant difference between mean values of the two algorithms. The alternative hypothesis considers a significant difference between the AfE values of both approaches.

All p-values reported in Table 5.2 are less than 0.05 (5 % significance level) which is a strong evidence against the null hypothesis. Therefore, such evidence indicates that SMS-TM results are statistically significant and that it has not occurred by coincidence (i.e. due to common noise contained in the process).

5.8 Conclusions

In this chapter, a nature-inspired algorithm called the States of Matter Search (SMS) has been presented for solving the template matching (TM) problem. The SMS algorithm is based on the simulation of the states of matter phenomenon. In SMS, individuals emulate molecules which interact to each other by using evolutionary operations that are based on the physical principles of the thermal-energy motion mechanism. Such operations allow the increase of the population diversity and avoid the concentration of particles around and inside a local minimum area. The approach combines the use of defined operators with a control strategy that modifies the parameter setting for each operation during the evolution process. The algorithm is devised by considering each state of matter at one different exploration–exploitation rate. Thus, the evolutionary process is divided into three stages which emulate the three states of matter: gas, liquid and solid. At each state, molecules (individuals) exhibit different behaviors. Beginning from the gas state (pure exploration), the algorithm modifies the intensities of exploration and exploitation until the solid state (pure exploitation) is reached.

The approach also incorporates a simple fitness calculation approach which is based on the Nearest Neighbor Interpolation (NNI) algorithm which aims to estimate the fitness value (NCC operation) for several candidate solutions (search locations). The method is able to save computational time by identifying which NCC values can be only estimated or must be calculated instead. As a result, the approach substantially reduces the number search positions (by using the SMS approach), but also avoids the NCC evaluation for most of them through the use of the NNI strategy. The presented method achieves the best balance over other TM algorithms in terms of the detection accuracy and the computational cost. As a result, the approach can substantially reduce the number of function evaluations, yet preserving the good search capabilities of SMS.

Since the resultant algorithm is designed to have a better exploration-exploitation balance than other evolutionary algorithms, a high probability for finding the true matching point (accurate detection point) is also expected regardless of the high multi-modality nature of the TM process.

The performance of the presented approach has been compared to other existing TM algorithms by considering different images that show a great variety of formats and complexities. Experimental results demonstrate a higher performance of the method in terms of the elapsed time and the number of NCC evaluations.

References

1. Brunelli, R.: Template Matching Techniques in Computer Vision: Theory and Practice. Wiley (2009). ISBN 978-0-470-51706-2
2. Hadi, G., Mojtaba, L., Hadi, S.Y.: An improved pattern matching technique for lossy/lossless compression of binary printed Farsi and Arabic textual images. Int. J. Intell. Comput. Cybernet. **2**(1), 120–147 (2009)
3. Krattenthaler, W., Mayer, K.J., Zeiler, M.: Point correlation: a reduced-cost template matching technique. In: Proceedings of the First IEEE International Conference on Image Processing, pp. 208–212 (1994)
4. Rosenfeld, A., VanderBrug, G.J.: Coarse-fine template matching. IEEE Trans. Syst. Man. Cybern. **SMC-7**(2), 104–107 (1977)
5. Tanimoto, S.L.: Template matching in pyramids. Comput. Graph. Image Process. **16**(4), 356–369 (1981)
6. Uenohara, M., Kanade, T.: Use of fourier and karhunen-loeve decomposition for fast pattern matching with a large set of templates. IEEE Trans. Pattern Anal. Mach. Intell. **19**(8), 891–898 (1997)
7. Dong, Na, Chun-Ho, Wu, Ip, Wai-Hung, Chen, Zeng-Qiang, Chan, Ching-Yuen, Yung, Kai-Leung: An improved species based genetic algorithm and its application in multiple template matching for embroidered pattern inspection. Expert Syst. Appl. **38**, 15172–15182 (2011)
8. Liu, Fang, Duana, Haibin, Deng, Yimin: A chaotic quantum-behaved particle swarm optimization based on lateral inhibition for image matching. Optik **123**, 1955–1960 (2012)
9. Chun-Ho, Wu, Wang, Da-Zhi, Ip, Andrew, Wang, Ding-Wei, Chan, Ching-Yuen, Wang, Hong-Feng: A particle swarm optimization approach for components placement inspection on printed circuit boards. J Intell. Manuf. **20**, 535–549 (2009)
10. Duan, H., Xu, C., Liu, S., Shao, S.: Template matching using chaotic imperialist competitive algorithm, Pattern Recogn. Lett. **31**, 1868–1875 (2010)
11. Chen, G., Low, C.P., Yang, Z.: Preserving and exploiting genetic diversity in evolutionary programming algorithms. IEEE Trans. Evol. Comput. **13**(3), 661–673 (2009)
12. Adra, S.F., Fleming, P.J.: Diversity management in evolutionary many-objective optimization. IEEE Trans. Evol. Comput. **15**(2), 183–195 (2011)
13. Tan, K.C., Chiam, S.C., Mamun, A.A., Goh, C.K.: Balancing exploration and exploitation with adaptive variation for evolutionary multi-objective optimization. Eur. J. Oper. Res. **197**, 701–713 (2009)
14. Jin, Y.: Comprehensive survey of fitness approximation in evolutionary computation. Soft. Comput. **9**, 3–12 (2005)
15. Jin, Yaochu: Surrogate-assisted evolutionary computation: Recent advances and future challenges. Swarm Evol. Comput. **1**, 61–70 (2011)
16. Branke, J., Schmidt, C.: Faster convergence by means of fitness estimation. Soft. Comput. **9**, 13–20 (2005)

17. Zhou, Z., Ong, Y., Nguyen, M., Lim, D.: A Study on Polynomial Regression and Gaussian Process Global Surrogate Model in Hierarchical Surrogate-Assisted Evolutionary Algorithm, IEEE Congress on Evolutionary Computation (ECiDUE'05). United Kingdom, Edinburgh, 2–5 Sept 2005
18. Ratle, A.: Kriging as a surrogate fitness landscape in evolutionary optimization. Artif. Intell. Eng. Des. Anal. Manuf. **15**, 37–49 (2001)
19. Lim, D., Jin, Y., Ong, Y., Sendhoff, B.: Generalizing surrogate-assisted evolutionary computation. IEEE Trans. Evol. Comput. **14**(3), 329–355 (2010)
20. Ong, Y., Lum, K., Nair, P.: Evolutionary algorithm with hermite radial basis function interpolants for computationally expensive adjoint solvers. Comput. Optim. Appl. **39**(1), 97–119 (2008)
21. Ceruti, G., Rubin, H.: Infodynamics: analogical analysis of states of matter and information. Inf. Sci. **177**, 969–987 (2007)
22. Chowdhury, D., Stauffer, D.: Principles of equilibrium statistical mechanics, 1st edn. Wiley-VCH (2000)
23. Cengel, Y.A., Boles, M.A.: Thermodynamics: an engineering approach, 5th edn. New York, McGraw-Hill (2005)
24. Frederick Bueche, Eugene Hecht, Schaum's Outline of College Physics, 11th Edition. New York, McGraw-Hill
25. David, S., Betts, R.E., Turner Introductory statistical mechanics, 1st edn. Addison Wesley (1992)
26. Wilcoxon, F.: Individual comparisons by ranking methods. Biometrics **1**, 80–83 (1945)
27. Garcia, S., Molina, D., Lozano, M., Herrera, F.: A study on the use of non-parametric tests for analyzing the evolutionary algorithms' behaviour: a case study on the CEC'2005 Special session on real parameter optimization. J. Heurist. (2008). doi:10.1007/s10732-008-9080-4

Chapter 6
Estimation of Multiple View Relations Considering Evolutionary Approaches

Abstract Many computer vision problems require estimating geometric relations between two different views that contain a considerable number of unwanted abnormal data. Despite several robust techniques with different criteria have been proposed to solve such a problem, the Random Sampling Consensus (RANSAC) algorithm is by far the most well-known method. In this chapter, a method for robustly estimating multiple view relations from point correspondences is presented. The approach combines the RANSAC method and the Clonal Selection algorithm. Upon such combination, the method adopts a different sampling strategy in comparison to RANSAC in order to generate putative solutions. Under the new mechanism, new candidate solutions are iteratively built by considering the quality of models that have been generated by previous candidate solutions, rather than relying over a pure random selection as it is the case with classic RANSAC. The rules for the generation of candidate solutions (samples) are motivated by the behavior of the immunological system in human beings. As a result, the approach can substantially reduce the number of iterations but still preserves the robust capabilities of RANSAC. The method is generic and its use is illustrated by the estimation of fundamental matrices and homographies for synthetic and real images. Experimental results validate the efficiency of the resultant method in terms of accuracy, speed, and robustness.

6.1 Introduction

The goal of estimating geometric relations between images is to find an appropriate global transformation to match images of the same scene, taken at different viewpoints. Such estimation can be applied either when an object moves in front of a static camera or when a static scene is captured by a moving camera or by multiple cameras from different viewpoints. This methodology has been widely adopted in many applications such as the case of series of images that can be stitched together to generate a panorama image [1–3]. Also, multiple image super-resolution

E. Cuevas et al., *Applications of Evolutionary Computation in Image Processing and Pattern Recognition*, Intelligent Systems Reference Library 100,
DOI 10.1007/978-3-319-26462-2_6

approaches can be applied in the overlapped region that is calculated according to the estimated geometry [4–6]. Another application is related to the motion of a moving object which can also be estimated using its geometric relations [7]. Likewise, a distributed camera network can be calibrated, with each camera's position, orientation and focal length being computed based on their correspondences [8–10]. Another example is the application of the RANSAC procedure for the localization and tracking of objects of interest in medical images [11, 12]. Moreover, the robot position can be controlled or approximated through the estimation of the homography [13–15].

In a generic modelling problem, it is important to identify as inliers those data that can be explained by the hypothetical model. Other points, e.g., those generated by matching errors are called *outliers*. The outliers are caused by external effects not related to the investigated model. Based on different criteria, several robust techniques have been proposed to identify points as inliers or outliers, being the random sampling consensus (RANSAC) algorithm [16] the most well-known method.

RANSAC adopts a simple hypothesize-and-evaluation process. Under such approach, a minimal subset of elements (correspondences) is sampled randomly, and a candidate model is hypothesized using this subset. Then, the candidate model is evaluated on the entire dataset separating all elements from the dataset into inliers and outliers, according to their degree of matching (error scale) to the candidate model. These steps are iterated until there is a high probability that an accurate model could be found during iterations. The model with the largest number of inliers is considered as the estimation result.

Although RANSAC algorithm is simple and powerful, it presents two main problems [17, 18]: the high consumption of iterations and the inflexible definition of its objective function. In the RANSAC algorithm, candidate models are generated by selecting data samples. Since such a strategy is completely random, a large number of iterations are required to explore a representative subset of noisy data and to find a reliable model that could contain the maximum number of inliers. In general terms, the number of iterations is strongly affected by the contamination of the dataset. The other crucial issue is the objective function to evaluate the correctness of a candidate model from contaminated data. In the RANSAC methodology, the best estimation result is the model that maximizes the number of inliers. Therefore, the objective function involves the count one by one of the number of inliers associated to a candidate model. Such an objective function is fixed and prone to obtain sub-optimal models under different circumstances [18].

Several variants have been proposed in order to enhance the performance of the RANSAC method. One example constitutes the approach MLESAC [19] which searches the best hypothesis by maximizing the likelihood via the RANSAC process. Such method assumes that inlier data would distribute as a Gaussian function while outliers are distributed randomly. Alternatively, instead of giving the error scale (i.e. the threshold to separate inliers from outliers) a priori, the SIMFIT method [20] proposes its prediction based on an iterative procedure. Other representative works, such as the projection-pursuit method [21] and TSSE (Two-Step Scale Estimator) [22], employ the mean shift technique to model the inlier distribution and obtain an inlier scale. Such approaches enables RANSAC to be data-driven; however, the whole process becomes quite time consuming.

Although all the proposed variants allow solving one of the two main RANSAC problems, the other challenge still remains. The fact is that the estimation process is approached as an optimization problem where the search strategy is a random walking algorithm and the objective function is fixed to the number of inliers that are associated to the candidate model. In order to overcome typical RANSAC problems, we propose to visualize the RANSAC operation as a generic optimization procedure. Under this point of view, a new efficient search strategy can be added for reducing the number of consumed iterations. Likewise, it can be defined as a new objective function which incorporates other elements that allow an accurate evaluation of the quality of a candidate model.

Two important difficulties in selecting a search strategy for RANSAC are the high multi-modality and the complex characteristics of the estimation process produced by the elevated contamination of the dataset. Under such circumstances, classical methods present a bad performance [23, 24], making way for recent new approaches that have been proposed to solve complex and ill-posed engineering problems. These methods include the application of modern optimization techniques such as evolutionary algorithms and swarm techniques [25, 26] which have delivered better solutions over those obtained by classical methods.

Clonal selection algorithm (CSA) [27] is one example of these approaches. The CSA is an evolutionary optimization algorithm that has been built on the basis of the human immune system [28]. In CSA, the antigen (Ag) represents the problem to be optimized and its constraints while the antibodies (Ab) are the candidate solutions. The antibody-antigen affinity indicates the matching quality between the solution and the problem (fitness function). The algorithm performs the selection of antibodies (candidate solutions) based on affinity either by matching against an antigen pattern or by evaluating the pattern via an objective function. As optimization approach, CSA has the ability to operate overhigh multi-modal and complex objective functions. It does not use derivatives or any of its related information as it employs probabilistic transition rules instead of deterministic ones. Despite to its simple and straightforward implementation, it has been extensively employed in the literature for solving several kinds of complex and ill-posed engineering problems [29–32].

This chapter presents a method for the robust estimation of multiple view relations from point correspondences. The approach combines the RANSAC method with the CSA. Upon such combination, the method adopts a different sampling strategy in comparison to RANSAC to generate putative solutions. Under the new mechanism, new candidate solutions are built iteratively by considering the quality of models that have been generated by previous candidate solutions, rather than relying over a pure random selection as it is the case of RANSAC. Likewise, a more accurate objective function is incorporated to accurately evaluate the quality of a candidate model. As a result, the approach can substantially reduce the number of iterations, yet preserving the robust capabilities of RANSAC. The method is generic and its use is illustrated by the estimation of fundamental matrices and homographies, considering synthetic and real images. Experimental results validate the efficiency of the resultant method in terms of accuracy, speed and robustness.

6.2 View Relations from Point Correspondences

The problem of image matching consists in finding a geometric transformation that maps one image of a scene to another image taken from a different point of view. In order to solve such a problem, it is necessary to conduct a feature correspondence process as a preprocessing step. The feature correspondence task aims to find the pixel coordinates in two different images that refer eventually to the same point in the world. This process can be divided into two steps: (I) feature detection and (II) correspondence map.

In the feature detection (I), a set of M feature points are acquired from each image, $\mathbf{E} = \{\mathbf{e}_1, \mathbf{e}_2, \ldots, \mathbf{e}_M\}$ for the first image and $\mathbf{E}' = \{\mathbf{e}'_1, \mathbf{e}'_2, \ldots, \mathbf{e}'_M\}$ for the second image. Such sets are obtained as a result of applying a feature detector algorithm. The function of a feature detector is to find interest points based on a descriptor which provides a unique and robust representation of an image feature. In the presented approach, the Speeded Up Robust Features (SURF) method has been employed [33, 34]. It has demonstrated to be effective for both, high and low resolution images [35]. During its operation, the SURF algorithm computes a 64-element descriptor vector to characterize each feature point form $\mathbf{E}$ and $\mathbf{E}'$.

Once the sets $\mathbf{E}$ and $\mathbf{E}'$ have been computed, the correspondence process (II) between the elements of $\mathbf{E}$ and $\mathbf{E}'$ is applied. During such process, descriptor vectors from the first image are compared to descriptors of the second image in order to build pairs of corresponding points. The similarity between descriptors is based on their Euclidian distance. Therefore, in the correspondence process, an element $\mathbf{e}_i$ of the first image is automatically assigned to the point $\mathbf{x}_i$. Then, the descriptor of $\mathbf{e}_i$ is compared to all the descriptors from $\mathbf{E}'$, so that the most similar element $\mathbf{e}'_j$ can be found $(\mathbf{e}'_j = \arg\min(dis(\mathbf{e}_i, \mathbf{e}'_b)), b \in \{1, 2, \ldots, M\})$. Under such conditions, the element $\mathbf{e}'_j$ is assigned to point $\mathbf{x}'_i$ and the correspondence pair $(\mathbf{x}_i, \mathbf{x}'_i)$ is generated.

Image conditions do not allow discriminating one point from another with complete certainty in spite of a proper deployment of the correspondence process. As a result, an erroneous matching about the correspondence of points that are located on different parts of different images may emerge.

In this section the geometric relations of points between two views are discussed, considering two cases: (a) the fundamental matrix [36] and (b) the homography [37].

In the discussion, it is assumed that there is a collection of pairs of corresponding points (correspondence map) that are found on two images, as follows:

$$\mathbf{U} = \left\{ (\mathbf{x}_1, \mathbf{x}'_1), (\mathbf{x}_2, \mathbf{x}'_2), \ldots, (\mathbf{x}_M, \mathbf{x}'_M) \right\}, \tag{6.1}$$

where $\mathbf{x}_i = (x_i, y_i, 1)^T$ and $\mathbf{x}'_i = (x'_i, y'_i, 1)^T$ are the positions of points in the first and second image, respectively.

(a) *Fundamental matrix*

Consider that the points $\mathbf{x}_i$ and $\mathbf{x}'_i$ are the projections of a three-dimensional point $\mathbf{X}_i = (X_i, Y_i, Z_i)$ in the first and second image, respectively (see Fig. 6.1). Such images have been captured considering two different positions which are related by a rotation $\mathbf{R}$ and translation $\mathbf{t}$ of the optical centers $(\mathbf{C}_1, \mathbf{C}_2)$. In this case, the collection of possible positions of the point $\mathbf{X}_i$ in the second image is restricted to the points lying on a straight line called epipolar line [38]. The epipolar line at the point $\mathbf{x}'_i$, in the second image, is the intersection of the epipolar plane passing through the optical centers and the point $\mathbf{X}_i$ within the plane of the second image. The epipolar geometry represents the intrinsic geometry between two-views. It is independent of the scene structure and only depends on the relative localization between the cameras $(\mathbf{R}, \mathbf{t})$. The fundamental matrix $\mathbf{F} \in \Re^{3\times3}$ is the algebraic representation of this intrinsic geometry. The fundamental matrix satisfies the epipolar constraint

$$(\mathbf{x}'_i)^T \mathbf{F} \mathbf{x}_i = 0. \tag{6.2}$$

To determine $\mathbf{F}$, it is sufficient to know the coordinates of at least eight pairs of corresponding points. The use of a greater number of points supports an improvement on computation accuracy by assuming that there are no outliers within the points. Once the matrix $\mathbf{F}$ and the point $\mathbf{x}_i$ in the first image are known, the epipolar line in the second image corresponding to point $\mathbf{x}'_i$, is given by $\mathbf{F}\mathbf{x}_i$ [39].

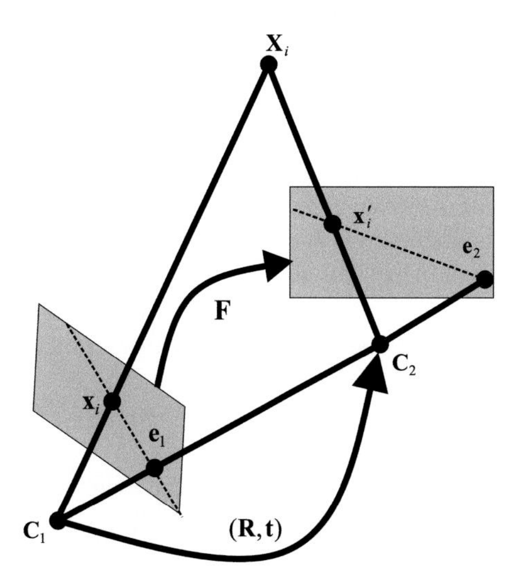

Fig. 6.1 Epipolar geometry between two-views

Similarly, $\mathbf{F}^T\mathbf{x}_i'$ specifies the epipolar line in the first image that corresponds to the point $\mathbf{x}_i'$ from the second image. By computing the fundamental matrix for each point from $\mathbf{U}$ in one of the images, the corresponding epipolar line that lies on the other image can be computed and its correspondence can be found. The quality of the estimated fundamental matrices is evaluated by considering the distance between the points and the epipolar lines to which they must belong, rather than the distance between each point, whose position is calculated through the matrix $\mathbf{F}$, and its corresponding observed position.

Considering the notation $\mathbf{F}\mathbf{x}_i = (\beta_1, \beta_2, \beta_3)^T$, the distance $d(\mathbf{x}_i', \mathbf{F}\mathbf{x}_i)$ between the point $\mathbf{x}_i'$ and the line $\mathbf{F}\mathbf{x}_i$ can be computed as follows:

$$d(\mathbf{x}_i', \mathbf{F}\mathbf{x}_i) = \frac{\mathbf{x}_i'\mathbf{F}\mathbf{x}_i}{\sqrt{\beta_1^2 + \beta_2^2}} \tag{6.3}$$

Likewise, denoting $\mathbf{F}^T\mathbf{x}_i' = (\beta_1', \beta_2', \beta_3')^T$, the other corresponding distance can be calculated as:

$$d(\mathbf{x}_i, \mathbf{F}^T\mathbf{x}_i') = \frac{(\mathbf{x}_i')^T\mathbf{F}\mathbf{x}_i}{\sqrt{(\beta_1')^2 + (\beta_2')^2}} \tag{6.4}$$

Therefore, the mismatch error EF_i^2 produced by the i-correspondence $(\mathbf{x}_i, \mathbf{x}_i')$ is defined by the sum of squared distances from the points to their corresponding epipolar lines as follows:

$$EF_i^2 = \left[d(\mathbf{x}_i', \mathbf{F}\mathbf{x}_i)\right]^2 + \left[d(\mathbf{x}_i, \mathbf{F}^T\mathbf{x}_i')\right]^2, \tag{6.5}$$

(b) *Homography*

Two perspective images can be geometrically linked through a plane Q of the scene by a homography $\mathbf{H} \in \Re^{3\times3}$ (see Fig. 6.2). This projective transformation $\mathbf{H}$ relates corresponding points of the plane projected into two images by $\mathbf{x}_i' = \mathbf{H}\mathbf{x}_i$ or $\mathbf{x}_i = \mathbf{H}^{-1}\mathbf{x}_i'$. The homography across two views can be computed by solving a linear system from a set of four point matches [40]. The quality of the estimated homography $\mathbf{H}$ is evaluated by considering the distance between the position of the point calculated with the help of the matrix $\mathbf{H}$ and the actually observed position. Therefore, the mismatch error EH_i^2 produced by the i-correspondence $(\mathbf{x}_i, \mathbf{x}_i')$ is defined as the sum of squared distances from the points to their estimated positions:

$$EH_i^2 = \left[d(\mathbf{x}_i', \mathbf{H}\mathbf{x}_i)\right]^2 + \left[d(\mathbf{x}_i, \mathbf{H}^{-1}\mathbf{x}_i')\right]^2, \tag{6.6}$$

A detailed description of the fundamental matrix $\mathbf{F}$ and the homography matrix $\mathbf{H}$ is provided by Hartley in [39].

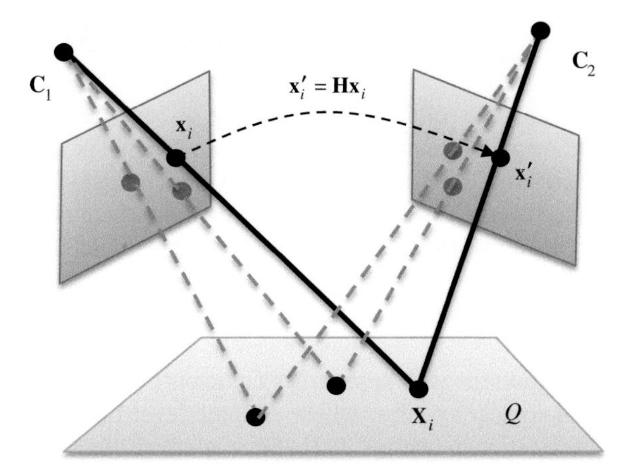

Fig. 6.2 Homography from a plane between two views

6.3 Random Sampling Consensus (RANSAC) Algorithm

The goal of RANSAC is to estimate the geometric transformation (fundamental matrix **F** or the homography **H**) from image correspondences over two views. Potentially there are a significant number of mismatches amongst the correspondences. Correct matches will obey the epipolar geometry or coincide with the homoghraphy transformation. Therefore, the aim is to obtain a set of *inliers* consistent with the epipolar geometry or with the homography transformation by using a robust technique. In this case *outliers* are points inconsistent with the epipolar geometry or with the homography transformation. Since the RANSAC algorithm has proven to be the most successful for solving the aforementioned problem, a summary of the RANSAC algorithm is given in the Appendix A.

In the RANSAC algorithm, the optimal hypothesis is found through a random-walking search strategy; therefore many attempts are necessary to investigate the space of possible samples in sufficient detail, and to find the sample for which the hypothesis has the greatest degree of agreement on the source data. The number of iterations and thus the elapsed time for the search can be reduced by choosing points according to some directed rules, rather than the random search. In contrast, evolutionary computation algorithms can be considered as a robust scheme [41] whose new candidate solutions are generated in accordance to information obtained from past candidate solutions.

In this chapter, we propose a different approach to solve the RANSAC matching problem that employs the Clonal Selection Algorithm (CSA) as a search strategy.

The goal is to demonstrate that a combination of RANSAC with an evolutionary strategy allows performing an efficient search among the correspondences to generate models of higher quality. It is also shown that the number of inliers that are found by the new method, through the use of a fixed number of samples, is greater than the number of inliers determined by the family of algorithms based purely on RANSAC.

6.4 Clonal Selection Algorithm (CSA)

An important difficulty in selecting an optimization strategy for the RANSAC method is the high multi-modality and complex characteristics of the estimation process that arise from the elevated contamination of the dataset. Under such circumstances, classical optimization methods have exhibited a bad performance [21, 22]. Different to classical methods, computational intelligent approaches have been proposed to solve complex and ill-posed engineering problems. These methods include the application of modern optimization techniques such as evolutionary methods [25, 26] that have recently been proposed with interesting results. The inspiration for such approaches commonly arises from our scientific understanding of biological systems, which at some level of abstraction, can be represented as optimization processes [42]. Such methods include the social behavior of bird flocking and fish schooling such as the Particle Swarm Optimization (PSO) algorithm [43], the emulation of the evolution in species such as the Genetic Algorithms (GA) [44, 45] and the behavior of the human immune system such as the Clonal Selection Algorithm (CSA) [27, 28], respectively.

Although PSO and GA are the most popular evolutionary methods for solving optimization problems, they exhibit serious flaws such as the difficulty to overcome local minima [46, 47]. The reason of this problem falls into the operators used for modifying the candidate solutions. In such algorithms, during their evolution, the position of each individual in the next iteration is updated yielding an attraction towards the position of the best particle seen so-far (in case of PSO) or interchanging information amongst the best candidate solutions (in case of GA). Just as the algorithm evolves, such behaviors produce that the entire population concentrates around the best particle or diverges without control, favoring the premature convergence to a local minima [48, 49].

In this chapter, the Clonal Selection Algorithm (CSA) is used to solve the problem of image matching. In the CSA, an antigen represents the problem to be optimized and its constraints (the estimation of geometric transformation, **T** or **H**). An antibody emulates a candidate solution of the problem (a single sample) whereas the affinity symbolizes an objective function that evaluates the similarity antibody-antigen (matching quality).

Different to PSO and GA, CSA exhibits interesting search capabilities such as its ability to avoid local minima in high modality environments [50].

6.4.1 Definitions

In an optimization problem, a candidate solution $\mathbf{v}_i$ (individual) consists of several decision variables $(v_1^i, \ldots, v_l^i)$ which are modified by an optimization strategy in order to obtain the optimal solution. CSA is an optimization method that operates considering binary encoded individuals. Therefore, each decision v_j^i ($j \in 1,\ldots,l$) variable must be coded into a binary number as $\mathbf{d}_j^i = encode(v_j^i)$. The concatenation of all encoded decision variables builds an antibody $\mathbf{a}_i = \left[\mathbf{d}_1^i \mathbf{d}_2^i \ldots \mathbf{d}_l^i\right]$. In order to obtain the value in the problem domain of a decision variable, the encoded binary information must be transformed into $v_j^i = decode(\mathbf{d}_j^i)$.

6.4.2 CSA Operators

Based on [32, 51], the CSA implements three different operators: the clonal proliferation operator ($T_{\mathrm{P}}^{\mathrm{C}}$), the affinity maturation operator (T_M^{A}) and the clonal selection operator ($T_{\mathrm{S}}^{\mathrm{C}}$). $\mathbf{A}(k)$ is the antibody population at time k that represents the set of n antibodies $\mathbf{a}_i$, such as $\mathbf{A}(k) = \{\mathbf{a}_1(k), \mathbf{a}_2(k), \ldots, \mathbf{a}_n(k)\}$. The evolution process of CSA can be described as follows:

$$\mathbf{A}(k) \longrightarrow T_{\mathrm{C}}^{\mathrm{P}}\mathbf{Y}(k) \longrightarrow T_{\mathrm{M}}^{\mathrm{A}}\mathbf{Z}(k) \cup \mathbf{A}(k) \longrightarrow T_{\mathrm{S}}^{\mathrm{C}}\mathbf{A}(k+1) \tag{6.7}$$

6.4.3 Clonal Proliferation Operator ($T_{\mathrm{P}}^{\mathrm{C}}$)

Define

$$\mathbf{Y}(k) = T_{\mathrm{P}}^{\mathrm{C}}(\mathbf{A}(k)) = \left[T_{\mathrm{P}}^{\mathrm{C}}(\mathbf{a}_1(k)), T_{\mathrm{P}}^{\mathrm{C}}(\mathbf{a}_1(k)), \ldots, T_{\mathrm{P}}^{\mathrm{C}}(\mathbf{a}_n(k))\right] \tag{6.8}$$

where $\mathbf{Y}(k) = T_{\mathrm{P}}^{\mathrm{C}}(\mathbf{A}(k)) = \mathbf{o}_i \cdot \mathbf{a}_i(k) i = 1, 2, \ldots, n,$, and $\mathbf{o}_i$ is a q_i-dimensional column vector that contains only ones. There are various methods for calculating q_i. In this chapter, it is calculated as follows:

$$q_i(k) = \mathrm{round}\left[N_c \cdot \frac{F(\mathbf{a}_i(k))}{\sum_{j=1}^{n} F(\mathbf{a}_j(k))}\right] \quad i = 1, 2, \ldots, n \tag{6.9}$$

where N_c is called the clonal size whereas the function round(w) gets the value to the least integer bigger than w. Therefore, the value of $q_i(k)$ is proportional to the fitness value of $\mathbf{a}_i(k)(F(\mathbf{a}_i(k)))$. Following the clonal proliferation, the population becomes

$$\mathbf{Y}(k) = \{\mathbf{Y}_1(k), \mathbf{Y}_2(k), \ldots, \mathbf{Y}_n(k)\} \tag{6.10}$$

where

$$\begin{aligned}\mathbf{Y}_i(k) &= \{\mathbf{y}_{ij}(k)\} = \{\mathbf{y}_{i1}(k), \mathbf{y}_{i2}(k), \ldots, \mathbf{y}_{iq_i}(k)\} \text{ and} \\ \mathbf{y}_{ij}(k) &= \mathbf{a}_i(k), \quad j = 1, 2, \ldots, q_i. \; i = 1, 2, \ldots, n\end{aligned} \tag{6.11}$$

6.4.4 *Affinity Maturation Operator* (T_M^{A})

The affinity maturation operation is performed by hypermutation. Random changes are introduced into the antibodies just as it happens in the immune system. Such changes may lead to an increase on the affinity. The hypermutation is performed by the operator T_M^{A} which is applied to the population $\mathbf{Y}(k)$ as it is obtained by clonal proliferation $\mathbf{Z}(k) = T_{\mathrm{M}}^{\mathrm{C}}(\mathbf{Y}(k))$.

The mutation rate is calculated using the following equation [52]:

$$\alpha = e^{(-\rho \cdot F(ab))}, \tag{6.12}$$

with α being the mutation rate, F being the objective function value of the antibody (ab) as it is normalized between [0, 1] and ρ being a fixed value. In [53], it is demonstrated the importance of including the factor ρ into Eq. (6.14) to improve the algorithm performance. The way ρ modifies the shape of the mutation rate is graphically explained by Fig. 6.3.

The number of mutations executed over a cloneis equal to $L \cdot \alpha$, considering L as the string length of the antibody. Therefore, the mutation operation is verified selecting (randomly) a determined bit of the string and then replacing it by its opposite value (i.e. 0–1 or 1–0).

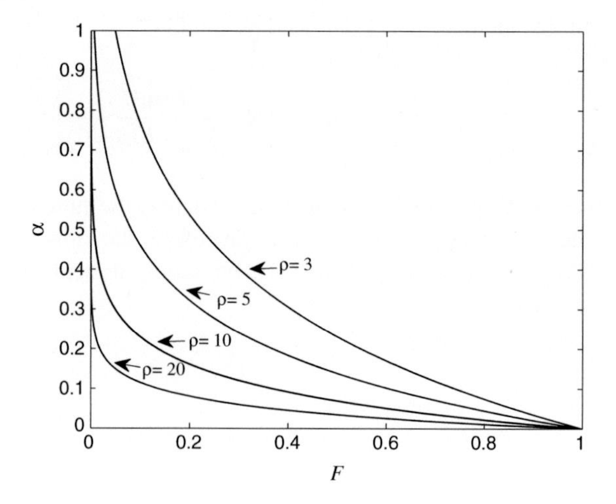

Fig. 6.3 Hypermutation rate versus fitness, considering some size steps

Following the affinity maturation operation, the population becomes:

$$\begin{aligned}
&\mathbf{Z}(k) = \{\mathbf{Z}_1(k), \mathbf{Z}_2(k), \ldots, \mathbf{Z}_n(k)\} \\
&\mathbf{Z}_i(k) = \{\mathbf{z}_{ij}(k)\} = \{\mathbf{z}_{i1}(k), \mathbf{z}_{i2}(k), \ldots, \mathbf{z}_{iq_1}(k)\}\text{and} \\
&\mathbf{z}_{ij}(k) = T_{\mathrm{M}}^{\mathrm{A}}(\mathbf{y}_{ij}(k)), \quad j = 1, 2, \ldots, q_1 \ \ i = 1, 2, \ldots, n
\end{aligned} \tag{6.13}$$

where T_M^A represent the maturation operator which is applied to the cloned antibody $\mathbf{y}_{ij}$.

6.4.5 *Clonal Selection Operator* $(T_{\mathrm{S}}^{\mathrm{C}})$

Finally, the elements of $\mathbf{A}(k)$ and $\mathbf{Z}_i(k)$ are merged into $\mathbf{A}(k+1)(\mathbf{A}(k+1) = \mathbf{Z}(k) \cup \mathbf{A}(k))$. Under such conditions, the clonal selection operation $T_{\mathrm{S}}^{\mathrm{C}}(\mathbf{A}(k+1))$ is defined as follows: From $\mathbf{A}(k+1)$, the n antibodies with the best fitness values are selected whereas the others are removed from $\mathbf{A}(k+1)$.

Since all individual are always cloned and muted by their respective operators, there is a low probability to be trapped into local minima [54]. Therefore, CSA can effectively handle challenging multimodal optimization problems [55–58]. Such fact clearly contrasts to well-known Genetic Algorithms (GA) [45] and Particle Swarm Optimization (PSO) [43] which usually tends to bias the whole population of chromosomes towards only the best candidate solution [59].

6.5 Method for Geometric Estimation Using CSA

In the presented method, the estimation process is approached as an optimization problem. Computational optimization generally involves two steps: (1) a search strategy to select candidate solutions and (2) an objective function that evaluates the quality of each selected candidate solution. Several computational algorithms are commonly available to perform the first step. On the other hand, the second step is unquestionably the most critical, because includes the designing of the objective function which is expected to embody the full performance complexity, biases, and restrictions related to the problem to be solved.

Considering that the problem consists in estimating the parameters of $\mathbf{F}$ or $\mathbf{H}$ through a set $\mathbf{U} = \{(\mathbf{x}_1, \mathbf{x}'_1), (\mathbf{x}_2, \mathbf{x}'_2), \ldots, (\mathbf{x}_M, \mathbf{x}'_M)\}$ of M different correspondences, RANSAC uses a random walking algorithm as a search strategy. Therefore, each candidate solution is randomly built. The idea of the presented method considers the use of CSA to generate samples (candidate solutions) based on information about their quality, rather than randomness. The quality of a sample, i.e., the fitness of an antibody $F(\mathbf{a}_i(k))$ is defined as the matching degree of the candidate transformations $\mathbf{F}_i$ or $\mathbf{H}_i$ that are constructed based on the correspondence numbers which in turn, are coded within $\mathbf{a}_i(k)$.

In the traditional RANSAC method, the only feasible objective function corresponds to the maximization of the number of inliers for a given permissible error (Eq. A.1). However, since the presented approach visualizes the estimation operation as a generic optimization procedure, different objective functions can be incorporated to accurately evaluate the quality of a candidate model yet avoiding any modification of the search strategy.

In order to appropriately design an objective function for estimating multiple view relations from point correspondences, the following alternatives can be considered: (I) to combine the number of inliers with the produced residual error, (II) to relate the number of inliers with regard to their residual error and, (III) to incorporate geometric restrictions among the selected points.

As a result of applying alternative (I), the fitness functions evaluate the matching degree of a candidate transformation (**F** or **H**) by considering two different objectives: the number of inliers and the approximation error. The idea is to find the candidate transformation that maximizes the number of inliers and simultaneously minimizes the approximation error. One possible representation of this class of objective functions is defined by the following expression:

$$\begin{aligned} F_{\mathrm{I}}(\mathbf{a}_i(k)) &= \sum_{j=1}^{M} \theta\left(e_j^2(h_i)\right) - \lambda \cdot e_j^2(h_i) \\ \theta\left(e_j^2(h_i)\right) &= \begin{cases} 0 & e_j^2(h_i) > Th \\ 0 & e_j^2(h_i) \leq Th \end{cases}, \end{aligned} \tag{6.14}$$

where $e_j^2(h_i)$ represents the quadratic errors EF_j^2 and EH_j^2 produced by the jth correspondence considering the candidate transformation $\mathbf{F}_i$ or $\mathbf{H}_i$, whereas λ is the penalty associated with the mismatch magnitude. Therefore, the best possible value delivered by F_I is M. This value is obtained when the candidate transformation $\mathbf{F}_i$ or $\mathbf{H}_i$ matches perfectly without presenting a residual error $(\sum_{j=1}^{M} e_j^2(h_i) = 0)$. In the original objective function (Eq. A.2) used by RANSAC, it is evaluated only the number of matching points which conflictingly depends on the permissible error (*Th*). A low *Th* value increases the accuracy of the model but degrades its generalization ability (number of matching points) to tolerate noisy data. By contrast, a high *Th* value improves the noise tolerance of the model but adversely drives the process to false detections. On the other hand, with the use of F_I, the obtained estimation represents the solution that presents the best balance between the approximation error (accuracy) and number of matching points (generalization ability).

In the case of alternative (II), the objective functions fuse the number of inliers and their residual error into a simple quotient. An example of this class of objective functions is defined by the following expression:

$$F_{\mathrm{II}}(\mathbf{a}_i(k)) = \frac{\sum_{j=1}^{M} \theta\left(e_j^2(h_i)\right)}{\sum_{j=1}^{M} e_j^2(h_i)} \tag{6.15}$$

Therefore, the maximization of F_{II} implies to obtain the candidate solution $\mathbf{a}_i(k)$ holding the highest number of inlier and the lowest residual error, simultaneously. One important difficulty associated to F_{II} emerges when the set of point correspondences $\mathbf{U}$ is composed by synthetic data (with no noise). Under such circumstances, F_{II} presents mathematical indeterminations, since the estimated model matches perfectly with no residual error whatsoever.

In the alternative (III), the objective functions incorporate additional restrictions to functions of type (I) and (II), since both are commonly used as basis. Under such scheme, new variants can be designed to consider either distances or angles between the points of candidate solutions in order to eliminate incidental correspondences between outliers.

In this chapter, the objective function that is defined by Eq. 6.16 has been used to construct the presented approach in order to assure that the overall algorithm remains simple.

6.5.1 Computational Procedure

In the presented approach, a candidate transformation (**F** or **H**) is encoded as an antibody. From the implementation point of view in the CSA-RANSAC method, a population $\mathbf{A}(k)\{\mathbf{a}_1(k), \mathbf{a}_2(k), \ldots, \mathbf{a}_n(k)\}$ of n candidate solutions (individuals) evolves from the initial point ($k = 1$) to a total $k_{\max}$ number iterations ($k = k_{\max}$). The algorithm begins by initializing the set of n candidate solutions with random values. The objective function F_I evaluates the matching quality of the candidate transformation. Guided by the values of this objective function, the set of encoded candidate solutions are modified by using the CSA operators so that they can improve their matching quality as the optimization process evolves. For the sake of clarity, Fig. 6.4 illustrates the computational procedure with a plain flow chart.

The first step in the algorithm is to generate an initial group of candidate transformations. The standard literature of evolutionary algorithms generally suggests the use of random solutions as the initial population, assuming the absence of knowledge about the problem [60]. This tendency has been reinforced, since several studies [61] have demonstrated that the use of solutions generated through some domain knowledge (i.e., non-random solutions) to set the initial population can only marginally improve its performance. From this point of view, in the presented algorithm, the random generation has been selected as the initialization method.

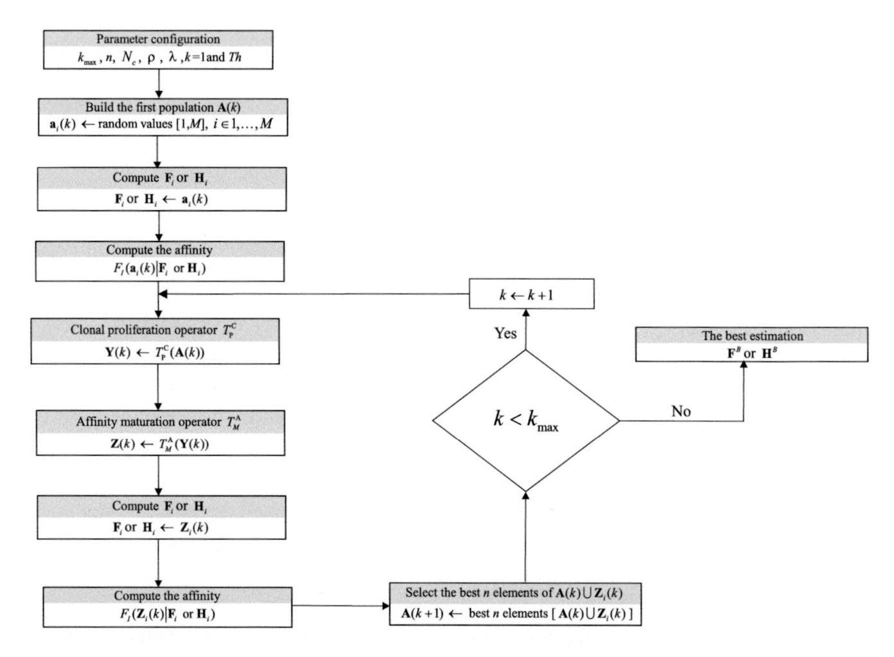

Fig. 6.4 Flow chart of the CSA-RANSAC computational procedure

The presented approach can be summarized as follows:

1.	Configuration	
	(a)	Calculate the necessary number of bits l for encoding the correspondence interval $[1, M]$
	(b)	The length L of the each antibody is $8 \times l$ in case of the fundamental matrix $\mathbf{F}$ and $4 \times l$ for the homography $\mathbf{H}$. Therefore, each antibody $\mathbf{a}_i(k)$ uses four or eight correspondence points as individuals for the optimization algorithm
	(c)	Set the number of antibodies n, the maximum number of iterations k_{max}, the clonal size N_c and the fixed constant ρ
2.	Initial population. For $i = 1,\ldots,n$	
	(a)	Build the first population $\mathbf{A}(1) = \{\mathbf{a}_1(1), \mathbf{a}_2(1), \ldots, \mathbf{a}_n(1)\}$ where each antibody $\mathbf{a}_i(1)$ consists of 8 (for $\mathbf{F}$) or 4 (for $\mathbf{H}$) random non-repeating indices from 1 to M
	(b)	Compute the fundamental matrix $\mathbf{F}_i$ or the homography $\mathbf{H}_i$ (hypothesis h_i) by using the indices from $\mathbf{a}_i(1)$
	(c)	Calculate the affinity (fitness value) $F(\mathbf{a}_i(1))$ as the matching quality of the constructed fundamental matrix $\mathbf{F}_i$ or the homography $\mathbf{H}_i$ considering the whole available data $\mathbf{U}$. Such fitness value is calculated by using Eq. 6.16
3.	Iterations $k = 1,\ldots,k_{max}$	
	(a)	Reproduce (T_P^C) the antibody population $\mathbf{A}(k)$ proportionally to the affinity of each element $F(\mathbf{a}_i(k))$ and generate a temporary population of clones $\mathbf{Y}(k)$

(continued)

(continued)

		Mutate (T_M^A) the population $\mathbf{Y}(k)$ of clones (copies of $\mathbf{A}(k)$) according to the affinity of each clone to the antigen (Eq. 6.16). A maturated antibody population $\mathbf{Z}(k)$ is thus generated. Since $\mathbf{Y}(k)$ are only copies of $\mathbf{A}(k)$, the fitness values of $\mathbf{Y}(k)$ are obtained from $\mathbf{A}(k)$
	(c)	Compute the fundamental matrix $\mathbf{F}_{i\cdot j}$ or the homography $\mathbf{H}_{i\cdot j}$ (hypothesis $h_{i\cdot j}$) by using the indices from the maturated antibody $\mathbf{z}_{ij}(k)$
	(d)	Calculate the affinity (fitness value) $F(\mathbf{z}_{ij}(k))$ as the matching quality of the constructed fundamental matrix $\mathbf{F}_{i\cdot j}$ or the homography $\mathbf{H}_{i\cdot j}$ considering the whole available data $\mathbf{U}$. Such fitness value is calculated by using the objective function described in Eq. 6.15
	(e)	Merge the elements of $\mathbf{A}(k)$ and $\mathbf{Z}_i(k)$ in a single population
	(f)	Select (T_S^C) the best n individuals according to their affinity from $\mathbf{Z}(k) \cup \mathbf{A}(k)$ to compose a new population $\mathbf{A}(k+1)$
4.	Estimation result	
	(a)	The best estimation of $\mathbf{F}^B$ or $\mathbf{H}^B$ consists of the parameters computed by using the indices from the best element $\mathbf{a}^B(k_{\max})$ of $\mathbf{A}(k_{\max})$ in terms of its affinity, so that $\mathbf{a}^B(k_{\max}) = \arg\max_{i=1,\ldots,n} F(\mathbf{a}_i(k_{\max}))$

Image matching refers to the task of finding a geometric transformation (fundamental matrix $\mathbf{F}$ or the homography $\mathbf{H}$) that maps one image of a scene to another image taken from a different point of view. Since our approach operates over the correspondence map, common problems of feature detection such as low resolution, blurred conditions, etc., must be associated to the capacities of the detection algorithm (it is used as a preprocessing step) and not to the characteristics of the presented method.

The presented approach combines the RANSAC method with the CSA adopting a different sampling strategy in comparison to RANSAC for generating putative solutions. Under the new mechanism, at each iteration, new candidate solutions are built by considering the quality of the models that have been generated by previous candidate solutions, rather than purely random as it is the case in RANSAC.

Since the approach visualizes the RANSAC operation as a generic optimization procedure, different objective functions can be incorporated to accurately evaluate the quality of a candidate model. Although several objective function can be tested, this chapter employs the expression in Eq. 6.16.

Table 6.1 presents the parameter values for the presented estimator. Such values have been obtained after extensive experimentation and have been kept for all experiments.

Table 6.1 CSA-RANSAC estimator parameters

n	$k_{\max}$	N_c	ρ	λ	Th
10	100	10	10	0.001	5

6.6 Experimental Results

A comprehensive set of experiments have been conducted to test the performance of the presented approach. We have applied the presented method to estimate fundamental matrices and homographies on real and synthetic data in order to compare its performance against other estimation algorithms such as the standard RANSAC [16], the MLESAC [19], the SIMFIT method [20], the projection-pursuit algorithm [21], the TSSE [22], the PSO algorithm (PSO-RANSAC) [43], the GA algorithm (GA-RANSAC) [45]. The first five approaches are RANSAC-based estimators whose results are broadly known. In all cases, the algorithms are tuned according to the value set which is originally proposed by their own references. However, the PSO and GA methods have been included only to validate the performance of the CSA as an optimization approach.

The PSO is used as a search strategy for RANSAC considering the following configuration: $P = 10$, $c_1 = 2$, $c_2 = 2$ and $Th = 5$ whereas the weight factor decreases linearly from 0.9 to 0.2. This set of parameters has been experimentally determined and delivers the best possible performance according to [62]. On the other hand, for the GA algorithm, the population size is set to 70, the crossover probability to 0.55, the mutation probability to 0.10 and number of elite individuals to 2. The roulette wheel selection and the 1-point crossover are applied. Such parameters achieve the best GA performance according to the experiments presented in [45], so they have been selected to conduct the comparisons.

In the performance study, three indexes are compared: the inlier number (*IN*), the error (E_s, E_r) and the number of function evaluations (*NFE*). The first two indexes assess the accuracy of the solution whereas the latter measures the computational cost.

The inlier number (*IN*) expresses the amount of elements contained in the set $\mathbf{I}$ of detected inliers. The error (E_s, E_r) provides a quality measure of the estimated relation. In case of synthetic data, the error is calculated as follows:

$$E_s = \left(\sum_{ij} \frac{d^2(\mathbf{x}_i^j, \hat{\mathbf{x}}_i^j)}{IN} \right)^{1/2}, \quad i \in \mathbf{I},\ j \in \{1,2\}, \tag{6.16}$$

where $\mathbf{x}_i^j$ is the inlier point calculated by the estimated relation in the *j*-view, $\hat{\mathbf{x}}_i^j$ is the inlier ground true point and $d(\cdot)$ is the Euclidian distance between the points. Therefore, E_s evaluates the fit of the estimated relation that is computed from the noisy data and is contrasted with the known ground truth.

In the case of real data, the error is assessed from the standard deviation of inliers. Thus, E_r is computed as follows:

$$E_r = \left(\sum_{i} \frac{e_i^2}{IN} \right)^{1/2}, \quad i \in \mathbf{I}, \tag{6.17}$$

where e_i^2 is the quadratic error produced by the ith inlier. In the context of this chapter, e_i^2 corresponds to EF_i^2 or EH_i^2 which represent the errors produced by the ith inlier considering the fundamental matrix **F** or homography **H**, respectively.

The number of function evaluations (*NFE*) specifies the total number of transformations that have been evaluated by the algorithm until the best estimation has been reached.

6.6.1 Fundamental Matrix Estimation with Synthetic Data

This section presents the experimental results that correspond to the estimation of fundamental matrix that considers synthetic data. In the experiment, the true fundamental matrix **F** was chosen as a random rotation matrix with $\theta_{\in}(0, 2\pi)$. 120 inliers have been generated by choosing a random point in the 320 × 240 region for the first view. Then, such a point is transformed by the true **F** and contaminated by normally distributed noise in order to construct its correspondence in the second view. The outliers have been produced by randomly selecting data points within the search space limits. In the test, the fraction of outliers varies from 0 to 100 %.

In the experiment, each execution of the algorithm involves 1000 iterations. Since the presented CSA-RANSAC and the PSO-RANSAC involve 10 individuals, they need to evolve up to 100 generations in order to fulfill the 1000 iterations. For illustrating the characteristics of each execution, Fig. 6.5 shows a test where the CSA-RANSAC has been applied to estimate a random **F** considering 10 % of additional outliers. Figure 6.5a, b exhibit the first and the second view, respectively. Considering the correspondence points shown by Fig. 6.5c, the CSA-RANSAC algorithm generates the estimation of **F** whose quality is visually exhibited by Fig. 6.5e, f, for both views. In Fig. 6.5e, the yellow squares indicate the position in the first view of a point from the second view as a result of the **F** transformation. Likewise, the green squares exhibit the position in the second view of a point from the first view as a result of the **F** transformation.

Figure 6.6 presents the performance for each algorithm. The results include averaged outcomes that are obtained through 50 different executions. In order to appropriately analyze such results, it is necessary to define the concept of a breakdown point [18]. The breakdown point is identified as the highest outlier ratio from which the algorithm abruptly degrades its capacity to find inliers. It can be seen from Fig. 6.6a that standard RANSAC has a breakdown point at 30 %, the MLESAC at 50 %, the SIMFIT method at 70 %, the projection-pursuit algorithm at 40 %, the TSSE at 40 %, the PSO-RANSAC at 70 %, and the GA-RANSAC at 70 %. In contrast to such methods, the presented approach, CSA-RANSAC, does not seem to have a prominent breakdown point, since its capacity to detect inliers smoothly degrades. It is also observed that the CSA-RANSAC algorithm presents the best performance in terms of the number of inliers (*IN*), as it is able to detect most of them. For the estimated **F**, the error E_s (Fig. 6.6b) is fairly comparable for

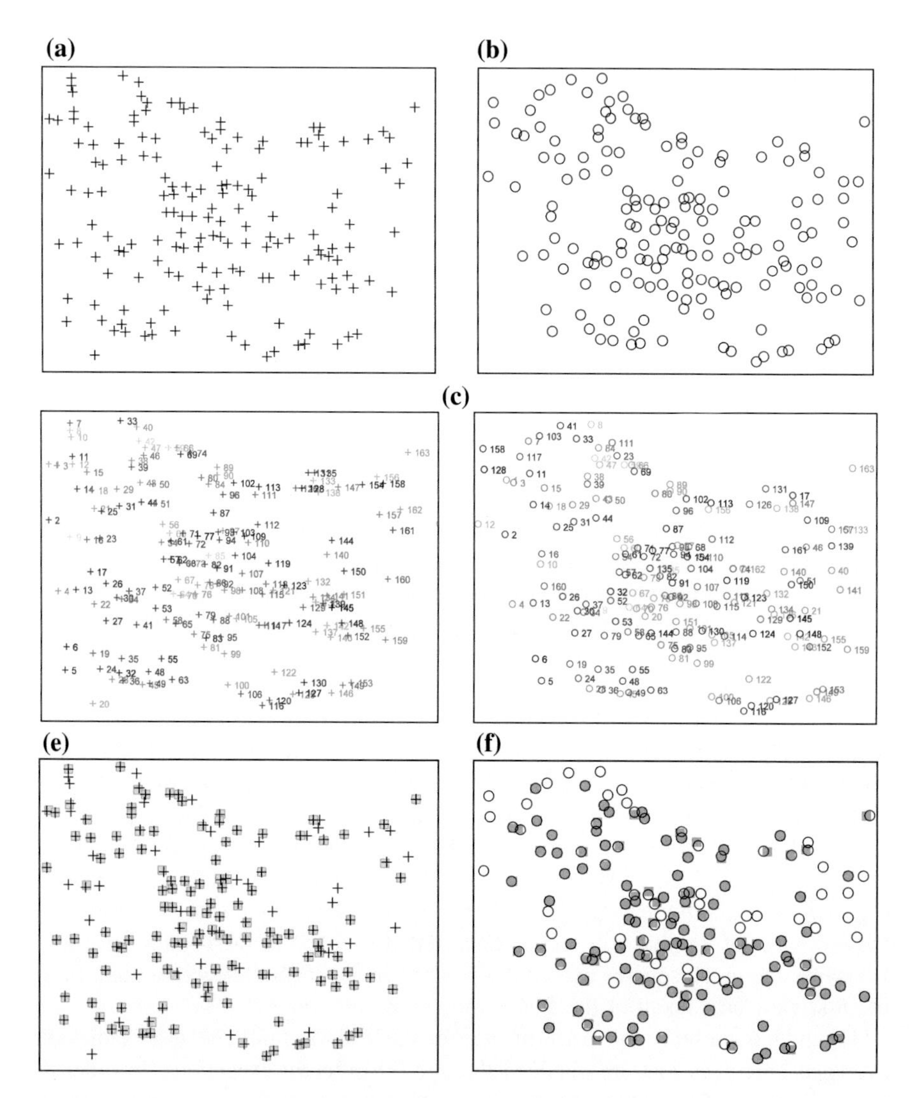

Fig. 6.5 A test example where the CSA-RANSAC has been applied to estimate a random transformation **F** considering 10 % of additional outliers. **a** First view, **b** second view, **c** correspondences, **e** corresponding point positions in the second view from the first view as a consequence of the **F** transformation, **f** corresponding point positions in the first view from the second view as a consequence of the **F** transformation

all methods until they reach their breakdown points. Nonetheless, the presented algorithm has performed better, being the only algorithm that consistently found the minimum error at all outlier ratios.

In terms of number of function evaluations (*NFE*), Fig. 6.6c shows that the standard RANSAC, the MLESAC, the projection-pursuit algorithm and the TSSE

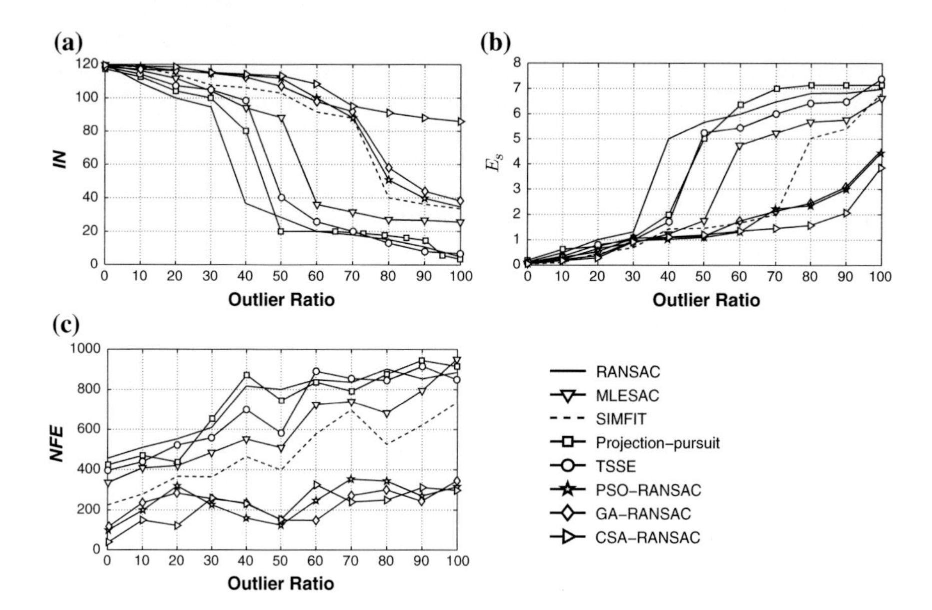

Fig. 6.6 Experimental results corresponding to the estimation of fundamental matrix **F** considering synthetic data

invest approximately the same number of iterations for reaching their best estimation of **F**. Since such methods use a random walking algorithm as a search strategy, the *NFE* significantly grows as the number of outliers increases. On the other hand, the PSO-RANSAC, the GA-RANSAC and the CSA-RANSAC (that use an optimization algorithm as a search strategy) maintain a considerably low *NFE* value with independence from the number of outliers.

From the experiment, it is evident that the use of an optimization approach can considerably reduce the *NFE* value. However, no optimization algorithm is suitable to find a good-enough estimation in view of the high multi-modality and complex characteristics of the estimation process which is produced by the elevated contamination of the dataset. Therefore, although the PSO-RANSAC and GA-RANSAC find its best estimated fundamental matrix **F** investing approximately the same number of evaluations as the CSA-RANSAC, such estimated matrix represents only a sub-optimal solution. This fact can be observed in Fig. 6.6b where it is clear that the PSO-RANSAC and the GA-RANSAC algorithms present higher E_s values in comparison to the CSA-RANSAC approach. The reason of this problem points to those operators used by PSO and GA for modifying the individual positions. In PSO, during their evolution, the position of each agent in the next iteration is updated yielding an attraction towards the position of the best particle seen so-far [63]. On the other hand, in GA, algorithmic features such as the high crossover rates, low mutation factors and high selection pressure tend to produce individuals in the proximity of the best elements in the population [64].

Such behaviors produce that, just as the algorithm evolves, the entire population concentrates around the best particle, favoring the premature convergence (reaching sub-optimal solutions) or damaging the exploration-exploitation balance [65].

6.6.2 Fundamental Matrix Estimation with Real Images

In this section, the experimental outcome of the estimation of fundamental matrix **F** considering real images is reported. Figures 6.7, 6.8, 6.9 and 6.10 show four images that have been used to compare the performance of the standard RANSAC [16], the MLESAC [19], the SIMFIT method [20], the projection-pursuit algorithm [21], the TSSE [22], the PSO algorithm (PSO-RANSAC) [43], the GA algorithm (GA-RANSAC) [45] and the presented CSA-RANSAC approach. For the experiment, the image correspondences are generated by using the SURF algorithm and matched according to their descriptor similarity.

The results consider 50 different executions for each algorithm over the four images. Table 6.2 presents the inlier number (*IN*), the error (E_r) and the number of

Fig. 6.7 Test image "Parking": **a** first view, **b** second view, **c** correspondence points and **d** inliers produced by CSA-RANSAC

Fig. 6.8 Test image "Garden": **a** first view, **b** second view, **c** correspondence points and **d** inliers produced by CSA-RANSAC

function evaluations (*NFE*). A close inspection reveals that the presented method is able to achieve the highest success rate (*IN* value) with the smallest error (E_r), yet requiring a small number of function evaluations (*NFE*) for most cases. Figures 6.6, 6.7, 6.8 and 6.9 also exhibit the results after applying the CSA-RANSAC estimator. Such results present the median case obtained throughout 50 runs.

In Fig. 6.7c labelled as "parking", it is presented the correspondence map that is generated by matching the SURF descriptors from Fig. 6.7a, b. For the sake of clarity, the figure shows both views are blended (Fig. 6.7a, b). In spite of the correspondence process, image conditions do not allow discriminating one point from another with complete certainty. As a result, several erroneous correspondence points have emerged. Figure 6.7d exhibits the detected inliers after applying the CSA-RANSAC estimator to the correspondence map 6.7c. From Fig. 6.7d, it is clear that most of the inliers ($\approx$223) have been detected whereas all the outliers have been suppressed. The evolutionary methods GA-RANSAC and PSO-RANSAC obtain an acceptable performance detecting 196 and 193 inliers, respectively.

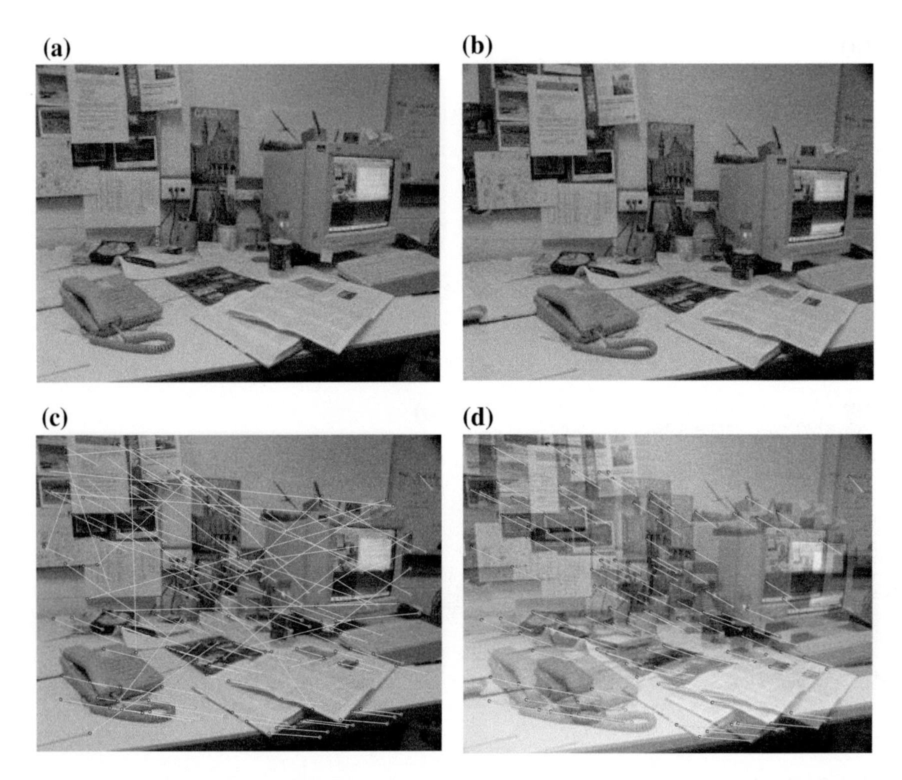

Fig. 6.9 Test image "Working place": **a** first view, **b** second view, **c** correspondence points and **d** inliers produced by CSA-RANSAC

However, such results have been obtained by investing more computational cost than CSA-RANSAC. Due to image conditions of Fig. 6.8a, b, more SURF feature points have been detected in comparison to Fig. 6.7. As a consequence, there exist a bigger percentage of erroneous correspondence points which contaminate the data set. This fact can be appreciated in Fig. 6.8c which shows a great amount of outlier points in the correspondence map. Figure 6.8d visualizes the detected inliers after applying the CSA-RANSAC estimator to the correspondence map 6.8c.

Despite of the great amount of outliers, the CSA-RANSAC method is able to detect most of the inliers ($\approx$419) and to suppress all the outliers. Different to images from Figs. 6.7, 6.8 and 6.9 presents a pair of images with lower resolution. As a consequence less SURF feature points are detected. After applying the CSA-RANSAC, an amount of approximately 120 inliers have been detected. In this case, in contrast to Figs. 6.7 and 6.8, the PSO-RANSAC presents a better performance than GA-RANSAC. Although this improvement is marginal, it has been reached using a low computational cost. Finally, Fig. 6.10 shows two images with

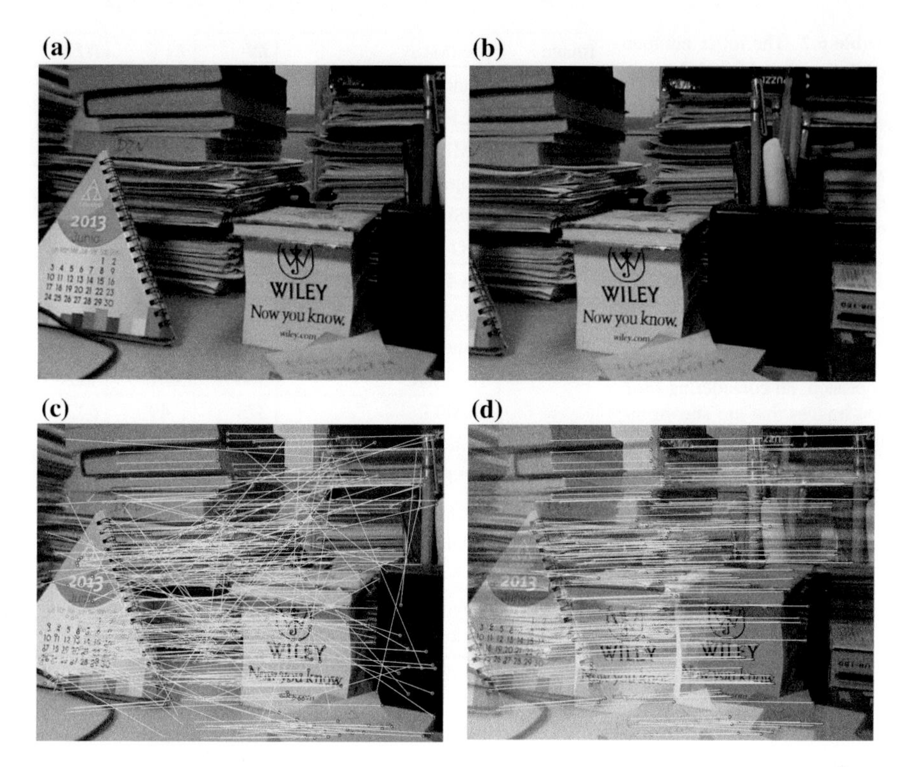

Fig. 6.10 Test image "Desk": **a** first view, **b** second view, **c** correspondence points and **d** inliers produced by CSA-RANSAC

objects close to the camera. The produced correspondence map is shown by Fig. 6.10c. As it is visualized, the great number of similar details produces a great number of outliers. Under such conditions, the CSA-RANSAC method is able to detect most of the inliers ($\approx$241) and to suppress all the outliers. In this case the method SIMFIT reaches a similar performance than the evolutionary techniques (PSO-RANSAC, GA-RANSAC and CSA-RANSAC) in terms of the number of detected inliers. However, its performance in terms of computational cost is almost the double.

6.6.3 Homography Estimation with Synthetic Data

This section reports on the experimental results corresponding to the estimation of homography matrix considering synthetic data. In the experiment, for the first view, 48 inliers have been generated through a rectangular pattern of 8×6 elements

Table 6.2 The inlier number (IN), the error (E_s) and the number of function evaluations (NFE) for standard RANSAC [16], the MLESAC [19], the SIMFIT method [20], the projection-pursuit algorithm [21], the TSSE [22], the PSO algorithm (PSO-RANSAC) [43], the GA algorithm (GA-RANSAC) [45] and the presented CSA-RANSAC approach, all considering four test images that are shown by Figs. 6.7, 6.8, 6.9 and 6.10

Image	Method	*IN*	E_s	*NFE*
(A)	Standard RANSAC	110	4.23	885
	MLESAC	163	3.64	651
	SIMFIT	178	2.53	584
	Projection-pursuit	148	3.83	716
	TSSE	137	3.78	843
	PSO-RANSAC	193	1.21	405
	GA-RANSAC	196	1.17	440
	CSA-RANSAC	223	0.78	378
(B)	Standard RANSAC	315	4.43	921
	MLESAC	350	3.61	712
	SIMFIT	373	2.49	632
	Projection-pursuit	321	3.58	823
	TSSE	311	3.65	864
	PSO-RANSAC	386	1.42	487
	GA-RANSAC	392	1.31	497
	CSA-RANSAC	419	0.69	402
(C)	Standard RANSAC	72	3.11	665
	MLESAC	86	2.67	421
	SIMFIT	94	2.02	337
	Projection-pursuit	75	2.73	412
	TSSE	81	2.86	364
	PSO-RANSAC	109	2.01	145
	GA-RANSAC	101	2.18	188
	CSA-RANSAC	120	0.31	98
(D)	Standard RANSAC	169	4.86	911
	MLESAC	189	3.12	843
	SIMFIT	210	2.87	725
	Projection-pursuit	175	3.21	885
	TSSE	181	3.45	795
	PSO-RANSAC	215	1.92	451
	GA-RANSAC	217	1.92	484
	CSA-RANSAC	241	0.47	312

within a 2-dimensional space of $[-300, 300]$. Such points are transformed by a random homography **H** and contaminated by normally distributed noise for constructing their correspondences in the second view. A set of outliers was added by selecting data points randomly within the space limits. In the test, the fraction of outliers varies from 0 to 100 %.

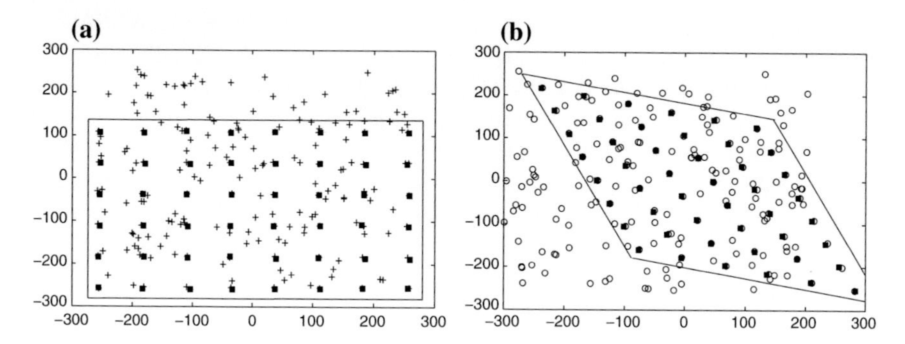

Fig. 6.11 A test example where the CSA-RANSAC has been applied to estimate a random transformation **H** considering only the 75 % of additional outliers. **a** the first view and **b** the second view, with *black squares* representing the detected inliers

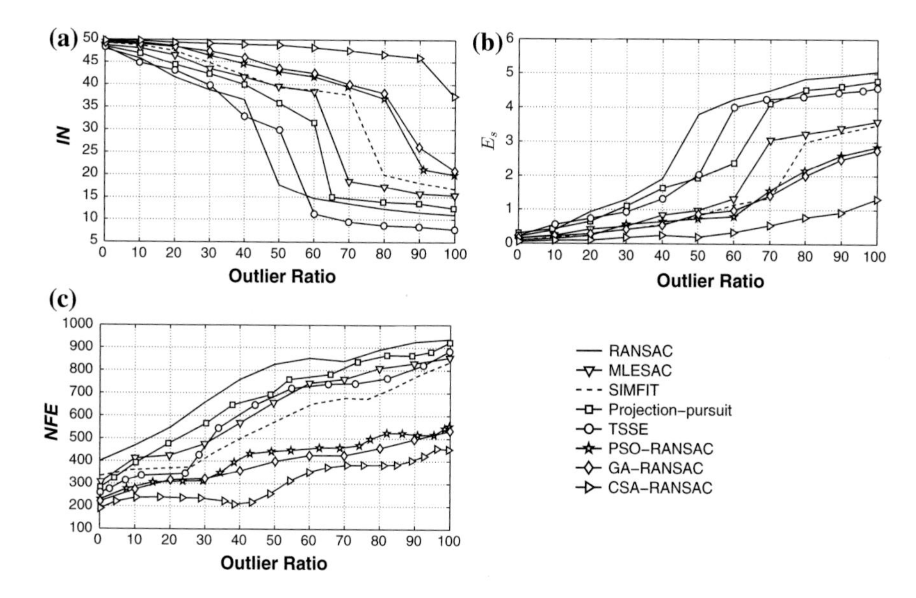

Fig. 6.12 Experimental results corresponding to the estimation of **H** considering synthetic data

In the experiment, each algorithm's execution requires 1000 iterations. In order to illustrate the characteristics of each execution, Fig. 6.11 shows a test where the CSA-RANSAC has been applied to estimate a random **H** considering only the 75 % of additional outliers. Figure 6.11 presents the performance results for each algorithm. Such results represent the averaged outcome after 50 different executions.

From Fig. 6.12a, it can be seen that the standard RANSAC has a breakdown point at 40 %, the MLESAC at 60 %, the SIMFIT method at 70 %, the projection-pursuit algorithm at 60 %, the TSSE at 50 %, the PSO-RANSAC at 80 %, the GA-RANSAC at 80 % and the presented approach, CSA-RANSAC at 90 %. It is also observed that the CSA-RANSAC algorithm presents the best performance in terms of the number of inliers (*IN*), as it is able to detect most of them. For the estimated **H**, the error E_s (Fig. 6.12b) is fairly comparable for all methods until they reach their breakdown points. Figure 6.12a–c reveal that the presented method is able to achieve the highest success rate (*IN* value) with the smallest error (E_s), yet requiring only a low number of function evaluations (*NFE*).

6.6.4 Homography Estimation with Real Images

In this section, the experimental results corresponding to the estimation of **H** considering real images are reported. Figures 6.13, 6.14 and 6.15 show four images that have been used to compare the performance of the standard RANSAC [16], the MLESAC [19], the SIMFIT method [20], the projection-pursuit algorithm

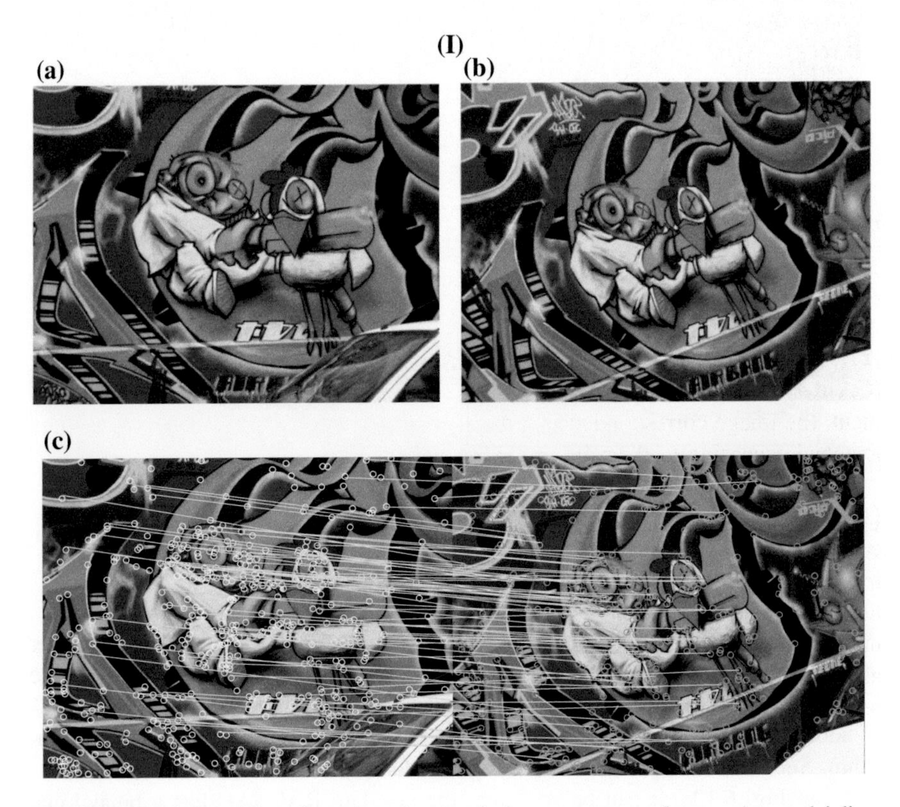

Fig. 6.13 Test image "I": **a** first view, **b** second view, **c** correspondence points and inliers produced by CSA-RANSAC

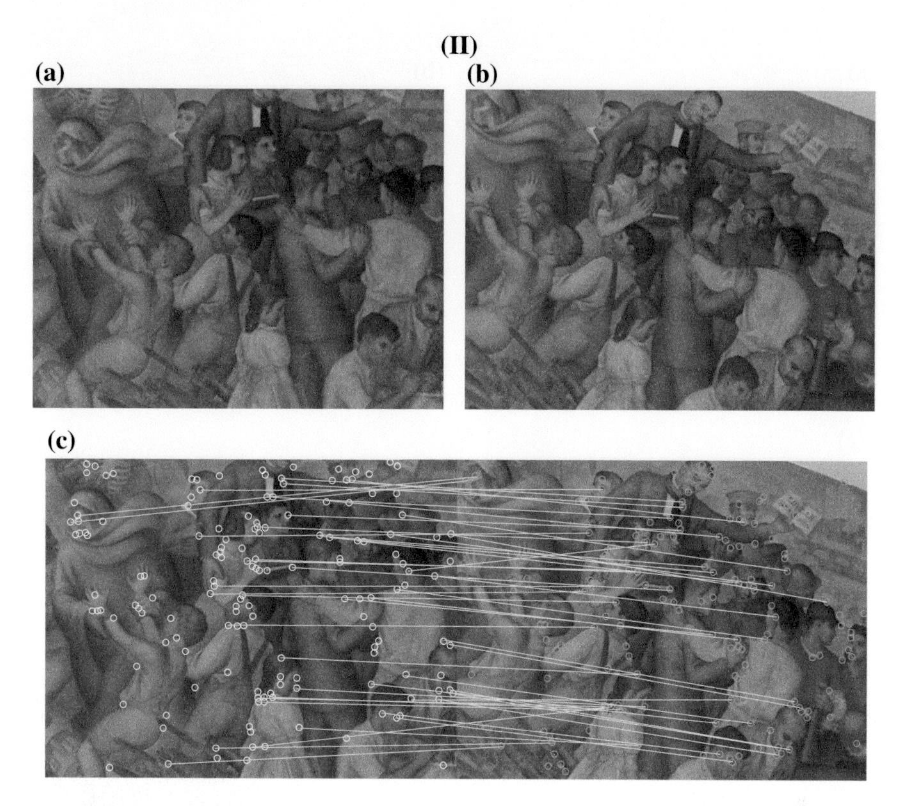

Fig. 6.14 Test image “II”: **a** first view, **b** second view, **c** correspondence points and inliers produced by CSA-RANSAC

[21], the TSSE [22], the PSO algorithm (PSO-RANSAC) [43], the GA algorithm (GA-RANSAC) [45] and the presented CSA-RANSAC approach. For the experiment, the image correspondences are generated by using the SURF algorithm and matched according to their descriptor similarity.

Table 6.3 presents the performance indexes: the inlier number (*IN*), the error (E_s) and the number of function evaluations (*NFE*). The results are analyzed by considering 50 different executions for each algorithm over the four images. An inspection from Table 6.3 reveals that the presented method is able to achieve the highest success rate (*IN* value) with the smallest error (E_s), yet requiring only a low number of function evaluations (*NFE*) for most cases. Figures 6.13, 6.14 and 6.15 also exhibit the results after applying the CSA-RANSAC estimator. Such results present the median case obtained throughout 50 runs.

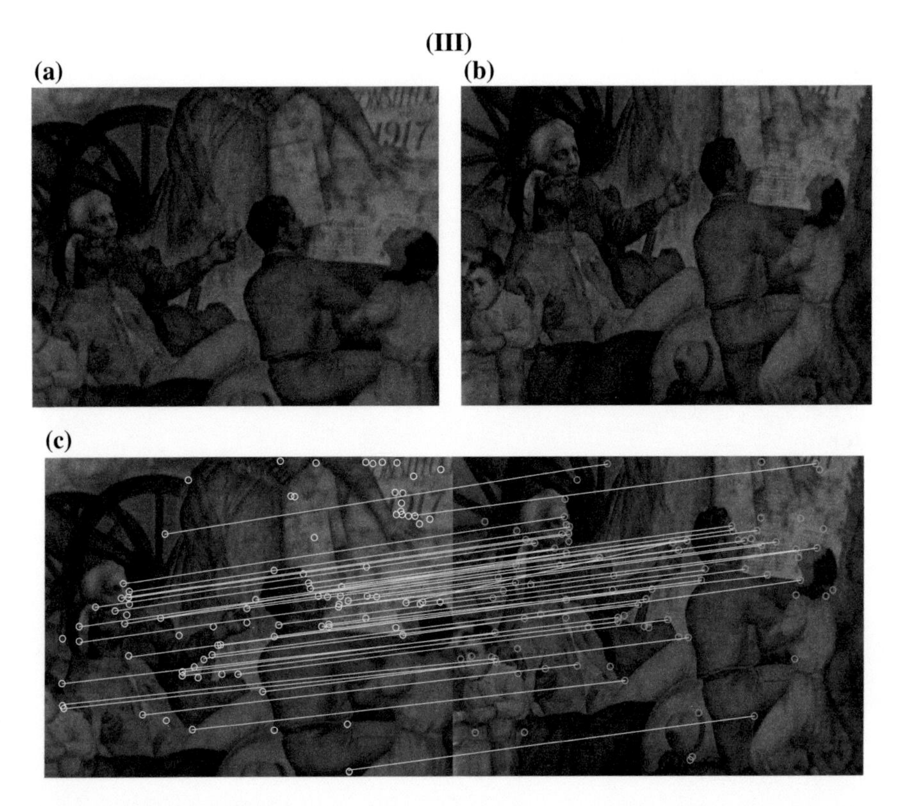

Fig. 6.15 Test image "III": **a** first view, **b** second view, **c** corresponding points and inliers produced by CSA-RANSAC

6.7 Conclusions

In this chapter, a method for robustly estimating multiple view relations from point correspondences has been presented. The approach combines the RANSAC method with the CSA. Upon such combination, the method adopts a different sampling strategy in comparison to RANSAC to generate putative solutions. Under the new mechanism, new candidate solutions are built iteratively by considering the quality of models generated by previous candidate solutions, rather than relying over a pure random selection as it is the case of RANSAC. Likewise, a more accurate objective function is incorporated to accurately evaluate the quality of a candidate model. As a result, the approach can substantially reduce the number of iterations still preserving the robust capabilities of RANSAC.

The resultant approach has been compared to other similar techniques proposed in the literature such as standard RANSAC [16], the MLESAC [19], the SIMFIT method [20], the projection-pursuit algorithm [21], the TSSE [22], the PSO

Table 6.3 The inlier number (IN), the error (E_s) and the number of function evaluations (NFE) for standard RANSAC [16], the MLESAC [19], the SIMFIT method [20], the projection-pursuit algorithm [21], the TSSE [22], the PSO algorithm (PSO-RANSAC) [43], the GA algorithm (GA-RANSAC) [45] and the presented CSA-RANSAC approach, considering the four test images shown by Figs. 6.13, 6.14 and 6.15

Image	Method	*IN*	E_r	*NFE*
(I)	Standard RANSAC	68	4.11	795
	MLESAC	82	3.21	721
	SIMFIT	88	2.33	697
	Projection-pursuit	75	3.11	775
	TSSE	78	3.51	758
	PSO-RANSAC	95	1.11	441
	GA-RANSAC	92	1.14	481
	CSA-RANSAC	110	0.47	378
(II)	Standard RANSAC	48	3.33	951
	MLESAC	65	3.11	802
	SIMFIT	71	2.32	765
	Projection-pursuit	56	2.87	821
	TSSE	50	2.46	911
	PSO-RANSAC	76	1.75	421
	GA-RANSAC	78	1.74	467
	CSA-RANSAC	82	0.52	389
(III)	Standard RANSAC	31	3.22	705
	MLESAC	39	2.87	657
	SIMFIT	41	2.65	617
	Projection-pursuit	35	2.56	559
	TSSE	37	2.79	632
	PSO-RANSAC	46	2.11	328
	GA-RANSAC	47	2.11	439
	CSA-RANSAC	55	0.53	287

algorithm (PSO-RANSAC) [43] the GA algorithm (GA-RANSAC) [43]. The efficiency of the algorithm has been evaluated in terms of accuracy and its computational cost. Experimental results after testing real and synthetic data provide evidence on the outstanding performance of the presented algorithm in comparison to previous methods.

Although the experimental results indicate that the presented method can yield better results on estimating fundamental matrices and homographies, it should be noticed that the aim of our chapter is not intended to beat all the RANSAC methods which have been proposed earlier, but to show that the use of evolutionary approaches can effectively serve as an attractive alternative to solve complex optimization problems, yet demanding fewer function evaluations.

References

1. Szeliski, R., Shum, H.-Y.: Creating full view panoramic image mosaics and texture-mapped models. In: Proceedings of the Computer Graphics, SIGGRAPH'97, pp. 251–258 (1997)
2. He, Y., Chung, R.: Image mosaicking for polyhedral scene and in particular singly visible surfaces. Pattern Recogn. **31**, 1200–1213 (2008)
3. Brown, M., Lowe, D.G.: Automatic panoramic image stitching using invariant features. Int. J. Comput. Vis. **74**, 59–73 (2007)
4. Akyol, A., Gökmen, M.: Super-resolution reconstruction of faces by enhanced global models of shape and texture. Pattern Recogn. **45**, 4103–4116 (2012)
5. Huang, H., He, H.T., Fan, X., Zhang, J.P.: Super-resolution of human face image using canonical correlation analysis. Pattern Recogn. **43**, 2532–2543 (2010)
6. Jia, K., Gong, S.G.: Hallucinating multiple occluded face images of different resoltuions. Pattern Recogn. Lett. **27**, 1768–1775 (2006)
7. Deniz, O., Bueno, G., Bermejo, E., Sukthankar, R.: Fast and accurate global motion compensation. Pattern Recogn. **44**(12), 2887–2901 (2011)
8. Montijano, E., Sagues, C.: Distributed multi-camera visual mapping using topological maps of planar regions. Pattern Recogn. **44**(7), 1528–1539 (2011)
9. Su, J., Chung, R., Jin, L.: Homography-based partitioning of curved surface for stereo correspondence establishment. Pattern Recogn. Lett. **28**(12), 1459–1471 (2007)
10. Santos, T.T., Morimoto, C.H.: Multiple camera people detection and tracking using support integration. Pattern Recogn. Lett. **32**(1), 47–55 (2011)
11. Uherčík, M., Kybic, J., Zhao, Y., Cachard, C., Liebgott, H.: Line filtering for surgical tool localization in 3D ultrasound images. Comput. Biol. Med. **43**(12), 2036–2045 (2013)
12. Zhao, Y., Cachard, C., Liebgott, H.: Automatic needle detection and tracking in 3D ultrasound using an ROI-based RANSAC and Kalman method. Ultrason. Imaging **35**(4), 283–306 (2013)
13. Zhang, H., Ostrowski, J.P.: Visual motion planning for mobile robots. IEEE Trans. Robot. Autom. **18**(2), 199–208 (2002)
14. Sagüés, C., Guerrero, J.J.: Visual correction for mobile robot homing. Robot Auton. Syst. **50**(1), 41–49 (2005)
15. López-Nicolás, G., Guerrero, J.J., Sagüés, C.: Visual control of vehicles using two-view geometry. Mechatronics **20**(2), 315–325 (2010)
16. Fischler, M.A., Bolles, R.C.: Random sample consensus: a paradigm for model fitting with applications to image analysis and automated cartography. Comm. ACM **24**, 381–395 (1981)
17. Matas, J., Chum, O.: Randomized RANSAC with $T_{d,d}$ test. Image Vis. Comput. **22**, 837–842 (2004)
18. Cheng, C.-M., Lai, S.-H.: A consensus sampling technique for fast and robust model fitting. Pattern Recogn. **42**, 1318–1329 (2009)
19. Torr, P.H.S.: MLESAC: a new robust estimator with application to estimating image geometry. Comput. Vis. Image Underst. **78**, 138–156 (2000)
20. Heinrich, S.B.: Efficient and robust model fitting with unknown noise scale. Image Vis. Comput. **31**, 735–747 (2013)
21. Subbarao, R., Meer, P.: Beyond RANSAC: user independent robust regression. In: Proceedings IEEE International Conference on Computer Vision and Pattern Recognition Workshop, no. 1, p. 101. 17–22 June 2006
22. Wang, H., Suter, D.: Robust adaptive-scale parametric model estimation for computer vision. IEEE Trans. Pattern Anal. Mach. Intell. **26**(11), 1459–1474 (2004)
23. Jun-hua, L., Ming, L.: An analysis on convergence and convergence rate estimate of elitist genetic algorithms in noisy environments. Optik—Int. J. Light Electron Opt. **124**(24), 6780–6785 (2013)
24. Mendel, E., Krohling, R.A., Campos, M.: Swarm algorithms with chaotic jumps applied to noisy optimization problems. Inf. Sci. **181**(20), 4494–4514 (2011)

25. Pan, H., Wang, L., Liu, B.: Particle swarm optimization for function optimization in noisy environment. Appl. Math. Comput. **181**, 908–919 (2006)
26. Beyer, H.-G.: Evolutionary algorithms in noisy environments: theoretical issues and guidelines for practice. Comput. Methods Appl. Mech. Eng. **186**, 239–267 (2000)
27. de Castro, L.N., von Zuben, F.J.: Learning and optimization using the clonal selection principle. IEEE Trans. Evol. Comput. **6**(3), 239–251 (2002)
28. Ada, G.L., Nossal, G.: The clonal selection theory. Sci. Am. **257**, 50–57 (1987)
29. CoelloCoello, C.A., Cortes, N.C.: Solving multiobjective optimization problems using an artificial immune system. Genet. Program Evolvable Mach. **6**, 163–190 (2005)
30. Campelo, F., Guimaraes, F.G., Igarashi, H., Ramirez, J.A.: A clonal selection algorithm for optimization in electromagnetics. IEEE Trans. Magn. **41**, 1736–1739 (2005)
31. Weisheng, D., Guangming, S., Li, Z.: Immune memory clonal selection algorithms for designing stack filters. Neurocomputing **70**, 777–784 (2007)
32. Cuevas, E., Osuna-Enciso, V., Wario, F., Zaldívar, D., Pérez-Cisneros, M.: Automatic multiple circle detection based on artificial immune systems. Expert Syst. Appl. **39**(1), 713–722 (2012)
33. Bay, H., Ess, A., Tuytelaars, T., Gool, L.V.: Surf: speeded up robust features. Comput. Vis. Image Underst. (CVIU) **110**(3), 346–359 (2008)
34. Harris, C., Stephens, M.: A combined corner and edge detector. In: Proceedings of the 4th Alvey Vision Conference, pp. 147–151 (1988)
35. Huynh, D.Q., Saini, A., Liu, W.: Evaluation of three local descriptors on low resolution images for robot navigation. In: Image and Vision Computing New Zealand, IVCNZ 2009. 24th International Conference, 113–118 (2009)
36. Faugeras, O.D.: What can be seen in three dimensions with an uncalibrated stereo? In: Sandini, G. (ed.) Proceedings of the 2nd European Conference on Computer Vision, Santa MargheritaLigure. Lecture Notes in Computer Science, vol. 588, pp. 563–578, Springer, Berlin (1992)
37. Hartley, R.I.: Estimation of relative camera positions for uncalibrated cameras. In Sandini, G. (ed.) Proceedings of the 2nd European Conference on Computer Vision, Santa MargheritaLigure. Lecture Notes in Computer Science, vol. 588, pp. 579–587, Springer, Berlin (1992)
38. Forsyth, D.A., Ponce, J.: Computer Vision a Modern Approach. Prentice Hall, Vilyams, Moscow, 2004 (2003)
39. Hartley, R.: In defense of the eight-point algorithm. IEEE Trans. Patt. Anal. Mach. Intell. **19**, 580–593 (1997)
40. Hartley, R.I., Zisserman A.: Multiple view geometry in computer vision (2nd ed.). Cambridge University Press, Cambridge (2004)
41. Meer, P.: Robust techniques in computer vision. In: Medioni, G., Kang, S.B. (eds.) Emerging Topics in Computer Vision, pp. 107–190. Prentice Hall, Boston (2004)
42. Nanda, S.J., Panda, G.: A survey on nature inspired metaheuristic algorithms for partitional clustering. Swarm Evol. Comput. **16**, 1–18 (2014)
43. Kennedy, J., Eberhart, R.C.: Particle swarm optimization. In: Proceedings of the 1995 IEEE International Conference on Neural Networks, vol. 4, pp. 1942–1948. (1995)
44. Holland, J.H.: Adaptation in Natural and Artificial Systems: An Introductory Analysis with Applications to Biology, Control, and Artificial Intelligence. MIT Press, Cambridge (1992)
45. Hamzaçebi, C.: Improving genetic algorithms' performance by local search for continuous function optimization. Appl. Math. Comput. **196**(1), 309–317 (2008)
46. Wang, Y., Li, B., Weise, T., Wang, J., Yuan, B., Tian, Q.: Self-adaptive learning based particle swarm optimization. Inf. Sci. **181**(20), 4515–4538 (2011)
47. Pandey, H.M., Chaudhary, A., Mehrotra, D.: A comparative review of approaches to prevent premature convergence in GA. Appl. Soft Comput. **24**, 1047–1077 (2014)
48. Wang, H., Sun, H., Li, C., Rahnamayan, S., Jeng-shyang, P.: Diversity enhanced particle swarm optimization with neighborhood. Inf. Sci. **223**, 119–135 (2013)

49. Hrstka, O., Kučerová, A.: Improvements of real coded genetic algorithms based on differential operators preventing premature convergence. Adv. Eng. Softw. **35**(3–4), 237–246 (2004)
50. Peng, Y., Lu, B.-L.: Hybrid learning clonal selection algorithm. Inf. Sci. **296**, 128–146 (2015)
51. Gong, M., Jiao, L., Zhang, L., Du, H.: Immune secondary response and clonal selection inspired optimizers. Prog. Nat. Sci. **19**, 237–253 (2009)
52. Goldsby, G.A., Kindt, T.J., Kuby, J., Osborne, B.A.: Immunology, 5th edn. Freeman, New York (2003)
53. Cutello, V., Narzisi, G., Nicosia, G., Pavone, M.: Clonal selection algorithms: a comparative case study using effective mutation potentials. In: Jacob, C. et al. (eds.) ICARIS 2005, LNCS 3627, pp. 13–28. (2005)
54. Gao, X., Wang, X., Ovaska, S.: Fusion of clonal selection algorithm and differential evolution method in training cascade-correlation neural network. doi:10.1016/j.neucom.2008.11.004
55. Yoo, J., Hajela, P.: Immune network simulations in multicriterion design. Struct. Optim. **18**(2–3), 85–94 (1999)
56. Wang, X., Gao, X.Z., Ovaska, S.J.: A hybrid optimization algorithm in power filter design. In: Proceedings of the 31st Annual Conference of the IEEE Industrial Electronics Society, Raleigh, NC, pp. 1335–1340, Nov 2005
57. Xu, X., Zhang, J.: An improved immune evolutionary algorithm for multimodal function optimization. In: Proceedings of the Third International Conference on Natural Computation, Haikou, China, pp. 641–646, Aug 2007
58. Tang, T., Qiu, J.: An improved multimodal artificial immune algorithm and its convergence analysis. In: Proceedings of the Sixth World Congress on Intelligent Control and Automation, Dalian, China, pp. 3335–3339, June 2006
59. Arani, B.O., Mirzabeygi, P., Panahi, M.S.: An improved PSO algorithm with a territorial diversity-preserving scheme and enhanced exploration–exploitation balance. Swarm Evol. Comput. **11**, 1–15 (2013)
60. Li, X., Xiao, N., Claramunt, C., Lin, H.: Initialization strategies to enhancing the performance of genetic algorithms for the p-median problem. Comput. Ind. Eng. **61**(4), 1024–1034 (2011)
61. Rahnamayan, S., Tizhoosh, H.R., Salama, M.M.A.: A novel population initialization method for accelerating evolutionary algorithms. Comput. Math Appl. **53**(10), 1605–1614 (2007)
62. Clerc, M., Kennedy, J.: The particle swarm—explosion, stability, and convergence in a multidimensional complex space. IEEE Trans. Evol. Comput. **6**(1), 58–73 (2002)
63. Wang, H., Sun, H., Li, C., Rahnamayan, S., Pan, J.S.: Diversity enhanced particle swarm optimization with neighborhood search. Inf. Sci. **223**, 119–135 (2013)
64. Fabio, F., Maurizio, R.: Comparison of artificial immune systems and genetic algorithms in electrical engineering optimization. Int. J. Comput. Math. Electr. Electron. Eng. **25**, 792–811 (2006)
65. Wang, H., Sun, H., Li, C., Rahnamayan, S., Jeng-shyang, P.: Diversity enhanced particle swarm optimization with neighborhood. Inf. Sci. **223**, 119–135 (2013)

Chapter 7
Circle Detection on Images Based on an Evolutionary Algorithm that Reduces the Number of Function Evaluations

Abstract Automatic circle detection in digital images is considered as an important and complex task for the computer vision community which has consequently devoted a tremendous amount of research to find an optimal circle detector. On the other hand, Evolutionary Algorithms (EA) are earning popularity as computational intelligence approaches for solving complex problems that are encountered in many engineering disciplines. They have exhibited robustness and suitability to locate global optimum as compared to deterministic (gradient based) optimization methods. Over the last decade, new algorithms based on EA have been applied to image processing. However, one critical difficulty in deploying EA to real-world problems is their high computational time associated to a large number of function evaluations which are required to deliver a satisfactory result. This chapter presents an algorithm for the automatic detection of circular shapes from complicated and noisy images with no consideration of the conventional Hough transform principles. The algorithm is based on a newly developed evolutionary algorithm called the Adaptive Population with Reduced Evaluations (APRE). The algorithm reduces the number of function evaluations through the use of two mechanisms: (1) adapting dynamically the size of the population and (2) incorporating a fitness calculation strategy which decides whether the calculation or estimation of the new generated individuals is feasible. As a result, the approach can substantially reduce the number of function evaluations, yet preserving the good search capabilities of an evolutionary approach. The algorithm uses the encoding of three pixels as candidate circles over the edge image. An objective function evaluates if such candidate circles are actually present in the edge image. Guided by the values of this objective function, the set of encoded candidate circles are evolved using the operators defined by APRE so that they can fit into the actual circles on the edge map of the image. Experimental results over several synthetic and natural images, with a varying range of complexity, validate the efficiency of the resultant technique with regard to accuracy, speed, and robustness.

E. Cuevas et al., *Applications of Evolutionary Computation in Image Processing and Pattern Recognition*, Intelligent Systems Reference Library 100,
DOI 10.1007/978-3-319-26462-2_7

7.1 Introduction

The problem of detecting circular features holds paramount importance in several engineering applications such as automatic inspection of manufactured products and components, aided vectorization of drawings, target detection, etc. [1, 2]. Circle detection in digital images has been commonly solved through the Circular Hough Transform (CHT) [3]. Unfortunately, this approach requires a large storage space that augments the computational complexity and yield a low processing speed. In order to overcome this problem, several approaches which modify the original CHT have been proposed. Onewell-known example is the Randomized Hough Transform (RHT) [4].

As an alternative to Hough Transform-based techniques, the problem of shape recognition has also been handled through evolutionary methods. In general, they have demonstrated to deliver better results than those based on the HT considering accuracy and robustness [5]. Evolutionary methods approach the detection task as an optimization problem whose solution involves the computational expensive evaluation of objective functions. Such fact strongly restricts their use in several image processing applications; despite of this EA methods have produced several robust circle detectors which use different evolutionary algorithms like Genetic algorithms (GA) [5], Harmony Search (HSA) [6], Electromagnetism-Like (EMO) [7], Differential Evolution (DE) [8] and Bacterial Foraging Optimization (BFOA) [9].

However, one particular difficulty in applying any EA to real-world problems, such as image processing, is its demand for a large number of fitness evaluations before reaching a satisfactory result. Fitness evaluations are not always straightforward as either an explicit fitness function does not exist or the fitness evaluation is computationally expensive.

The problem of excessively long fitness function calculations has already been faced in the field of evolutionary algorithms (EA) and is better known as evolution control or as fitness estimation [10]. For such an approach, the idea is to replace the costly objective function evaluation for some individuals by alternative models which are based on an approximate model of the fitness landscape. The individuals to be evaluated and those to be estimated are determined following some fixed criteria which depend on the specific properties of the approximate model [11]. The models involved at the estimation can be dynamically built during the actual EA execution, since EA repeatedly sample the search space at different points [12]. There are several alternative models which have been used in combination with popular EAs. Some examples include polynomial functions [13], kriging schemas [14], multi-layer Perceptrons [15] and radial basis-function networks [16]. In the practice, it is very difficult the construction of successful models which can globally deal with the high dimensionality, ill distribution and limited number of training samples. Experimental studies [17] have demonstrated that if an alternative model is used for fitness evaluation, it is very likely that the evolutionary algorithm will converge to a false optimum. A false optimum is an optimum of the alternative model, which does not coincide with the optimum of the original fitness function. Under such conditions, the use the alternative fitness models degrade the search effectiveness of the original EAs, producing frequently inaccurate solutions [18].

In an EA, the population size has a direct influence on the solution quality and its computational cost [19]. Traditionally, population size is set in advance to a pre-specified value and remains fixed through the entire execution of the algorithm. If the population size is too small then the EA may converge too quickly affecting severely the solution quality [20]. On the other hand, if it is too large then the EA may present a prohibitive computational cost [19]. Therefore, an appropriate population size allows maintaining a trade-off between computational cost and effectiveness of the algorithm. In order to solve such a problem, several approaches have been proposed for dynamically adapting the population size. These methods are grouped into three categories [21]: (i) methods that increment or decrement the number of individuals according to a fixed function; (ii) methods in which the number of individuals is modified according to the performance of the average fitness value and (iii) algorithms based on the population diversity.

In order to use either a fitness estimation strategy or an adaptive population size approach, it is necessary but not sufficient to tackle the problem of reducing the number of function evaluations. Using a fitness estimation strategy during the evolution process with no adaptation of the population size to improve the population diversity, makes the algorithm defenseless against the convergence to a false minimum and may result in poor exploratory characteristics of the algorithm [18]. On the other hand, the adaptation of the population size omitting the fitness estimation strategy leads to increase the computational cost [20]. Therefore, it does seem reasonable to incorporate both approaches into a single algorithm.

Since most of the EAs have been primarily designed to completely evaluate all involved individuals, techniques for reducing the evaluation number are usually incorporated into the original EAs in order to estimate fitness values or to reduce the number of individuals being evaluated [22]. However, the use of alternative fitness models degrades the search effectiveness of the original EAs, producing frequently inaccurate solutions [23].

This chapter presents an algorithm for the automatic detection of circular shapes from complicated and noisy images without considering of the conventional Hough transform principles. The algorithm is based on a newly developed evolutionary algorithm called the Adaptive Population with Reduced Evaluations (APRE). The presented method reduces the number of function evaluations through the use of two mechanisms: (1) adapting dynamically the size of the population and (2) incorporating a fitness calculation strategy which decides when it is feasible to calculate or only to estimate new generated individuals.

The APRE method begins with an initial population which is to be considered as a memory during the evolution process. To each memory element, a normalized fitness value, called quality factor is assigned to indicate the solution capacity that is provided by the element. Only a variable sub-set of memory elements is considered to be evolved. Like all EA-based methods, the algorithm generates new individuals considering two operators: exploration and exploitation. Both operations are applied to improve the quality of the solutions by: (1) searching through the unexplored solution space to identify promising areas that contain better solutions than those found so far, and (2) successive refinement of the best found solutions. Once the new individuals are generated, the memory is accordingly updated. At such stage, the new individuals compete against the memory elements to build the final memory configuration. In order to save computational time, the approach incorporates a fitness estimation strategy that decides which individuals can be estimated or actually evaluated. The proposed fitness calculation strategy estimates the fitness value of new individuals using memory elements located in neighboring positions which have been visited during the evolution process. In the strategy, new individuals that are located near to memory element whose quality factor is high, have a great probability to be evaluated by using the true objective function. Similarly, it is also evaluated those new particles lying in regions of the search space with no previous evaluations. The remaining search positions are only estimated by assigning the same fitness value that the nearest location element on the memory. The use of such a fitness estimation method contributes to save computational time, since the fitness value of only very few individuals are actually evaluated whereas the rest is just estimated.

Different to other approaches that use an already existent EA as framework, the APRE method has been completely designed to substantially reduce the computational cost, yet preserving good search effectiveness.

In order to detect circular shapes, the detector is implemented by encoding three pixels as candidate circles over the edge image. An objective function evaluates if such candidate circles are actually present in the edge image. Guided by the values of this objective function, the set of encoded candidate circles are evolved using the operators defined by APRE so that they can fit into the actual circles on the edge map of the image. Comparisons to several state-of-the-art evolutionary-based methods and the Randomized Hough Transform (RHT) approach on multiple images demonstrate a better performance of the resultant method in terms of accuracy, speed and robustness.

7.2 The Adaptive Population with Reduced Evaluations (APRE) Algorithm

In the algorithm, a population of candidate solutions to an optimization problem is evolved toward better solutions. The algorithm begins with an initial population which will be used as a memory during the evolution process. To each memory element, it is assigned a normalized fitness value called quality factor that indicates the solution capacity provided by the element. As a search strategy, the presented algorithm implements two operations: “exploration” and “exploitation”, both necessary in all EAs [24]. Exploration is the operation of visiting entirely new points of a search space, whilst exploitation is the process of refining those points of a search space within the neighborhood of previously visited locations in order to improve their solution quality. Pure exploration degrades the precision of the evolutionary process but increases its capacity to find new potentially solutions [25]. On the other hand, pure exploitation allows refining existent solutions but adversely drives the process to fall in local optimal solutions [26]. Therefore, the ability of an EA to find a global optimal solution depends on its capacity to find a good trade-off between the exploitation of found-so-far elements and the exploration of the search space.

The APRE algorithm is an iterative process in which several actions are executed. First, the number of memory elements to be evolved is computed. Such number is automatically modified at each iteration. Then, a set of new individuals is generated as a consequence of the execution of the exploration operation. For each new individual, its fitness value is estimated or evaluated according to a decision taken by a fitness estimation strategy. Afterwards, the memory is updated. In this stage, the new individuals produced by the exploration operation compete against the memory elements to build the final memory configuration. Finally, a sample of the best elements contained in the final memory configuration is undergone to the exploitation operation. Thus, the complete process can be divided in six phases: Initialization, selecting the population to be evolved, exploration, fitness estimation strategy, memory updating and exploitation.

7.2.1 Initialization

Like in EA, the APRE algorithm is an iterative method whose first step is to randomly initialize the population $\mathbf{M}(k)$ which will be used as a memory during the evolution process. The algorithm begins by initializing ($k = 0$) a set of N_p elements ($\mathbf{M}(k) = \{\mathbf{m}_1, \mathbf{m}_2, \ldots, \mathbf{m}_{N_p}\}$). Each element $\mathbf{m}_i$ is an n-dimensional vector containing the parameter values to be optimized. Such values are randomly and uniformly distributed between the pre-specified lower initial parameter bound p_j^{low} and the upper initial parameter bound p_j^{high}, just as it described by the following expression:

$$m_{i,j} = p_j^{low} + \text{rand}(0,1) \cdot (p_j^{high} - p_j^{low}); \quad \text{for } j = 1, 2, \ldots, n; \text{ and } i = 1, 2, \ldots, N_p, \tag{7.1}$$

being j and i the parameter and element indexes, respectively. Hence, $m_{i,j}$ is the j-th parameter of the i-the element.

Each element $\mathbf{m}_i$ has two associated characteristics: a fitness value $J(\mathbf{m}_i)$ and a quality factor $Q(\mathbf{m}_i)$. The fitness value $J(\mathbf{m}_i)$ assigned to each element $\mathbf{m}_i$ can be calculated by using the true objective function $f(\mathbf{m}_i)$ or only estimated by using the presented fitness strategy $F(\mathbf{m}_i)$. In addition to the fitness value, it is also assigned to $\mathbf{m}_i$ a normalized fitness value called quality factor $Q(\mathbf{m}_i)$ ($Q(\cdot) \in [0, 1]$) which is computed as follows:

$$Q(\mathbf{m}_i) = \frac{J(\mathbf{m}_i) - worst_{\mathbf{M}}}{best_{\mathbf{M}} - worst_{\mathbf{M}}}, \tag{7.2}$$

where $J(\mathbf{m}_i)$ is the fitness value obtained by evaluation $f(\cdot)$ or by estimation $F(\cdot)$ of the memory element $\mathbf{m}_i$. The values $worst_{\mathbf{M}}$ and $best_{\mathbf{M}}$ are defined as follows (considering a maximization problem):

$$best_{\mathbf{M}} = \max_{k \in \{1,2,\ldots,N_p\}} (J(\mathbf{m}_k)) \text{ and } worst_{\mathbf{M}} = \min_{k \in \{1,2,\ldots,N_p\}} (J(\mathbf{m}_k)) \tag{7.3}$$

Since the mechanism by which an EA accumulates information regarding the objective function is an exact evaluation of the quality of each potential solution, initially, all the elements of $\mathbf{M}(k)$ are evaluated without considering the fitness estimation strategy presented in this chapter. This fact is only allowed at this initial stage.

7.2.2 Selecting the Population to Be Evolved

At each k iteration, it must be selected which and how many elements from $\mathbf{M}(k)$ will be considered to build the population $\mathbf{P}^k$ to be evolved. Such selected elements will be undergone by the exploration and exploitation operators in order to generate a set of new individuals. Therefore, two things need to be defined: the number of elements N_e^k to be selected and the strategy of selection.

7.2.3 The Number of Elements N_e^k to Be Selected

One of the mechanisms used by the APRE algorithm for reducing the number of function evaluations is to modify dynamically the size of the population to be evolved. The idea is to operate with the minimal number of individuals that guarantee the correct efficiency of the algorithm. Hence, the method aims to vary the population size in an adaptive way during the execution of each iteration. At the beginning of the process, a predetermined number N_e^0 of elements are considered to build the first population; then, it will be incremented or decremented depending on the algorithm´s performance. The adaptation mechanism is based on the lifetime of the individuals and on their solution quality. In order to compute the lifetime of each individual, it is assigned a counter $c_i(i \in (1, 2, 3, \ldots, N_p))$ to each element $\mathbf{m}_i$ of $\mathbf{M}(k)$. When the initial population $\mathbf{M}(k)$ is created, all the counters are set to zero. Since the memory $\mathbf{M}(k)$ is updated at each generation, some elements prevail and others will be substituted by new individuals. Therefore, the counter of the surviving elements is incremented by one whereas the counter of new added elements is set to zero.

Another important requirement to calculate the number of elements to be evolved is the solution quality provided by each individual. The idea is to identify two classes of elements, those that provide good solutions and those that can be considered as bad solutions. In order to classify each element, the average fitness value J_A produced by all the elements of $\mathbf{M}(k)$ is calculated as:

$$J_A = \frac{1}{N_p} \sum_{i=1}^{N_p} J(\mathbf{m}_i), \tag{7.4}$$

where $J(\cdot)$ represents the fitness value corresponding to $\mathbf{m}_i$. These values are evaluated either by the true objective function $f(\mathbf{m}_i)$ or by the fitness estimation strategy $F(\mathbf{m}_i)$. Considering the average fitness value, two groups are built: the set $\mathbf{G}$ constituted by the elements of $\mathbf{M}(k)$ whose fitness values are greater than J_A and the set $\mathbf{B}$ which groups the elements of $\mathbf{M}(k)$ whose fitness values are equal or lower than J_A. Therefore, the number of individuals of the current population that will be incremented or decremented at each generation is calculated by the following model:

$$A = \text{floor}\left(\frac{|\mathbf{G}| \cdot \sum_{l \in \mathbf{G}} c_l - |\mathbf{B}| \cdot \sum_{q \in \mathbf{B}} c_q}{s} \right), \tag{7.5}$$

where the floor$(\cdot)$ function maps a real number to the previous integer. $|\mathbf{G}|$ and $|\mathbf{B}|$ represent the number of elements of $\mathbf{G}$ and $\mathbf{B}$, respectively whereas $\sum_{l \in \mathbf{G}} c_l$ and $\sum_{q \in \mathbf{B}} c_q$ indicate the sum of the counters that correspond to the elements of $\mathbf{G}$ and $\mathbf{B}$, respectively. The factor s is a term used for fine tuning. A small value of s implies a better algorithm's performance at the price of an increment in the computational cost. On the other hand, a big value of s involves a low computational cost at the price of a decrement in the performance algorithm. Therefore, the s value must reflex a compromise between performance and computational cost. In our experiments such compromise has been found with $s = 10$.

Therefore, the number the elements that define the population to be evolved is computed according to the following model:

$$N_e^k = N_e^{k-1} + A \tag{7.6}$$

Since the value of A can be positive or negative, the size of the population $\mathbf{P}^k$ may be higher or lesser than $\mathbf{P}^{k-1}$. The computational procedure that implements this method is presented in Algorithm 7.1, in form of pseudo code.

Algorithm 7.1 Selection of the number of individuals N_e^k to be evolved

1: **Input**: Current population $\mathbf{M}(k)$, counters $c_1, \ldots, c_{N_p}$, the past number of individuals N_e^{k-1} and the constant factor s.

2: $J_A \leftarrow \frac{1}{N_p}\sum_{i=1}^{N_p} J(\mathbf{m}_i)$

3: $\mathbf{G} \leftarrow$ FindIndividualsOverJA($\mathbf{M}(k)$, J_A)

4: $\mathbf{B} \leftarrow$ FindIndividualsUnderJA($\mathbf{M}(k)$, J_A)

5: $c_l \leftarrow$ FindCountersOfG($\mathbf{G}$) (Where $l \in \mathbf{G}$)

6: $c_q \leftarrow$ FindCountersOfB($\mathbf{B}$) (Where $q \in \mathbf{B}$)

7: $A \leftarrow \text{floor}\left(\frac{|\mathbf{G}|\cdot\sum_{l\in\mathbf{G}} c_l - |\mathbf{B}|\cdot\sum_{q\in\mathbf{B}} c_q}{s}\right)$

8: $N_e^k \leftarrow N_e^{k-1} + A$

9: **Output**: The number N_e^k

7.2.4 Selection Strategy for Building $\mathbf{P}^k$

Once the number of individuals has been defined, the next step is the selection of N_e^k elements from $\mathbf{M}(k)$ for building $\mathbf{P}^k$. A new population **MO** which contains the same elements that $\mathbf{M}(k)$ is generated, but sorted according to their fitness values. Thus, **MO** presents in its first positions the elements whose fitness values are better than those located in the last positions. Then, **MO** is divided in two parts: **X** and **Y**. The section **X** corresponds to the first N_e^k elements of **MO** whereas the rest of the elements constitute the part **Y**. Figure 7.1 shows this process.

In order to promote diversity, in the selection strategy, the 80 % of the N_e^k individuals of $\mathbf{P}^k$ are taken from the first elements of **X** and named as Fe as shown

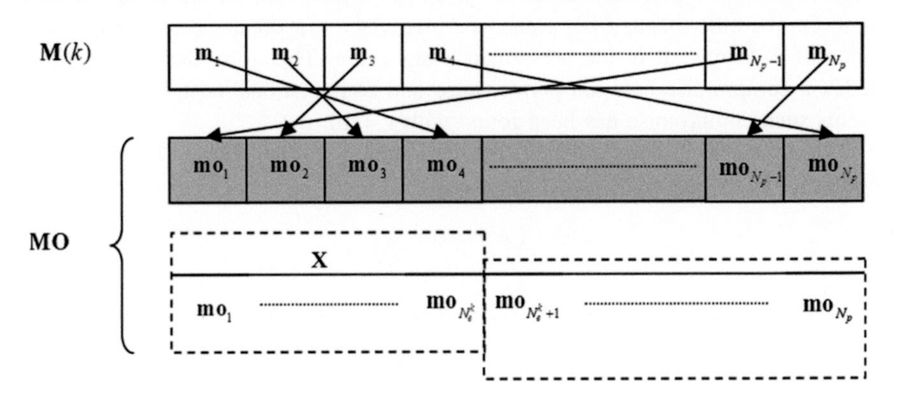

Fig. 7.1 Necessary post-processing implemented by the selection strategy

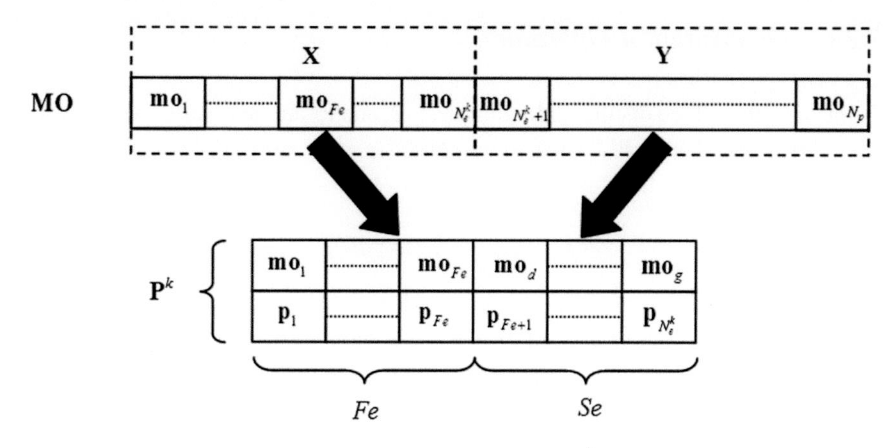

Fig. 7.2 Employed selection strategy to build the population $\mathbf{P}^k$, where $d, g \in \mathbf{Y}$

in Fig. 7.2, where $Fe = \text{floor}(0.8 * N_e^k)$. The remaining 20 % of the individuals are randomly selected from section **Y**. Hence, the last set of *Se* elements (where $Se = N_e^k - Fe$) is chosen considering that all elements of **Y** have the same possibility of being selected. Figure 7.2 shows a description of the selection strategy. The computational procedure that implements this method is presented in Algorithm 7.2, in form of pseudo code.

Algorithm 7.2 Selection strategy for building $\mathbf{P}^k$

1: **Input**: Current population $\mathbf{M}(k)$ and the number of individuals N_e^k.
2: **MO** ← SortElementsFitness($\mathbf{M}(k)$)
3: [**X**, **Y**] ← DivideMO(**MO**, N_e^k)
4: $Fe \leftarrow \text{floor}(0.8 * N_e^k)$
5: $Se \leftarrow N_e^k - Fe$
6: $(\mathbf{p}_1^k, \ldots, \mathbf{p}_{Fe}^k) \leftarrow$ SelectElementsOfX(**X**, *Fe*)
7: $(\mathbf{p}_{Fe+1}^k, \ldots, \mathbf{p}_{Fe+Se}^k) \leftarrow$ SelectRandomElementsOfY(**Y**, *Se*)
8: **Output**: Population $\mathbf{P}^k$ to be evolve

7.2.5 Exploration Operation

The first main operation applied to the population $\mathbf{P}^k$ is the exploration operation. Considering $\mathbf{P}^k$ as the input population, APRE mutates $\mathbf{P}^k$ to produce a temporal population $\mathbf{T}^k$ of N_e^k vectors. In the exploration operation two different mutation models are used: the mutation employed by the Differential Evolution algorithm (DE) [27] and the trigonometric mutation operator [28].

7.2.6 DE Mutation Operator

In this mutation, three distinct individuals *r1*, *r2*, and *r3* are randomly selected from the current population $\mathbf{P}^k$. Then, it is created a new value $h_{i,j}$ considering the following model:

$$h_{i,j} = p_{r1,j} + F(p_{r2,j} - p_{r3,j}) \tag{7.7}$$

where *r1*, *r2*, and *r3* are randomly selected individuals such that they satisfy: $r1 \neq r2 \neq r3 \neq i$; $i = 1$ to N_e^k (population size), and $j = 1$ to n (number of decision variable). Hence, $p_{i,j}$ is the *j*-th parameter of the *i*-th individual of $\mathbf{P}^k$. The scale factor, F (0, 1+), is a positive real number that controls the rate at which the population evolves.

7.2.7 Trigonometric Mutation Operator

The trigonometric mutation operation is performed according to the following formulation:

$$\begin{aligned} h_{i,j} &= p_{Av}(j) + F_1(p_{r1,j} - p_{r2,j}) + F_2(p_{r2,j} - p_{r3,j}) + F_3(p_{r3,j} - p_{r1,j}) \\ p_{Av}(j) &= \frac{p_{r1,j} + p_{r2,j} + p_{r3,j}}{3} \\ F_1 &= (d_{r2} - d_{r1}), F_2 = (d_{r3} - d_{r2}), F_1 = (d_{r1} - d_{r3}), \\ d_{r1} &= \frac{|J(\mathbf{p}_{r1})|}{d_T}, d_{r2} = \frac{|J(\mathbf{p}_{r2})|}{d_T}, d_{r3} = \frac{|J(\mathbf{p}_{r3})|}{d_T} \\ d_T &= |J(\mathbf{p}_{r1})| + |J(\mathbf{p}_{r2})| + |J(\mathbf{p}_{r3})|, \end{aligned} \tag{7.8}$$

where $\mathbf{p}_{r1}$, $\mathbf{p}_{r2}$ and $\mathbf{p}_{r3}$ represent the individuals *r1*, *r2*, and *r3* randomly selected from the current population $\mathbf{P}^k$ whereas $J(\cdot)$ represents the fitness value (calculated or estimated) corresponding to $\mathbf{p}_i$. Under this formulation, the individual $p_{Av}(j)$ to be perturbed is the average value of three randomly selected vectors (*r1*, *r2*, and *r3*). The perturbation to be imposed over such individual is implemented by the sum of three weighted vector differentials. F_1, F_2, and F_3 are the weights applied to these vector differentials. Notice that the trigonometric mutation is a greedy operator since it biases the $p_{Av}(j)$ strongly in the direction where the best one of three individuals is lying.

7.2.7.1 Computational Procedure

Considering $\mathbf{P}^k$ as the input population, all its N_e^k individuals are sequentially processed in cycles beginning by the first individual $\mathbf{p}_1$. Therefore, in the cycle

i (where it is processed the individual i), three distinct individuals *r1*, *r2*, and *r3* are randomly selected from the current population considering that they satisfy the following conditions $r1 \neq r2 \neq r3 \neq i$. Then, it is processed each dimension of $\mathbf{p}_i$ beginning by the first parameter 1 until the last dimension n has been reached. At each processing cycle, the parameter $p_{i,j}$ considered as a parent, creates an offspring $t_{i,j}$ in two steps. In the first step, from the selected individuals *r1*, *r2*, and *r3*, a donor vector $h_{i,j}$ is created by means of two different mutation models. In order to select which mutation model is applied, a uniform random number is generated within the range [0, 1]. If such number is less than a threshold *MR*, the donor vector $h_{i,j}$ is generated by the DE mutation operator; otherwise, it is produced by the trigonometric mutation operator. Such process can be modeled as follows:

$$h_{i,j} = \begin{cases} \text{By using Eq. (7)} & \text{with probability } MR \\ \text{By using Eq. (8)} & \text{with probability } (1 - MR) \end{cases} \tag{7.9}$$

In the second step, the final value of the offspring $t_{i,j}$ is determined. Such decision is stochastic; hence a second uniform random number is generated within the range [0, 1]. If this random number is less than *CH*, $t_{i,j} = h_{i,j}$; otherwise, $t_{i,j} = p_{i,j}$. This operation can be formulated as following:

$$t_{i,j} = \begin{cases} h_{i,j} & \text{with probability } CH \\ p_{i,j} & \text{with probability } (1 - CH) \end{cases} \tag{7.10}$$

The complete computational procedure is presented in Algorithm 7.3, in form of pseudo code.

Algorithm 7.3 Exploration operation of APRE algorithm

1: **Input**: Current population $\mathbf{P}^k$
2: **for** i=1 to N_e^k **do**
3: $(\mathbf{p}_{r1}, \mathbf{p}_{r2}, \mathbf{p}_{r3}) \leftarrow$ SelectElements() %Considering that $r1 \neq r2 \neq r3 \neq i$
4: **for** j=1 to n **do**
5: **if** (rand(0,1)<=*MR*) **then**
6: $h_{i,j} \leftarrow$ DEMutation $(\mathbf{p}_{r1}, \mathbf{p}_{r2}, \mathbf{p}_{r3})$ % Eq. 7.7
7: **else**
8: $h_{i,j} \leftarrow$ TrigonometricMutation $(\mathbf{p}_{r1}, \mathbf{p}_{r2}, \mathbf{p}_{r3})$ % Eq. 7.8
9: **end if**
10: **if** (rand(0,1)<=*CH*) **then**
11: $t_{i,j} \leftarrow h_{i,j}$
12: **else**
13: $t_{i,j} \leftarrow p_{i,j}$
14: **end if**
15: **end for**
16: **end for**
17: **Output**: Population $\mathbf{T}^k$

7.2.8 Fitness Estimation Strategy

Once the population $\mathbf{T}^k$ has been generated by the exploration operation, it is necessary to calculate the fitness value provided by each individual. In order to reduce the number of function evaluations, it is introduced a fitness estimation strategy that decides which individuals can be estimated or actually evaluated. The idea of such a strategy is to find the global optimum of a given function considering only a very few number of function evaluations.

In this chapter, we explore a local approximation scheme that estimates the fitness values based on previously evaluated neighboring individuals, stored in the memory $\mathbf{M}(k)$ during the evolution process. The strategy decides if an individual $\mathbf{t}_i$ is calculated or estimated based on two criteria. The first one considers the distance between $\mathbf{t}_i$ and the nearest element $\mathbf{m}^{ne}$ contained in $\mathbf{M}(k)$ (where $\mathbf{m}^n \in (\mathbf{m}_1, \mathbf{m}_2, \ldots, \mathbf{m}_{N_p})$) whereas the second one examines the quality factor provided by the nearest element $\mathbf{m}^{ne}(Q(\mathbf{m}^{ne}))$.

In the model, individuals of $\mathbf{T}^k$ that are near the elements of $\mathbf{M}(k)$ holding the best quality values have a high probability to be evaluated. Such individuals are important, since they will have a stronger influence on the evolution process than other individuals. In contrast, individuals of $\mathbf{T}^k$ that are also near the elements of $\mathbf{M}$ (k) but with a bad quality value maintain a very low probability to be evaluated. Thus, most of such individuals will be only estimated, assigning it the same fitness value that the nearest element of $\mathbf{M}(k)$. On the other hand, those individuals in regions of the search space with few previous evaluations (individuals of $\mathbf{T}^k$ located farther than a distance D) are also evaluated. The fitness values of these individuals are uncertain since there is no close reference (close points contained in $\mathbf{M}(k)$).

Therefore, the fitness estimation strategy follows two rules in order to evaluate or estimate the fitness values:

1. If the new individual $\mathbf{t}_i$ is located closer than a distance D with respect to the nearest element $\mathbf{m}^{ne}$ stored in $\mathbf{M}$, then a uniform random number is generated within the range [0, 1]. If such number is less than $Q(\mathbf{m}^{ne})$, $\mathbf{t}_i$ is evaluated by the true objective function $(f(\mathbf{t}_i))$. Otherwise, its fitness value is estimated assigning it the same fitness value that $\mathbf{m}^{ne}(F(\mathbf{t}_i) = J(\mathbf{m}^{ne}))$. Figure 7.3a, b draw the rule procedure.
2. If the new individual $\mathbf{t}_i$ is located longer than a distance D with respect to the nearest individual location $\mathbf{m}^{ne}$ stored in $\mathbf{M}$, then the fitness value of $\mathbf{t}_i$ is evaluated using the true objective function $(f(\mathbf{t}_i))$. Figure 7.3c outlines the rule procedure.

From the rules, the distance D controls the trade-off between the evaluation and estimation of new individuals. Unsuitable values of D result in a lower convergence rate, longer computation time, larger function evaluation number, convergence to a local maximum or unreliability of solutions. Therefore, the D value is computed considering the following equation:

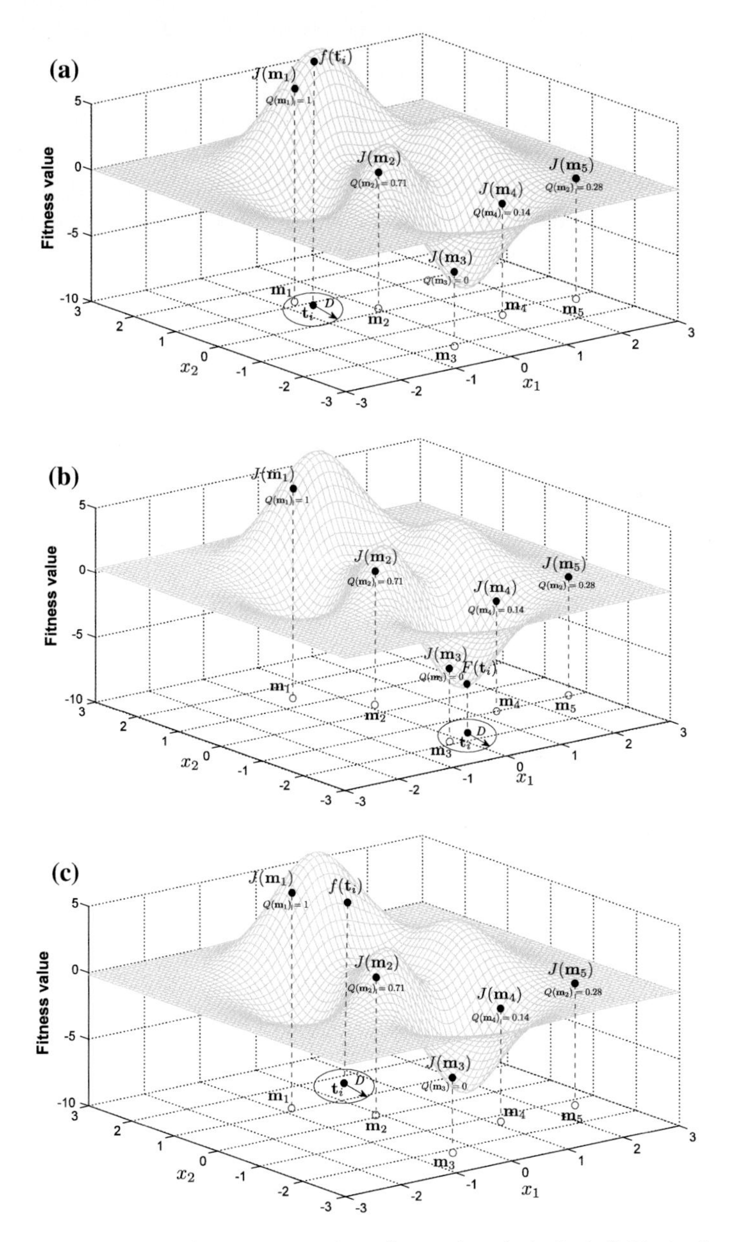

Fig. 7.3 The fitness estimation strategy. **a** According to the rule 1, the individual $\mathbf{t}_i$ has a high probability to be evaluated $f(\mathbf{t}_1)$, since it is located closer than a distance D with respect to the nearest element $\mathbf{m}^{ne} = \mathbf{m}_1$ whose quality factor $Q(\mathbf{m}_1)$ corresponds to the best value. **b** According to the rule 1, the individual $\mathbf{t}_i$ has a high probability to be estimated $F(\mathbf{t}_1)$ (assigning it the same fitness value that $\mathbf{m}^{ne}$ $(F(\mathbf{t}_i) = J(\mathbf{m}_3)))$, since it is located closer than a distance D with respect to the nearest element $\mathbf{m}^{ne} = \mathbf{m}_3$ whose quality factor $Q(\mathbf{m}_3)$ corresponds to the worst value. **c** According to the rule 2, the individual $\mathbf{t}_i$ is evaluated, as there is no close reference in its neighborhood

$$D = \frac{\sum_{j=1}^{n} (p_j^{high} - p_j^{low})}{50 \cdot n} \tag{7.11}$$

where p_j^{low} and p_j^{high} represent the pre-specified lower bound and the upper bound of the j-parameter respectively, within an n-dimensional space. Both rules show that the fitness estimation strategy is simple and straightforward. Figure 7.3 illustrates the procedure of fitness computation for a new candidate solution $\mathbf{t}_i$ considering the two different rules. In the problem the objective function f is maximized with respect to two parameters (x_1, x_2). In all figures (Figs. 7.3a–c) the memory **M** (k) contains five different elements $(\mathbf{m}_1, \mathbf{m}_2, \mathbf{m}_3, \mathbf{m}_4, \mathbf{m}_5)$ with their corresponding fitness values $(J(\mathbf{m}_1), J(\mathbf{m}_2), J(\mathbf{m}_3), J(\mathbf{m}_4), J(\mathbf{m}_5))$ and quality factors $(Q(\mathbf{m}_1), Q(\mathbf{m}_2), Q(\mathbf{m}_3), Q(\mathbf{m}_4), Q(\mathbf{m}_5))$. Figures 7.3a, b show the fitness evaluation $(f(x_1, x_2))$ or estimation $(F(x_1, x_2))$ of the new individual $\mathbf{t}_i$ following the rule 1. Figure 7.3a represent the case when $\mathbf{m}^{ne}$ holds a good quality factor whereas Fig. 7.3b when $\mathbf{m}^{ne}$ maintains a bad quality factor. Finally, Fig. 7.3c presents the fitness evaluation of $\mathbf{t}_i$ considering the conditions of rule 2. The procedure that implements the fitness estimation strategy is presented in Algorithm 7.4, in form of pseudo code.

Algorithm 7.4 Fitness estimation strategy

```
1:   Input: Population T^k and memory M(k)
2:   for i=1 to N_e^k do
3:     m^ne ← FindNearestElementOfM(t_i)
4:     distance ← FindTheDistance(t_i, m^ne)
5:       if (distance<D) then
6:           if (rand(0,1)<=Q(m^ne)) then
7:    J(t_i) ← f(t_i)  % Evaluation
8:    else                                        Rule 1
9:    J(t_i) ← J(m^ne)  % Estimation
10:          end if
11:  else
12:   J(t_i) ← f(t_i)  % Evaluation                Rule 2
13:      end if
14:  end for
15:  Output: fitness values of T^k
```

7.2.9 Memory Updating

Once the operations of exploration and fitness estimation have been applied, it is necessary to update the memory $\mathbf{M}(k)$. In the APRE algorithm, the memory $\mathbf{M}(k)$ is updated considering the following procedure:

1. The elements of $\mathbf{M}(k)$ and $\mathbf{T}^k$ are merged into $\mathbf{M}_U$ ($\mathbf{M}_U = \mathbf{M}(k) \cup \mathbf{T}^k$).
2. From the resulting elements of $\mathbf{M}_U$, it is selected the N_p best elements according to their fitness values to build the new memory $\mathbf{M}(k+1)$.
3. The counters $c_1, c_2, \ldots, c_{N_p}$ must be updated. Thus, the counter of the surviving elements is incremented by 1 whereas the counter of modified elements is set to zero.

7.2.10 Exploitation Operation

The second main operation applied by the APRE algorithm is the exploitation operation. Exploitation, in the context of EA, is the process of refining the solution quality of existent promising solutions within a small neighborhood. In order to implement such a process, a new memory **ME** is generated, which contains the same elements that $\mathbf{M}(k+1)$, but sorted according to their fitness values. Thus, **ME** presents in its first positions the elements whose fitness values are better than those located in the last positions. Then, the 10 % of the N_p (N_e) individuals are taken from the first elements of **ME** to build the set **E** ($\mathbf{E} = \{\mathbf{me}_1, \mathbf{me}_2, \ldots, \mathbf{me}_{N_e}\}$ where $N_e = \text{ceil}(0.1 \cdot N_p)$).

To each element $\mathbf{me}_i$ of **E** is assigned a probability p_i which express the likelihood of the element $\mathbf{me}_i$ to be exploited. Such a probability is computed as follows:

$$p_i = \frac{N_e + 1 - i}{N_e} \tag{7.12}$$

Therefore, the first elements of **E** have a better probability to be exploited than the last ones. In order to decide if the element $\mathbf{me}_i$ must be exploited, a uniform random number is generated within the range [0, 1]. If such a number is less than p_i, then the element $\mathbf{me}_i$ will be modified by the exploitation operation. Otherwise, it remains without changes.

If the exploitation operation over $\mathbf{me}_i$ is verified, the position of $\mathbf{me}_i$ is perturbed considering a small neighborhood. The idea is to test if it is possible to refine the solution provided by $\mathbf{me}_i$ modifying slightly its position. In order to improve the exploitation process, the algorithm starts perturbing the original position within the interval $[-D, D]$ (where D is the distance defined in Eq. 7.11) and then gradually is reduced as the process evolves. Thus, the perturbation over a generic element $\mathbf{me}_i$ is modeled as follows:

$$me_{i,j}^{new} = me_{i,j} + \left[D \frac{ng - k}{ng} \right] (2 \cdot \text{rand}(0,1) - 1) \tag{7.13}$$

where k is the current iteration and ng is the total number of iterations from which consist the evolution process. Once $\mathbf{me}_i^{new}$ has been calculated, its fitness value is computed by using the true objective function ($J(\mathbf{me}_i^{new}) = f(\mathbf{me}_i^{new})$). If $\mathbf{me}_i^{new}$ is better than $\mathbf{me}_i$ according to their fitness values, the value of $\mathbf{me}_i$ in the original memory $\mathbf{M}(k+1)$ is updated with $\mathbf{me}_i^{new}$, otherwise the memory $\mathbf{M}(k+1)$ remains without changes. The procedure that implements the exploitation operation is presented in Algorithm 7.5, in form of pseudo code.

Algorithm 7.5 Exploitation operation of APRE

1: **Input**: New memory $\mathbf{M}(k+1)$, current iteration k
2: $\mathbf{ME} \leftarrow$ SortElementsFitness($\mathbf{M}(k+1)$)
3: $N_e \leftarrow \text{ceil}(0.1 \cdot N_p)$
4: $\mathbf{E} \leftarrow$ SelectTheFirstElements($\mathbf{ME}$, N_e)
5: **for** i=1 to N_e **do**
6: $p_i \leftarrow (N_e+1-i)/\ N_e$
7: **if** (rand(0,1)<= p_i) **then**
8: **for** j=1 to n**do**
9: $me_{i,j}^{new} \leftarrow me_{i,j} + \left[D \frac{ng-k}{ng} \right] (2 \cdot \text{rand}(0,1) - 1)$
10: **end for**
11: $J(\mathbf{me}_i^{new}) \leftarrow f(\mathbf{me}_i^{new})$
12: **if** ($J(\mathbf{me}_i^{new}) > J(\mathbf{me}_i)$) **then**
13: $\mathbf{M}(k+1) \leftarrow$ MemoryIsUpdated($\mathbf{me}_i^{new}$)
14: **end if**
15: **end if**
16: **end for**
17: **Output**: Memory $\mathbf{M}(k+1)$

In order to demonstrate the exploitation operation, Fig. 7.4a illustrates a simple example. It is assumed a memory $\mathbf{M}(k+1)$ of ten different 2-dimensional elements (N_p = 10). Figure 7.4b shows the previous configuration of the proposed example before the exploitation operation takes place. Since only the 10 % of the best elements of $\mathbf{M}(k+1)$ will build the set $\mathbf{E}$, $\mathbf{m}_5$ is the single element that constitutes $\mathbf{E}$ ($\mathbf{me}_1 = \mathbf{m}_5$). Therefore, according to Eq. 7.12, the probability p_1 assigned to $\mathbf{me}_1$

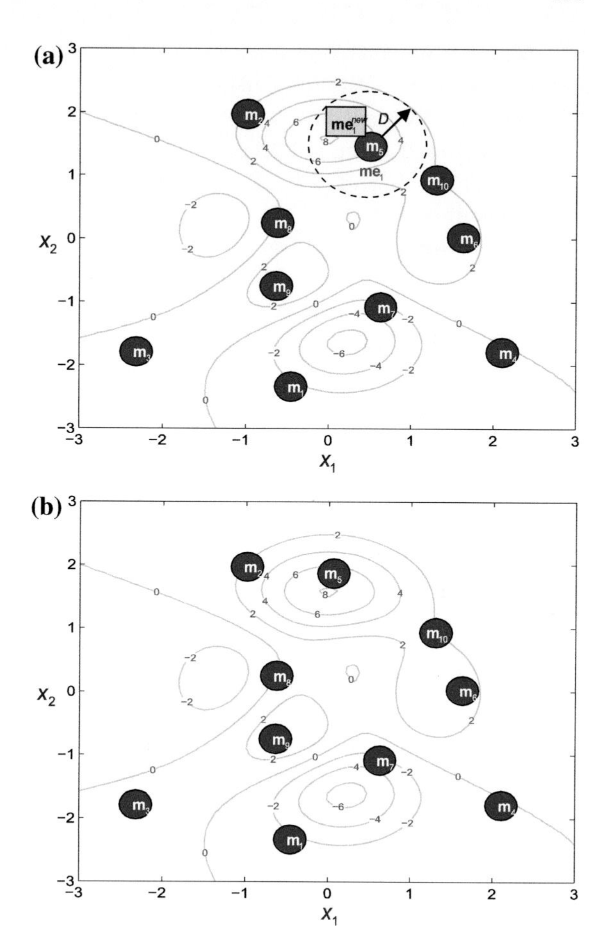

Fig. 7.4 Example of the mating operation: **a** function example, **b** initial configuration before the exploitation operation and **c** configuration after the operation

is 1. Under such circumstances, the element $\mathbf{me}_1$ is perturbed considering the Eq. 7.13, generating the new position $\mathbf{me}_1^{new}$. As $\mathbf{me}_1^{new}$ is better than $\mathbf{me}_1$ according to their fitness values, the value of $\mathbf{m}_5$ in the original memory $\mathbf{M}(k+1)$ is updated with $\mathbf{me}_i^{new}$. Figure 7.4c shows the final configuration of $\mathbf{M}(k+1)$ after the exploitation operation has been achieved.

7.2.11 Computational Procedure

The computational procedure for the presented algorithm can be summarized as follows:

Computational procedure of APRE

1: **Input**: N_p,N_e^0, MR, CH and Max$k$(where Max$k$ is the maximum number of iterations).

2: $\mathbf{M}(1) \leftarrow$ InitializeM(N_p)
3: $c_1, \ldots, c_{N_p} \leftarrow$ ClearCounters()
4: **for** k = 1 to Max$k$**do**
5: **Algorithm 1** (Sect. 7.2.2.1)
6: **Algorithm 2** (Sect. 7.2.2.2)
7: **Algorithm 3** (Sect. 7.2.3)
8: **Algorithm 4** (Sect. 7.2.4)
9: $\mathbf{M}(k+1) \leftarrow$ UpdateM($\mathbf{M}(k)$) (Sect. 7.2.5)
10: $c_1, \ldots, c_{N_p} \leftarrow$ UpdateCounters($c_1, \ldots, c_{N_p}$)
11: $\mathbf{Q} \leftarrow$ CalculateQualityFactor($\mathbf{M}(k)$)
12: **Algorithm 7.5**
13: **end for**
14: *Solution* $\leftarrow$ FindBestElement($\mathbf{M}(k)$)
15: **Output**: *Solution*

The APRE algorithm is an iterative process in which several actions are executed. After initialization (lines 2–3), it is computed the number of memory elements to be evolved. Such number is automatically modified at each iteration (lines 5–6). Then, a set of new individuals is generated as a consequence of the execution of the exploration operation (line 7). For each new individual, its fitness value is estimated or evaluated according to a decision taken by a fitness estimation strategy (line 8). Afterwards, the memory is updated. In this stage, the new individuals produced by the exploration operation compete against the memory elements to build the final memory configuration (lines 9–11). Finally, a sample of the best elements contained in the final memory configuration is undergone to the exploitation operation (line 12). This cycle is repeated until the maximum number the iterations Max$k$ has been reached.

7.3 Implementation of APRE-Based Circle Detector

7.3.1 Individual Representation

In order to detect circle shapes, candidate images must be preprocessed first by the well-known Canny algorithm which yields a single-pixel edge-only image. Then, the (x_i, y_i) coordinates for each edge pixel $\mathbf{e}_i$ are stored inside the edge vector $\mathbf{E} = \{\mathbf{e}_1, \mathbf{e}_2, \ldots, \mathbf{e}_{zn}\}$, with zn being the total number of edge pixels. Each circle C uses three edge points as individuals in the optimization algorithm. In order to construct such individuals, three indexes $\mathbf{e}_i$, $\mathbf{e}_j$ and $\mathbf{e}_k$, are selected from vector $\mathbf{E}$, considering the circle's contour that connects them. Therefore, the circle $C = \{\mathbf{e}_i, \mathbf{e}_j, \mathbf{e}_k\}$ that crosses over such points may be considered as a potential solution for the detection problem. Considering the configuration of the edge points shown by Fig. 7.5, the circle center (x_0, y_0) and the radius r of C can be computed as follows:

$$(x - x_0)^2 + (y - y_0)^2 = r^2 \tag{7.14}$$

considering

$$\mathbf{A} = \begin{bmatrix} x_j^2 + y_j^2 - (x_i^2 + y_i^2) & 2 \cdot (y_j - y_i) \\ x_k^2 + y_k^2 - (x_i^2 + y_i^2) & 2 \cdot (y_k - y_i) \end{bmatrix}$$
$$\mathbf{B} = \begin{bmatrix} 2 \cdot (x_j - x_i) & x_j^2 + y_j^2 - (x_i^2 + y_i^2) \\ 2 \cdot (x_k - x_i) & x_k^2 + y_k^2 - (x_i^2 + y_i^2) \end{bmatrix}, \tag{7.15}$$

$$x_0 = \frac{\det(\mathbf{A})}{4((x_j - x_i)(y_k - y_i) - (x_k - x_i)(y_j - y_i))},$$
$$y_0 = \frac{\det(\mathbf{B})}{4((x_j - x_i)(y_k - y_i) - (x_k - x_i)(y_j - y_i))}, \tag{7.16}$$

and

$$r = \sqrt{(x_0 - x_d)^2 + (y_0 - y_d)^2}, \tag{7.17}$$

being det(.) the determinant and $d \in \{i, j, k\}$. Figure 7.5 illustrates the parameters defined by Eqs. 7.14–7.17.

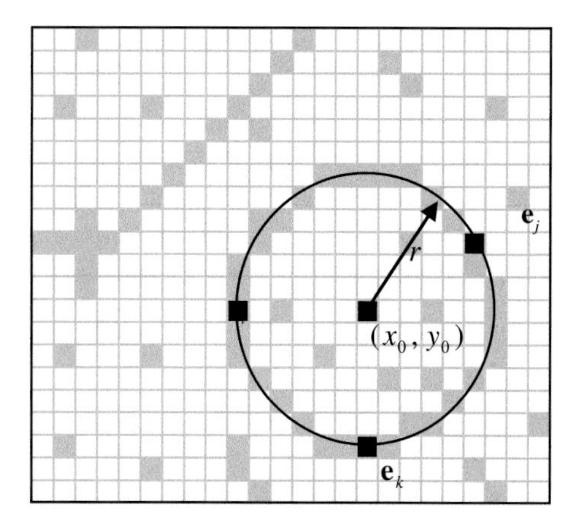

Fig. 7.5 Circle candidate (individual) built from the combination of points $\mathbf{e}_i$, $\mathbf{e}_j$ and $\mathbf{e}_k$

7.3.2 Objective Function

In order to calculate the error produced by a candidate solution *C*, a set of test points is calculated as a virtual shape which, in turn, must be validated, i.e. if it really exists in the edge image. The test set is represented by $\mathbf{A} = \{\mathbf{a}_1, \mathbf{a}_2, \ldots, \mathbf{a}_{sn}\}$, where *sn* is the number of points over which the existence of an edge point, corresponding to *C*, should be validated. In our approach, the set **A** is generated by the Midpoint Circle Algorithm (MCA) [29]. The MCA is a searching method which seeks the required points for drawing a circle digitally. Therefore MCA calculates the necessary number of test points *sn* to totally draw the complete circle. Such a method is considered the fastest because MCA avoids computing square-root calculations by comparing the pixel separation distances among them.

The objective function *J*(*C*) represents the matching error produced between the pixels **A** of the circle candidate *C* (individual) and the pixels that actually exist in the edge image, yielding:

$$J(C) = \frac{\sum_{v=1}^{sn} G(\mathbf{a}_v)}{sn}, \tag{7.18}$$

where $G(\mathbf{a}_v)$ is a function that verifies the pixel existence in $\mathbf{a}_v$, with $\mathbf{a}_v \in \mathbf{A}$ and *sn* being the number of pixels lying on the perimeter corresponding to *C* currently under testing. Hence, function $G(\mathbf{a}_v)$ is defined as:

$$G(\mathbf{a}_v) = \begin{cases} 1 & \text{if the pixel } (\mathbf{a}_v) \text{ is an edge point} \\ 0 & \text{otherwise} \end{cases} \tag{7.19}$$

A value of *J*(*C*) near to one implies a better response from the "circularity" operator. Figure 7.6 shows the procedure to evaluate a candidate solution *C* by using the objective function *J*(*C*). Figure 7.6a shows the original edge map **E**, while Fig. 7.6b presents the virtual shape **A** representing the particle $C = \{\mathbf{e}_i, \mathbf{e}_j, \mathbf{e}_k\}$. In Fig. 7.6c, the virtual shape **A** is compared to the edge image, point by point, in order to find coincidences between virtual and edge points. The particle has been built from points $\mathbf{e}_i$, $\mathbf{e}_j$ and $\mathbf{e}_k$ which are shown by Fig. 7.6a. The virtual shape**A**, obtained by MCA, gathers 56 points ($sn = 56$) with only 17 of them existing in both images (shown as white points in Fig. 7.6c) and yielding: $\sum_{v=1}^{sn} G(\mathbf{a}_v) = 17$, therefore $J(C) \approx 0.30$.

7.3.3 The Multiple Circle Detection Procedure

In order to detect multiple circles, the APRE-detector is iteratively applied. At each iteration, two actions are developed. In the first one, a new circle is detected as a

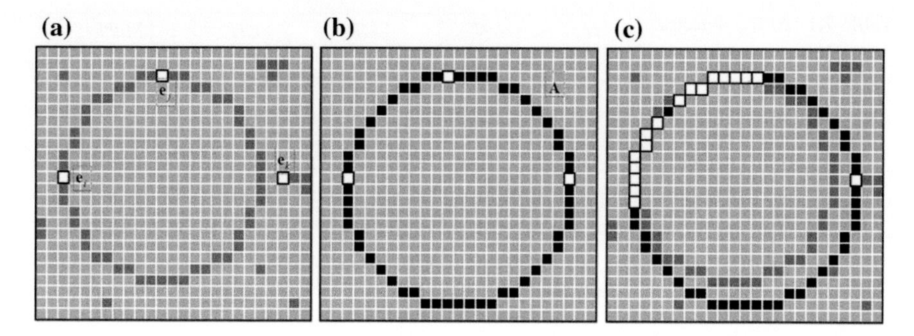

Fig. 7.6 Procedure to evaluate the objective function $J(C)$: The image **a** presents the original edge map while **b** present the virtual shape **A** corresponding to C. The image **c** shows the coincidences between both images by means of white pixels whereas the virtual shape is depicted in black

consequence of the execution of the APRE algorithm. The detected circle corresponds to the candidate solution C with the best found $J(C)$ value. In the second one, the detected circle is removed from the original edge map. The processed edge map without the removed circle represents the input image for the next iteration. Such process is executed over the sequence of images until the $J(C)$ value would be lower than a determined threshold that is considered as permissible.

7.4 Results on Multi-circle Detection

In order to achieve the performance analysis, the presented approach is compared to the GA-based algorithm [5], the BFAO detector [9] and the RHT method [4] over an image set.

The GA-based algorithm follows the proposal of Ayala-Ramirez et al. [5], which considers the population size as 70, the crossover probability as 0.55, the mutation probability as 0.10, the number of elite individuals as 2 and 200 generations. The roulette wheel selection and the 1-point crossover operator are both applied. The parameter setup and the fitness function follow the configuration suggested in [5]. The BFAO algorithm follows the implementation from [9] considering the experimental parameters as: $S = 50$, $N_c = 350$, $N_s = 4, N_{\mathrm{ed}} = 1, P_{\mathrm{ed}} = 0.25$, $d_{\mathrm{attract}} = 0.1$, $w_{\mathrm{attract}} = 0.2$, $w_{\mathrm{repellant}} = 10$, $h_{\mathrm{repellant}} = 0.1$, $\lambda = 400$ and $\psi = 6$. Such values are found to be the best configuration set according to [9]. Both, the GA-based algorithm and the BAFO method use the same objective function that is defined by Eq. 7.18. Likewise, the RHT method has been implemented as it is described in [4]. Finally, Table 7.1 presents the parameters for the APRE algorithm used in this work. They have been kept for all test images after being experimentally defined.

Images rarely contain perfectly-shaped circles. Therefore, with the purpose of testing accuracy for a single-circle, the detection is challenged by a ground-truth

Table 7.1 APRE detector parameters

N_p	N_e^0	MR	CH	Max$k$
50	20	0.6	0.8	200

circle which is determined from the original edge map. The parameters $(x_{true}, y_{true}, r_{true})$ representing the testing circle are computed using the Eqs. (7.6)–(7.9) for three circumference points over the manually-drawn circle. Considering the centre and the radius of the detected circle are defined as (x_D, y_D) and r_D, the Error Score (Es) can be accordingly calculated as:

$$Es = \eta \cdot (|x_{true} - x_D| + |y_{true} - y_D|) + \mu \cdot |r_{true} - r_D| \tag{7.20}$$

The central point difference $(|x_{true} - x_D| + |y_{true} - y_D|)$ represents the centre shift for the detected circle as it is compared to a benchmark circle. The radio mismatch $(|r_{true} - r_D|)$ accounts for the difference between their radii. η and μ represent two weighting parameters which are to be applied separately to the central point difference and to the radio mismatch for the final error Es. At this work, they are chosen as $\eta = 0.05$ and $\mu = 0.1$. Such a choice ensures that the radius difference would be strongly weighted in comparison to the difference of central circular positions between the manually detected and the machine-detected circles. Here we assume that if Es is found to be less than 1, then the algorithm gets a success, otherwise, we say that it has failed to detect the edge-circle. Note that for $\eta = 0.05$ and $\mu = 0.1$; $Es < 1$ means the maximum difference of radius tolerated is 10 while the maximum mismatch in the location of the center can be 20 (in number of pixels). In order to appropriately compare the detection results, the Detection Rate (DR) is introduced as a performance index. DR is defined as the percentage of reaching detection success after a certain number of trials. For "success" it does mean that the compared algorithm is able to detect all circles contained in the image, under the restriction that each circle must hold the condition $Es < 1$. Therefore, if at least one circle does not fulfil the condition of $Es < 1$, the complete detection procedure is considered as a failure.

In order to use an error metric for multiple-circle detection, the averaged Es produced from each circle in the image is considered. Such criterion, defined as the Multiple Error (ME), is calculated as follows:

$$\mathrm{ME} = \left(\frac{1}{NC}\right) \cdot \sum_{R=1}^{NC} \mathrm{Es}_R, \tag{7.21}$$

where $\boldsymbol{NC}$ represents the number of circles within the image according to a human expert.

Figure 7.7 shows three synthetic images and the resulting images after applying the GA-based algorithm [5], the BFOA method [9] and the presented approach. Figure 7.8 presents experimental results considering three natural images. The performance is analyzed by considering 35 different executions for each algorithm.

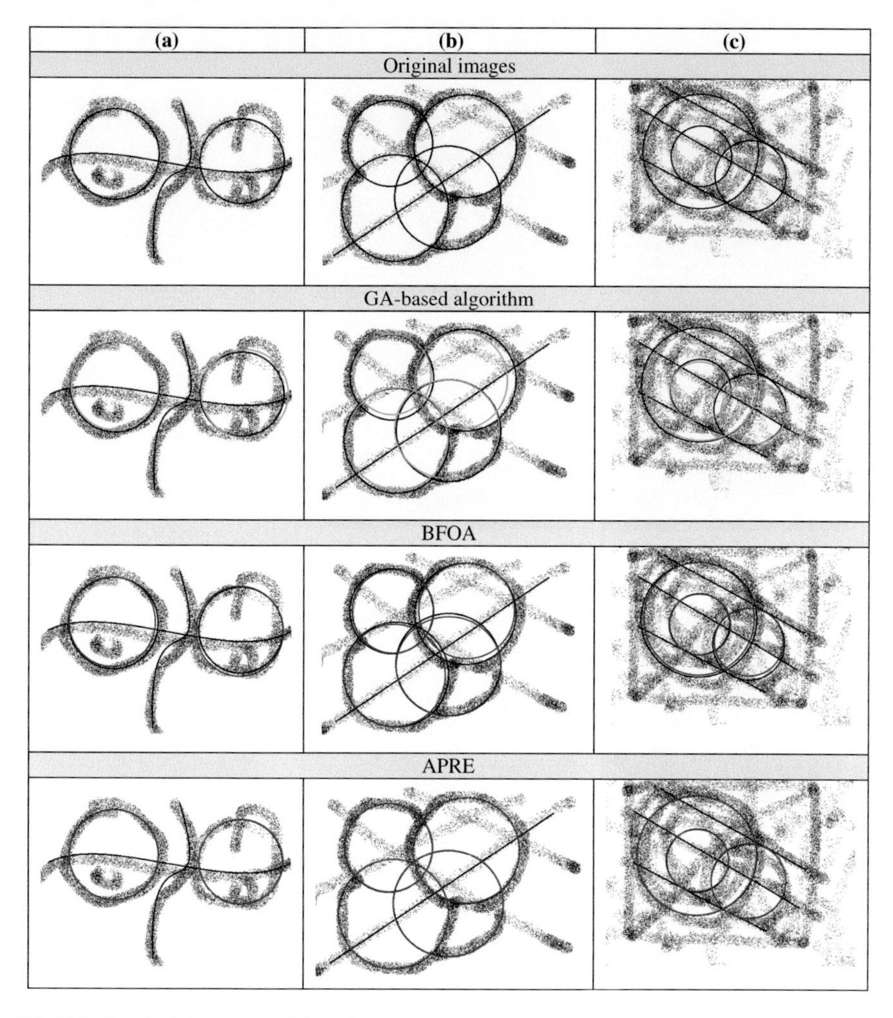

Fig. 7.7 Synthetic images and their detected circles for: GA-based algorithm, the BFOA method and the presented APRE algorithm

Table 7.2 shows the averaged execution time, the average number of function evaluations, the detection rate in percentage and the averaged multiple error (ME), considering six test images (shown by Figs. 7.7 and 7.8). Close inspection reveals that the presented method is able to achieve the highest success rate still keeping the smallest error and demanding less computational time and a lower number of function evaluations for all cases.

In order to statistically analyze the results in Table 7.2, a non-parametric significance proof known as the Wilcoxon's rank test [30–32] for 35 independent samples has been conducted. Such proof allows assessing result differences among two related methods. The analysis is performed considering a 5 % significance level

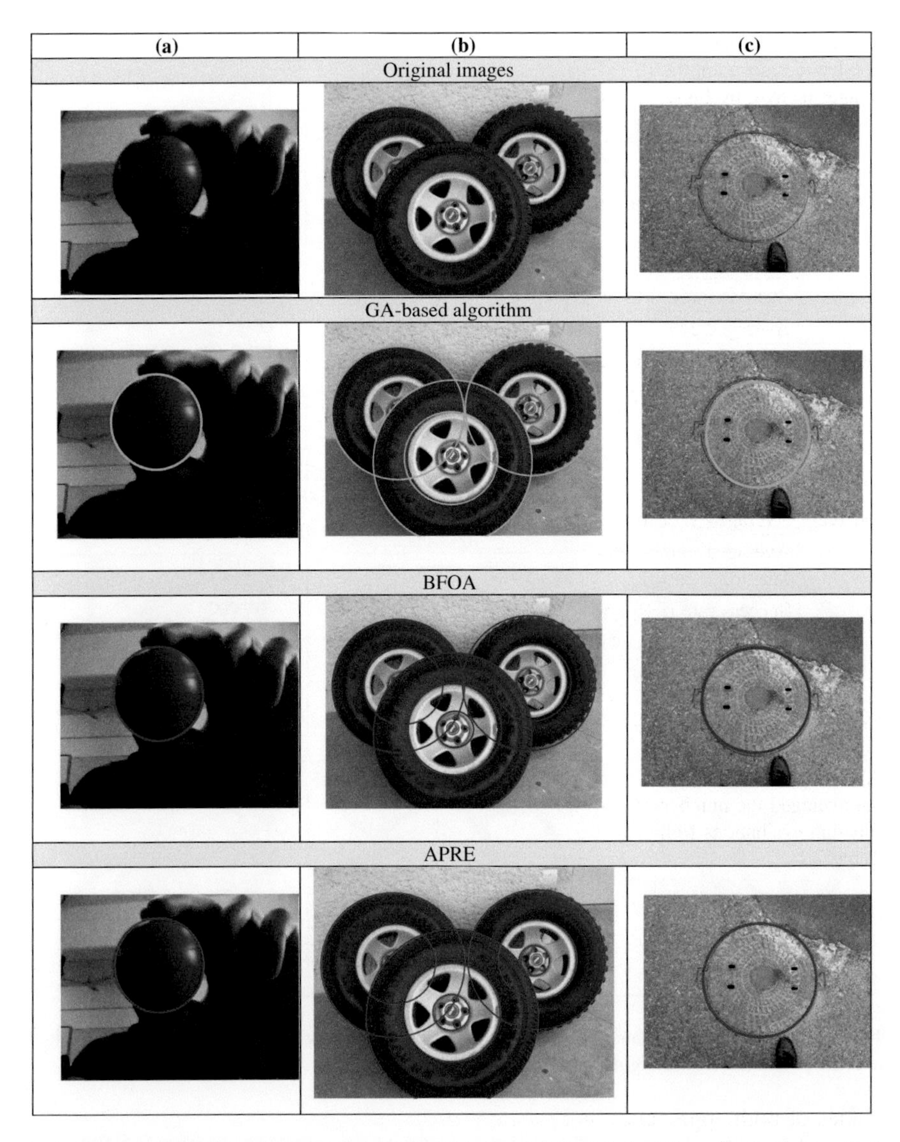

Fig. 7.8 Real-life images and their detected circles for: GA-based algorithm, the BFOA method and the presented APRE algorithm

over the number of function evaluations and a multiple error (ME) data. Tables 7.3 and 7.4 report the *p*-values produced by Wilcoxon's test for a pair-wise comparison of the number of function evaluations and the multiple error (ME), considering two groups gathered as APRE versus GA and APRE versus BFOA. As a null hypothesis, it is assumed that there is no difference between the values of the two algorithms. The alternative hypothesis considers an existent difference between the

Table 7.2 The averaged execution-time, detection rate and the averaged multiple error for the GA-based algorithm, the BFOA method and the presented APRE algorithm, considering six test images (shown by Figs. 7.8 and 7.9)

	Performance indexes	Synthetic images			Natural images		
		(a)	(b)	(c)	(a)	(b)	**(c)**
GA	Averaged execution time	2.23	3.15	4.21	5.11	6.33	7.62
	Averaged number of function evaluations	14,000	14,000	14,000	14,000	14,000	14,000
	Success rate (DR) (%)	88	79	74	90	83	84
	Averaged ME	0.41	0.51	0.48	0.45	0.81	0.92
BFOA	Averaged execution time	1.71	2.80	3.18	3.45	4.11	5.36
	Averaged number of function evaluations	17,500	17,500	17,500	17,500	17,500	17,500
	Success rate (DR) (%)	99	92	88	96	89	92
	Averaged ME	0.33	0.37	0.41	0.41	0.77	0.37
APRE	Averaged execution time	0.21	0.36	0.20	1.10	1.61	1.95
	Averaged number of function evaluations	2321	2756	3191	4251	3768	3834
	Success rate (DR) (%)	100	100	100	100	100	100
	Averaged ME	0.22	0.26	0.15	0.25	0.37	0.41

Table 7.3 p-values produced by Wilcoxon's test comparing APRE to GA and BFOA over the averaged the number of function evaluations from Table 7.2

Image	p-Value	
	APRE versus GA	APRE versus BFOA
Synthetic images		
(a)	3.3124e−006	5.7628e−006
(b)	5.3562e−007	6.8354e−007
(c)	4.1153e−006	1.1246e−005
Natural images		
(a)	4.5724e−006	4.5234e−006
(b)	6.7186e−006	5.3751e−006
(c)	8.7691e−007	6.2876e−006

values of both approaches. All p-values reported in the Tables 7.3 and 7.4 are less than 0.05 (5 % significance level) which is a strong evidence against the null hypothesis, indicating that the best APRE mean values for the performance are statistically significant which has not occurred by chance.

Figure 7.9 demonstrates the relative performance of APRE in comparison with the RHT algorithm as it is described in [4]. All images belonging to the test are complicated and contain different noise conditions. The performance analysis is achieved by considering 35 different executions for each algorithm over the three images. The results, exhibited in Fig. 7.9, present the median-run solution (when the runs were ranked according to their final ME value) obtained throughout the 35 runs. On the other hand, Table 7.5 reports the corresponding averaged execution

Table 7.4 p-values produced by Wilcoxon's test comparing APRE to GA and BFOA over the averaged ME from Table 7.2

Image	p-Value	
	APRE versus GA	APRE versus BFOA
Synthetic images		
(a)	1.7345e−004	1.5294e−004
(b)	1.6721e−004	1.4832e−004
(c)	1.0463e−004	1.9734e−004
Natural images		
(a)	1.5563e−004	1.6451e−004
(b)	1.2748e−004	1.5621e−004
(c)	1.0463e−004	1.7213e−004

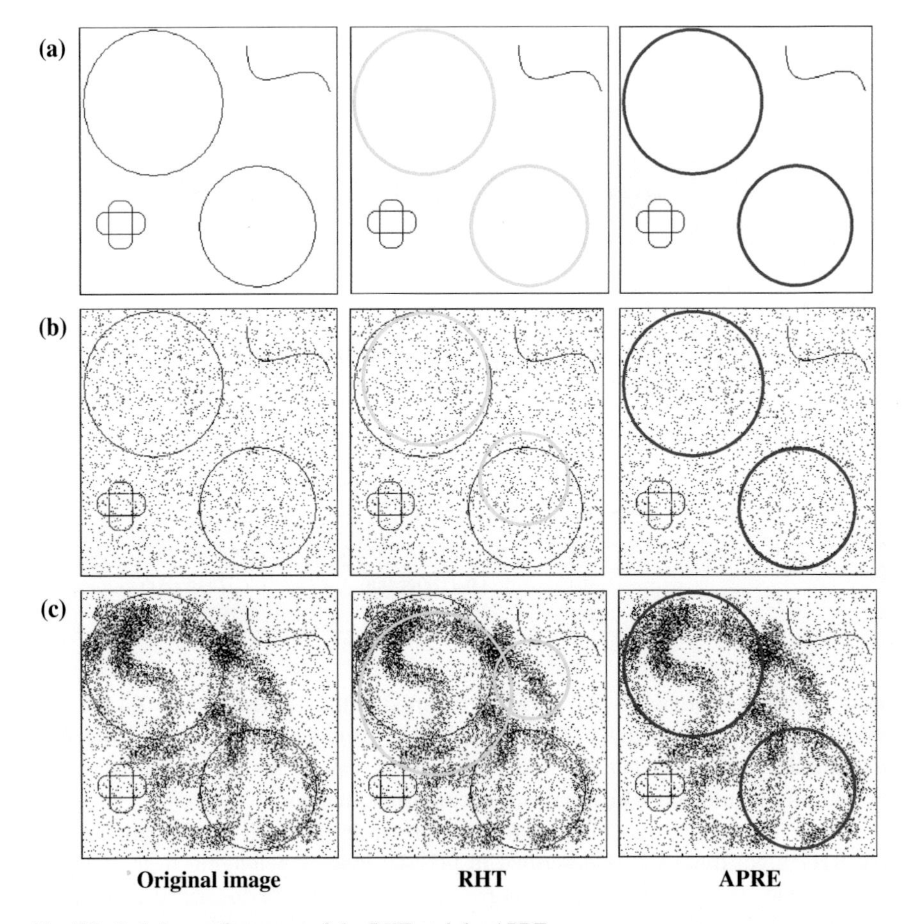

Fig. 7.9 Relative performance of the RHT and the APRE

Table 7.5 Average time, detection rate and averaged error for APRE and RHT, considering three test images

Image	Average time		Success rate (DR) (%)		Average ME	
	RHT	APRE	RHT	APRE	RHT	APRE
(A)	7.82	0.20	100	100	0.19	0.12
(B)	8.65	0.12	64	100	0.47	0.11
(C)	10.65	0.15	11	100	1.21	0.13

time, detection rate (in %), and average multiple error (using (7.10)) for APRE and RHT algorithms over the set of images. Table 7.5 shows a decrease in performance of the RHT algorithm as noise conditions change. Yet the APRE algorithm holds its performance under the same circumstances.

7.5 Conclusions

In this chapter, an evolutionary algorithm called the Adaptive Population with Reduced Evaluations (APRE) is presented to solve the problem of circle detection. The algorithm reduces the number of function evaluations through the use of two mechanisms: (1) adapting dynamically the size of the population and (2) incorporating a fitness calculation strategy which decides when it is feasible to calculate or only estimate new generated individuals.

The algorithm begins with an initial population which will be used as a memory during the evolution process. To each memory element, it is assigned a normalized fitness value called quality factor that indicates the solution capacity provided by the element. From the memory, only a variable sub-set of elements is considered to be evolved. Like other population based methods, the presented algorithm generates new individuals considering two operators: exploration and exploitation. Such operations are applied to improve the quality of the solutions by: (1) searching the unexplored solution space to identify promising areas containing better solutions than those found so far, and (2) successive refinement of the best found solutions. Once the new individuals are generated, the memory is updated. In such stage, the new individuals compete against the memory elements to build the final memory configuration.

In order to save computational time, the approach incorporates a fitness estimation strategy that decides which individuals can be estimated or actually evaluated. As a result, the approach can substantially reduce the number of function evaluations, yet preserving its good search capabilities. The presented fitness calculation strategy estimates the fitness value of new individuals using memory elements located in neighboring positions that have been visited during the evolution process. In the strategy, those new individuals, close to a memory element whose quality factor is high, have a great probability to be evaluated by using the true objective function. Similarly, it is also evaluated those new particles lying in regions of the search space with no previous evaluations. The remaining search positions are estimated assigning them the same fitness value that the nearest location of the memory element. By the use of such fitness estimation method, the fitness value of only very few individuals are actually evaluated whereas the rest is just estimated.

Different to other approaches that use an already existent EA as framework, the APRE method has been completely designed to substantially reduce the computational cost, but preserving good search effectiveness.

To detect the circular shapes, the detector is implemented by using the encoding of three pixels as candidate circles over the edge image. An objective function evaluates if such candidate circles are actually present in the edge image. Guided by the values of this objective function, the set of encoded candidate circles are evolved using the operators defined by APRE so that they can fit to the actual circles on the edge map of the image.

In order to test the circle detection accuracy, a score function is used (Eq. 7.21). It can objectively evaluate the mismatch between a manually detected circle and a machine-detected shape. We demonstrated that the APRE method outperforms both the evolutionary methods (GA and BFOA) and Hough Transform-based techniques (RHT) in terms of speed and accuracy, considering a statistically significant framework (Wilconxon test). Results show that the APRE algorithm is able to significantly reduce the computational overhead as a consequence of decrementing the number of function evaluations.

References

1. Baia, X., Yangb, X., Jan-Latecki, L.: Detection and recognition of contour parts based on shape similarity. Pattern Recogn. **41**, 2189–2199 (2008)
2. Schindlera, K., Suterb, D.: Object detection by global contour shape. Pattern Recogn. **41**, 3736–3748 (2008)
3. Atherton, T.J., Kerbyson, D.J.: Using phase to represent radius in the coherent circle Hough transform. In: Proceeings of IEE Colloquium on the Hough Transform, IEE, London (1993)
4. Xu, L., Oja, E., Kultanen, P.: A new curve detection method: randomized Hough transform (RHT). Pattern Recogn. Lett. **11**(5), 331–338 (1990)
5. Ayala-Ramirez, V., Garcia-Capulin, C.H., Perez-Garcia, A., Sanchez-Yanez, R.E.: Circle detection on images using genetic algorithms. Pattern Recogn. Lett. **27**, 652–657 (2006)
6. Cuevas, E., Ortega-Sánchez, N., Zaldivar, D., Pérez-Cisneros, M.: Circle detection by harmony search optimization. J. Intell. Rob. Syst. **66**(3), 359–376 (2012)
7. Cuevas, E., Oliva, D., Zaldivar, D., Pérez-Cisneros, M., Sossa, Humberto: Circle detection sing electro-magnetism optimization. Inf. Sci. **182**(1), 40–55 (2012)
8. Cuevas, E., Zaldivar, D., Pérez-Cisneros, M., Ramírez-Ortegón, M.: Circle detection using discrete differential evolution optimization. Pattern Anal. Appl. **14**(1), 93–107 (2011)
9. Dasgupta, S., Das, S., Biswas, A., Abraham, A.: Automatic circle detection on digital images with an adaptive bacterial foraging algorithm. Soft. Comput. **14**(11), 1151–1164 (2010)
10. Jin, Y.: Comprehensive survey of fitness approximation in evolutionary computation. Soft. Comput. **9**, 3–12 (2005)
11. Jin, Yaochu: Surrogate-assisted evolutionary computation: recent advances and future challenges. Swarm and Evolutionary Computation **1**, 61–70 (2011)

12. Branke, J., Schmidt, C.: Faster convergence by means of fitness estimation. Soft. Comput. **9**, 13–20 (2005)
13. Zhou, Z., Ong, Y., Nguyen, M., Lim, D.: A study on polynomial regression and gaussian process global surrogate model in hierarchical surrogate-assisted evolutionary algorithm. In: IEEE Congress on Evolutionary Computation (ECiDUE '05), Edinburgh, United Kingdom, 2–5 Sept 2005
14. Ratle, A.: Kriging as a surrogate fitness landscape in evolutionary optimization. Artif. Intell. Eng. Des. Anal. Manuf. **15**, 37–49 (2001)
15. Lim, D., Jin, Y., Ong, Y., Sendhoff, B.: Generalizing surrogate-assisted evolutionary computation. IEEE Trans. Evol. Comput. **14**(3), 329–355 (2010)
16. Ong, Y., Lum, K., Nair, P.: Evolutionary algorithm with hermite radial basis function interpolants for computationally expensive adjoint solvers. Comput. Optim. Appl. **39**(1), 97–119 (2008)
17. Jin, Y., Olhofer, M., Sendhoff, B.: A framework for evolutionary optimization with approximate fitness functions. IEEE Trans. Evol. Comput. **6**(5), 481–494 (2002)
18. Ong, Y.S., Nair, P.B., Keane, A.J.: Evolutionary optimization of computationally expensive problems via surrogate modeling. AIAA J. **41**(4), 687–696 (2003)
19. DeBao, Chen, ChunXia, Zhao: Particle swarm optimization with adaptive population size and its application. Appl. Soft Comput. **9**, 39–48 (2009)
20. Zhu, Wu, Tang, Yang, Fang, Jian-an, Zhang, Wenbing: Adaptive population tuning scheme for differential evolution. Inf. Sci. **223**, 164–191 (2013)
21. Brest, J., Maučec, M.S.: Population size reduction for the differential evolution algorithm. ApplIntell **29**, 228–247 (2008)
22. Oh, S., Jin, Y., Jeon, M.: Approximate models for constraint functions in evolutionary constrained optimization. Int. J. Innovative Comput. Info. Control **7**(10) (2011)
23. Luoa, C., Shao-Liang, Z., Wanga, C., Jiang, Z.: A metamodel-assisted evolutionary algorithm for expensive optimization. J. Comput. Appl. Math. (2011). doi:10.1016/j.cam.2011.05.047
24. Tan, K.C., Chiam, S.C., Mamun, A.A., Goh, C.K.: Balancing exploration and exploitation with adaptive variation for evolutionary multi-objective optimization. Eur. J. Oper. Res. **197**, 701–713 (2009)
25. Alba, E., Dorronsoro, B.: The exploration/exploitation tradeoff in dynamic cellular genetic algorithms. IEEE Trans. Evol. Comput. **9**(3), 126–142 (2005)
26. Ostadmohammadi, B., Mirzabeygi, P., Panahi, M.: An improved PSO algorithm with a territorial diversity-preserving scheme and enhanced exploration–exploitation balance, Swarm and Evolutionary Computation (In Press)
27. Storn, R., Price, K.: Differential evolution-a simple and efficient adaptive scheme for global optimisation over continuous spaces. Tech. Rep. TR-95–012, ICSI, Berkeley, Calif (1995)
28. Angira, R., Santosh, A.: Optimization of dynamic systems: a trigonometric differential evolution approach. Comput. Chem. Eng. **31**, 1055–1063 (2007)
29. Bresenham, J.E.: A linear algorithm for incremental digital display of circular arcs. Commun. ACM **20**, 100–106 (1977)
30. Wilcoxon, F.: Individual comparisons by ranking methods. Biometrics **1**, 80–83 (1945)
31. Garcia, S., Molina, D., Lozano, M., Herrera, F.: A study on the use of non-parametric tests for analyzing the evolutionary algorithms' behaviour: a case study on the CEC '2005 Special session on real parameter optimization. J. Heurist (2008). doi:10.1007/s10732-008-9080-4
32. Santamaría, J., Cordón, O., Damas, S., García-Torres, J.M., Quirin, A.: Performance evaluation of memetic approaches in 3D reconstruction of forensic objects. Soft Comput. in press (2008). doi:10.1007/s00500-008-0351-7

Chapter 8
Otsu and Kapur Segmentation Based on Harmony Search Optimization

Abstract Segmentation is one of the most important tasks in image processing that endeavors to classify pixels into two or more groups according to their intensity levels and a threshold value. Since traditional image processing techniques exhibit several difficulties when they are employed to segment images, the use of evolutionary algorithms has been extended to segmentation tasks in last years. The Harmony Search Algorithm (HSA) is an evolutionary method which is inspired in musicians improvising new harmonies while playing. Different to other evolutionary algorithms, HSA exhibits interesting search capabilities still keeping a low computational overhead. In this chapter, a multilevel thresholding (MT) algorithm based on the HSA is presented. The approach combines the good search capabilities of HSA with objective functions suggested by the popular MT methods of Otsu and Kapur. The presented algorithm takes random samples from a feasible search space inside the image histogram. Such samples build each harmony (candidate solution) in the HSA context whereas its quality is evaluated considering the objective function that is employed by the Otsu's or Kapur's method. Guided by these objective values, the set of candidate solutions are evolved through HSA operators until an optimal solution is found. The approach generates a multilevel segmentation algorithm which can effectively identify threshold values of a digital image in a reduced number of iterations. Experimental results show a high performance of the presented method for the segmentation of digital images.

8.1 Introduction

Segmentation is one of the most important tasks in image processing that endeavors to identify whether a pixel intensity corresponds to a predefined class. Thresholding is the easiest method for segmentation as it works taking a threshold (*th*) value so that pixels whose intensity value is higher than *th* are labeled as first class while the rest correspond to a second class label. When the image is segmented into two classes, the task is called bi-level thresholding (BT) and requires only one *th* value.

E. Cuevas et al., *Applications of Evolutionary Computation in Image Processing and Pattern Recognition*, Intelligent Systems Reference Library 100,
DOI 10.1007/978-3-319-26462-2_8

On the other hand, when pixels are separated into more than two classes, the task is named as MT and demands more than one *th* value [1].

In recent years image processing has been applied to different areas as engineering, medicine, agriculture, etc. Since most of such implementations use a TH methodology, several techniques had been studied. Generally, TH methods are divided into parametric and nonparametric [2–5]. Parametric approaches need to estimate values of a probability density function to model each class. The estimation process is time consuming and computationally expensive. On the other hand, the TH nonparametric employs several criteria such as the between-class variance, the entropy and the error rate [6–8] in order to verify the quality of a *th* value. These metrics could also be used as optimization functions since they result as an attractive option due their robustness and accuracy.

There exist two classical thresholding methods. The first, proposed by Otsu in [6] that maximizes the variance between classes while the second method, submitted by Kapur in [7], uses the maximization of the entropy to measure the homogeneity among classes. Their efficiency and accuracy have been already proved for a bi-level segmentation [4]. Although they can be expanded for MT, their computational complexity increases exponentially when a new threshold is incorporated [5].

As an alternative to classical methods, the MT problem has also been handled through evolutionary optimization methods. In general, they have demonstrated to deliver better results than those based on classical techniques in terms of accuracy, speed and robustness. Numerous evolutionary approaches have been reported in the literature.

Hammouche et al. in [9] provides an interesting survey of how different evolutionary algorithms are used to solve the Kaptur's and Otsu's problems. The study uses four classical evolutionary algorithms to test their efficiency in MT. Such methods include Differential Evolution (DE) [10], Simulated Annealing (SA) [11] and Tabu Search (TS) [12].

Genetic Algorithms (GA) [13], inspired on the biological evolution, have been also used for solving segmentation tasks. One interesting example is presented in [14], where a GA-based algorithm is combined with Gaussian models for multilevel thresholding. Other similar works, such as that of Yin [15] proposes an improved GA for optimal thresholding. In the approach, it is used as a learning strategy to increase the speed of convergence.

Evolutionary approaches inspired on swarm behavior, such as Particle Swarm Optimization (PSO) [16] and Artificial Bee Colony (ABC) [17], have been employed to face the segmentation problem. In [18], both methods are used to find the optimal multilevel threshold points by using the Kapur's entropy as fitness function.

Finally, in [19], the optimal segmentation threshold values are determined by using the Bacterial Foraging Algorithm (BFA). Such method aims to maximize the Kapur's and Otsu's objective functions by considering a set of operators that are based on the social foraging behavior of the bacteria *Eschericha Colli*.

On the other hand, the Harmony Search Algorithm (HSA) introduced by Geem et al. [20] is an evolutionary optimization algorithm which is based on the metaphor of the improvisation process that occurs when a musician searches for a better state of harmony. The HSA generates a new candidate solution from all existing solutions. The solution vector is analogous to the harmony in music while the local and global search schemes are analogous to musician's improvisations. In comparison to other meta-heuristics methods in the literature, HSA imposes fewer mathematical requirements as it can be easily adapted for solving several sorts of engineering optimization challenges [21]. Furthermore, numerical comparisons have demonstrated that the convergence for the HSA is faster than GA [22] which attracts further attention. It has been successfully applied to solve a wide range of practical optimization problems such as discrete and continuous structural optimization [23], parameter estimation of the nonlinear Muskingum model [24], design optimization of water distribution networks [25], vehicle routing [26], combined heat and power economic dispatch [27], design of steel frames [28], and image processing [29], among others. Although the standard HSA presents good search characteristics, several modifications to the original HSA have been proposed in the literature in order to enhance its own features [30].

In this chapter, a multi-thresholding segmentation algorithm is presented. The approach, called the Harmony Search Multi-thresholding Algorithm (HSMA), combines the original harmony search algorithm (HSA) and the Otsu's and Kapur's methodologies. The presented algorithm takes random samples from a feasible search space inside the image histogram. Such samples build each harmony (candidate solution) in the HSA context whereas its quality is evaluated considering the objective function that is employed by the Otsu's or the Kapur's method. Guided by these objective values, the set of candidate solutions are evolved using the HSA operators until the optimal solution is found. The approach generates a multilevel segmentation algorithm which can effectively identify the threshold values of a digital image within a reduced number of iterations. Experimental results over several complex images have validated the efficiency of the presented technique regarding accuracy, speed and robustness.

8.2 Harmony Search Algorithm

8.2.1 The Harmony Search Algorithm

In the basic HSA, each solution is called a "harmony" and is represented by an n-dimension real vector. An initial population of harmony vectors are randomly generated and stored within a Harmony Memory (HM). A new candidate harmony is thus generated from the elements in the HM by using a memory consideration operation either by a random re-initialization or a pitch adjustment operation. Finally, the HM is updated by comparing the new candidate harmony and the worst harmony vector in the HM. The worst harmony vector is replaced by the new candidate vector when the latter delivers a better solution in the HM. The above process is repeated until a certain termination criterion is met. The basic HS algorithm consists of three main phases: HM initialization, improvisation of new harmony vectors and updating of the HM. The following discussion addresses details about each stage.

8.2.1.1 Initializing the Problem and the Algorithm Parameters

In general, the global optimization problem can be summarized as follows:

$$\begin{array}{ll} \text{minimize}\, f(\mathbf{x}), & \mathbf{x} = (x(1), x(2), \ldots, x(n)) \in \mathbf{R}^n \\ \text{subject to:} & x(j) \in [l(j), u(j)] \quad j = 1, 2, \ldots, n, \end{array} \tag{8.1}$$

where $f(\mathbf{x})$ is the objective function, $\mathbf{x} = (x(1), x(2), \ldots, x(n))$ is the set of design variables, n is the number of design variables, and $l(j)$ and $u(j)$ are the lower and upper bounds for the design variable $x(j)$, respectively. The parameters for HSA are the harmony memory size, i.e. the number of solution vectors lying on the harmony memory (HM), the harmony-memory consideration rate (*HMCR*), the pitch adjusting rate (*PAR*), the distance bandwidth (*BW*) and the number of improvisations (*NI*) which represents the total number of iterations. The performance of HSA is strongly influenced by values assigned to such parameters, which in turn, depend on the application domain [31].

8.2.1.2 Harmony Memory Initialization

In this stage, initial vector components at HM, i.e. *HMS* vectors, are configured. Let $\mathbf{x}_i = \{x_i(1), x_i(2), \ldots, x_i(n)\}$ represent the i-th randomly-generated harmony vector: $x_i(j) = l(j) + (u(j) - l(j)) \cdot \text{rand}(0,1)$ for $j = 1, 2, \ldots, n$ and $i = 1, 2, \ldots, HMS$, where rand (0, 1) is a uniform random number between 0 and 1, the upper and lower bounds of the search space are defined by $l(j)$ and $u(j)$ respectively. Then, the HM matrix is filled with the *HMS* harmony vectors as follows:

$$\text{HM} = \begin{bmatrix} \mathbf{x}_1 \\ \mathbf{x}_2 \\ \vdots \\ \mathbf{x}_{HMS} \end{bmatrix} \tag{8.2}$$

8.2.1.3 Improvisation of New Harmony Vectors

In this phase, a new harmony vector $\mathbf{x}_{new}$ is built by applying the following three operators: memory consideration, random re-initialization and pitch adjustment. Generating a new harmony is known as 'improvisation'. In the memory consideration step, the value of the first decision variable $x_{new}(1)$ for the new vector is chosen randomly from any of the values already existing in the current HM i.e. from the set $\{x_1(1), x_2(1), \ldots, x_{HMS}(1)\}$. For this operation, a uniform random number r_1 is generated within the range [0, 1]. If r_1 is less than *HMCR*, the decision variable $x_{new}(1)$ is generated through memory considerations; otherwise, $x_{new}(1)$ is obtained from a random re-initialization between the search bounds $[l(1), u(1)]$. Values of the other decision variables $x_{new}(2), x_{new}(3), \ldots, x_{new}(n)$ are also chosen accordingly. Therefore, both operations, memory consideration and random re-initialization, can be modeled as follows:

$$x_{new}(j) = \begin{cases} x_i(j) \in \{x_1(j), x_2(j), \ldots, x_{HMS}(j) & \text{with probablity } HMCR \\ l(j) + (u(j) - l(j)) \cdot \text{rand})(0,1) & \text{with probablity } 1-HMCR \end{cases} \quad (8.3)$$

Every component obtained by memory consideration is further examined to determine whether it should be pitch-adjusted. For this operation, the Pitch-Adjusting Rate (*PAR*) is defined as to assign the frequency of the adjustment and the Bandwidth factor (*BW*) to control the local search around the selected elements of the HM. Hence, the pitch adjusting decision is calculated as follows:

$$x_{new}(j) = \begin{cases} x_{new}(j) = x_{new}(j) \pm \text{rand}(0,1) \cdot BW & \text{with probablity } PAR \\ x_{new}(j) & \text{with probablity } (1 - PAR) \end{cases} \quad (8.4)$$

Pitch adjusting is responsible for generating new potential harmonies by slightly modifying original variable positions. Such operation can be considered similar to the mutation process in evolutionary algorithms. Therefore, the decision variable is either perturbed by a random number between 0 and *BW* or left unaltered. In order to protect the pitch adjusting operation, it is important to assure that points lying outside the feasible range [l, u] must be re-assigned i.e. truncated to the maximum or minimum value of the interval.

8.2.1.4 Updating the Harmony Memory

After a new harmony vector x_{new} is generated, the harmony memory is updated by the survival of the fit competition between x_{new} and the worst harmony vector x_w in the HM. Therefore x_{new} will replace x_w and become a new member of the HM in case the fitness value of x_{new} is better than the fitness value of x_w.

8.2.1.5 Computational Procedure

The computational procedure of the basic HSA can be summarized as follows [20]:

Step 1: Set the parameters *HMS*, *HMCR*, *PAR*, *BW* and *NI*.
Step 2: Initialize the HM and calculate the objective function value of each harmony vector.
Step 3: Improvise a new harmony $\mathbf{x}_{new}$ as follows:

for ($j = 1$ to n) do
  if ($r_1 < HMCR$) then
  $x_{new}(j) = x_a(j)$ where $a \in (1, 2, \ldots, HMS)$
    if ($r_2 < PAR$) then
    $x_{new}(j) = x_{new}(j) \pm r_3 \cdot BW$ where $r_1, r_2, r_3 \in \text{rand}(0,1)$
    end if
    if $x_{new}(j) < l(j)$
    $x_{new}(j) = l(j)$
    end if
    if $x_{new}(j) > u(j)$
    $x_{new}(j) = u(j)$
    end if
  else
  $x_{new}(j) = l(j) + r \cdot (u(j) - l(j))$, where $r \in \text{rand}(0,1)$
  end if
end for

Step 4: Update the *HM* as $\mathbf{x}_w = \mathbf{x}_{new}$ if $f(\mathbf{x}_{new}) < f(\mathbf{x}_w)$
Step 5: If *NI* is completed, the best harmony vector $\mathbf{x}_b$ in the HM is returned; otherwise go back to step 3.

This procedure is implemented for minimization. If the intention is to maximize the objective function, a sign modification of Step 4 ($\mathbf{x}_w = \mathbf{x}_{new}$ if $f(\mathbf{x}_{new}) > f(\mathbf{x}_w)$) is required. In this chapter the HSA is used for maximization proposes.

8.3 Image Multilevel Thresholding (MT)

Thresholding is a process in which the pixels of a gray scale image are divided in sets or classes depending on their intensity level (L). For this classification it is necessary to select a threshold value (th) and follows the simple rule of Eq. (8.5).

$$\begin{aligned} C_1 &\leftarrow p \quad \text{if} \quad 0 \leq p < th \\ C_2 &\leftarrow p \quad \text{if} \quad th \leq p < L-1 \end{aligned} \tag{8.5}$$

where p is one of the $m \times n$ pixels of the gray scale image I_g that can be represented in L gray scale levels $L = \{0, 1, 2, \ldots, L-1\}$. C_1 and C_2 are the classes in which the pixel p can be located, while th is the threshold. The rule in Eq. (8.5) corresponds to a bi-level thresholding and can be easily extended for multiple sets:

$$\begin{aligned} C_1 &\leftarrow p \quad \text{if} \quad 0 \leq p < th_1 \\ C_2 &\leftarrow p \quad \text{if} \quad th_1 \leq p < th_2 \\ C_i &\leftarrow p \quad \text{if} \quad th_i \leq p < th_{i+1} \\ C_n &\leftarrow p \quad \text{if} \quad th_n \leq p < L-1 \end{aligned} \tag{8.6}$$

where $\{th_1 \quad th_2 \quad \ldots \quad th_i \quad th_{i+1} \quad th_k\}$ represent different thresholds. The problem for both bi-level and MT is to select the th values that correctly identify the classes. Although, Otsu's and Kapur's methods are well-known approaches for determining such values, both propose a different objective function which must be maximized in order to find optimal threshold values, just as it is discussed below.

8.3.1 Between—Class Variance (Otsu's Method)

This is a nonparametric technique for thresholding proposed by Otsu [6] that employs the maximum variance value of the different classes as a criterion to segment the image. Taking the L intensity levels from a gray scale image or from each component of a RGB (red, green, blue) image, the probability distribution of the intensity values is computed as follows:

$$Ph_i^c = \frac{h_i^c}{NP}, \quad \sum_{i=1}^{NP} Ph_i^c = 1, \quad c = \begin{cases} 1,2,3 & \text{if} \quad \text{RGB Image} \\ 1 & \text{if} \quad \text{Gray scale Image} \end{cases}, \tag{8.7}$$

where i is a specific intensity level ($0 \leq i \leq L-1$), c is the component of the image which depends if the image is gray scale or RGB whereas NP is the total number of pixels in the image. h_i^c (histogram) is the number of pixels that corresponds to the i intensity level in c. The histogram is normalized within a probability distribution Ph_i^c. For the simplest segmentation (bi-level) two classes are defined as:

$$C_1 \frac{Ph_1^c}{\omega_0^c(th)}, \ldots, \frac{Ph_{th}^c}{\omega_0^c(th)} \quad \text{and} \quad C_2 = \frac{Ph_{th+1}^c}{\omega_1^c(th)}, \ldots, \frac{Ph_L^c}{\omega_1^c(th)}, \tag{8.8}$$

where $\omega_0(th)$ and $\omega_1(th)$ are probabilities distributions for C_1 and C_2, as it is shown by Eq. (8.9).

$$\omega_0^c(th) = \sum_{i=1}^{th} Ph_i^c, \quad \omega_1^c(th) \sum_{i=th+1}^{L} Ph_i^c \tag{8.9}$$

It is necessary to compute mean levels μ_0^c and μ_1^c that define the classes using Eq. (8.10). Once those values are calculated, the Otsu variance between classes σ^{2^c} is calculated using Eq. (8.11) as follows:

$$\mu_0^c = \sum_{i=1}^{th} \frac{iPh_i^c}{\omega_0^c(th)}, \quad \mu_1^c = \sum_{i=th+1}^{L} \frac{iPh_i^c}{\omega_1^c(th)} \tag{8.10}$$

$$\sigma^{2^c} = \sigma_1^c + \sigma_2^c \tag{8.11}$$

Notice that for both equations, Eqs. (8.10) and (8.11), c depends on the type of image. In Eq. (8.11) the number two is part of the Otsu´s variance operator and does not represent an exponent in the mathematical sense. Moreover σ_1^c and σ_2^c in Eq. (8.11) are the variances of C_1 and C_2 which are defined as:

$$\sigma_1^c = \omega_0^c(\mu_0^c + \mu_T^c)^2, \quad \sigma_2^c = \omega_1^c(\mu_1^c + \mu_T^c)^2, \tag{8.12}$$

where $\mu_T^c = \omega_0^c\mu_0^c + \omega_1^c\mu_1^c$ and $\omega_0^c + \omega_1^c = 1$. Based on the values σ_1^c and σ_2^c, Eq. (8.13) presents the objective function.

$$J(th) = \max(\sigma^{2^c}(th)), \quad 0 \le th \le L-1, \tag{8.13}$$

where $\sigma^{2^c}(th)$ is the Otsu´s variance for a given th value. Therefore, the optimization problem is reduced to find the intensity level (th) that maximizes Eq. (8.13).

Otsu's method is applied for a single component of an image. In case of RGB images, it is necessary to apply separation into single component images. The previous description of such bi-level method can be extended for the identification of multiple thresholds. Considering k thresholds it is possible separate the original image into k classes using Eq. (8.6), then it is necessary to compute the k variances and their respective elements. The objective function $J(th)$ in Eq. (8.13) can thus be rewritten for multiple thresholds as follows:

$$J(\mathbf{TH}) = \max(\sigma^{2^c}(\mathbf{TH})), \quad 0 \le th_i \le L-1, \quad i = 1, 2, \ldots, k, \tag{8.14}$$

where $\mathbf{TH} = [th_1, th_2, \ldots, th_{k-1}]$, is a vector containing multiple thresholds and the variances are computed through Eq. (8.15) as follows.

$$\sigma^{2^c} = \sum_{i=1}^{k} \sigma_i^c = \sum_{i=1}^{k} \omega_i^c (\mu_i^c - \mu_T^c)^2, \tag{8.15}$$

here, i represents and specific class, ω_i^c and μ_j^c are respectively the probability of occurrence and the mean of a class. In MT, such values are obtained as:

$$\begin{aligned}
\omega_0^c(th) &= \sum_{i=1}^{th_1} Ph_i^c \\
\omega_1^c(th) &= \sum_{i=th_1+1}^{th_2} Ph_i^c \\
&\vdots \\
\omega_{k-1}^c(th) &= \sum_{i=th_k+1}^{L} Ph_i^c
\end{aligned} \tag{8.16}$$

and, for the mean values:

$$\begin{aligned}
\mu_0^c &= \sum_{i=1}^{th_1} \frac{iPh_i^c}{\omega_0^c(th_1)} \\
\mu_1^c &= \sum_{i=th_1+1}^{th_2} \frac{iPh_i^c}{\omega_0^c(th_2)} \\
&\vdots \\
\mu_{k-1}^c &= \sum_{i=th_1+1}^{L} \frac{iPh_i^c}{\omega_k^c(th_k)}
\end{aligned} \tag{8.17}$$

Similar to the bi-level case, for the MT using the Otsu's method c corresponds to the image components, RGB $c = 1, 2, 3$ and gray scale $c = 1$.

8.3.2 Entropy Criterion Method (Kapur's Method)

Another nonparametric method that is used to determine the optimal threshold values has been proposed by Kapur et al. [7]. It is based on the entropy and the probability distribution of the image histogram. The method aims to find the optimal *th* that maximizes the overall entropy. The entropy of an image measures the compactness and separability among classes. In this sense, when the optimal *th* value appropriately separates the classes, the entropy has the maximum value. For the bi-level example, the objective function of the Kapur's problem can be defined as:

$$J(th) = H_1^c + H_2^c, \quad c = \begin{cases} 1, 2, 3 & \text{if} \quad \text{RGB Image} \\ 1 & \text{if} \quad \text{Gray scale image} \end{cases}, \tag{8.18}$$

where the entropies H_1 and H_2 are computed by the following model:

$$H_1^c = \sum_{i=1}^{th} \frac{Ph_i^c}{\omega_0^c} \ln\left(\frac{Ph_i^c}{\omega_0^c}\right), \quad H_2^c = \sum_{i=th+1}^{L} \frac{Ph_i^c}{\omega_1^c} \ln\left(\frac{Ph_i^c}{\omega_1^c}\right), \tag{8.19}$$

Ph_i^c is the probability distribution of the intensity levels which is obtained using Eq. (8.7). $\omega_0(th)$ and $\omega_1(th)$ are probabilities distributions for C_1 and C_2. $\ln(\cdot)$ stands for the natural logarithm. Similar to the Otsu's method, the entropy-based approach can be extended for multiple threshold values; for such a case, it is necessary to divide the image into *k* classes using the similar number of thresholds. Under such conditions, the new objective function is defined as:

$$J(\mathbf{TH}) = \max\left(\sum_{i=1}^{k} H_i^c\right), \quad c = \begin{cases} 1, 2, 3 & \text{if} \quad \text{RGB Image} \\ 1 & \text{if} \quad \text{Gray scale image} \end{cases}, \tag{8.20}$$

where $\mathbf{TH} = [th_1, th_2, \ldots, th_{k-1}]$, is a vector that contains the multiple thresholds. Each entropy is computed separately with its respective *th* value, so Eq. (8.21) is expanded for *k* entropies.

$$\begin{aligned} \mathrm{H}_1^c &= \sum_{i=1}^{th_1} \frac{Ph_i^c}{\omega_0^c} \ln\left(\frac{Ph_i^c}{\omega_0^c}\right), \\ \mathrm{H}_2^c &= \sum_{i=th_1+1}^{th_2} \frac{Ph_i^c}{\omega_1^c} \ln\left(\frac{Ph_i^c}{\omega_1^c}\right), \\ &\vdots \qquad \vdots \\ \mathrm{H}_k^c &= \sum_{i=th_k+1}^{L} \frac{Ph_i^c}{\omega_{k-1}^c} \ln\left(\frac{Ph_i^c}{\omega_{k-1}^c}\right) \end{aligned} \tag{8.21}$$

The values of the probability occurrence $(\omega_0^c, \omega_1^c, \ldots, \omega_{k-1}^c)$ of the k classes are obtained using Eq. (8.16) and the probability distribution Ph_i^c with Eq. (8.10). Finally, it is necessary to use Eq. (8.6) to separate the pixels into the corresponding classes.

8.4 Multilevel Thresholding Using Harmony Search Algorithm (HSMA)

8.4.1 Harmony Representation

Each harmony (candidate solution) uses k different elements as decision variables within the optimization algorithm. Such decision variables represent a different threshold point th that is used for the segmentation. Therefore, the complete population is represented as:

$$\mathbf{HM} = \left[\mathbf{x}_1^c, \mathbf{x}_2^c, \ldots, \mathbf{x}_{HMS}^c\right]^T, \quad \mathbf{x}_i^c = \left[th_1^c, th_2^c, \ldots, th_k^c\right], \tag{8.22}$$

where T refers to the transpose operator, HMS is the size of the harmony memory, $\mathbf{x}_i$ is the i-th element of HM and $c = 1, 2, 3$ is set for RGB images while $c = 1$ is chosen for gray scale images. For this problem, the boundaries of the search space are set to $l = 0$ and $u = 255$, which correspond to image intensity levels.

8.4.2 HMA Implementation

The presented segmentation algorithm has been implemented considering two different objective functions: Otsu and Kapur. Therefore, the HSA has been coupled with the Otsu and Kapur functions, producing two different segmentation algorithms. The implementation of both algorithms can be summarized into the following steps:

Step 1: Read the image I and if it is RGB separate it into I_R, I_G and I_B. If I is gray scale store it into I_{Gr}. $c = 1, 2, 3$ for RGB images or $c = 1$ for gray scale images.

Step 2: Obtain histograms: for RGB images h^R, h^G, h^B and for gray scale images h^{Gr}.

Step 3: Calculate the probability distribution using Eq. (8.7) and obtain the histograms.

Step 4: Initialize the HSA parameters: HMS, k, $HMCR$, PAR, BW, NI, and the limits l and u.

Step 5: Initialize a HM $\mathbf{x}_i^c$ of HMS random particles with k dimensions.

Step 6: Compute the values ω_i^c and μ_i^c. Evaluate each element of $\mathbf{HM}$ in the objective function $J(\mathbf{HM})$ Eq. (8.14) or Eq. (8.20) depending on the thresholding method (Otsu or Kapur respectively).

Step 7: Improvise a new harmony $\mathbf{x}_{new}^c$ as follows:

for $(j = 1 \text{ to } n)$ do
  if $(r_1 < HCMR)$ then
    $x_{new}^c(j) = x_a^c(j)$ where $a \in (1, 2, \ldots, HMS)$
    if $(r_2 < PAR)$ then
      $x_{new}^c(j) = x_a^c(j) \pm r_3 \cdot BW$ where $r_1, r_2, r_3 \in \text{rand}(0,1)$
    end if
    if $x_{new}^c(j) < l(j)$
      $x_{new}^c(j) = l(j)$
    end if
    if $x_{new}^c(j) > u(j)$
      $x_{new}^c(j) = u(j)$
    end if
  else
    $x_{new}^c(j) = l(j) + r \cdot (u(j) - l(j))$ whew $r \in \text{rand}(0,1)$
  end if
end for

Step 8: Update the HM as $\mathbf{x}_{worst}^c = \mathbf{x}_{new}^c$ if $f(\mathbf{x}_{new}^c) > f(\mathbf{x}_{worst}^c)$

Step 9: If NI is completed or the stop criteria is satisfied, then jump to step 10; otherwise go back to step 6.

Step 10: Select the harmony that has the best x_{best}^c objective function value.

Step 11: Apply the thresholds values contained in x_{best}^c to the image I Eq. (8.6).

8.4.3 Parameter Setting

The performance of HSA is strongly influenced by values assigned to parameters HM, *HMCR*, *PAR*, *BW* and *NI*. Determining the most appropriate parameter values for an arbitrary problem is a complex issue, since such parameters interact to each other in a highly nonlinear manner, and no mathematical models of such interaction currently exist. The common method to find the best set of parameter values is to fix each parameter value to a random number within the parameter limits and then HSA is executed. If the final result is not satisfactory, then a new set of parameter values is defined, and the evolutionary algorithm is executed again. Evidently, this process can be very expensive (in terms of computational time), since many different trials may be required before reaching a set of satisfactory parameter values. Additionally, the set of values chosen by the user are not necessarily the best possible, but only the best from the arbitrary number of trials performed by the user. In order to reduce the number of experiments in this chapter, it has been used the factorial design method proposed in [32, 33] to systematically identify the best parameters of HSA. The factorial design method [34] is a statistical technique that evaluates at the same time all process variables in order to determine which ones really exert significant influence on the final response. All variables are called factors and the different values chosen to study the factors are called levels. The factors to be considered in the factorial design are the HSA parameters, the harmony memory (HM), the harmony-memory consideration rate (*HMCR*), the pitch adjusting rate (*PAR*), the distance bandwidth (*BW*) and the number of improvisations (*NI*) whereas the response is the best fitness value obtained as a consequence of the HSA execution. Table 8.1 show the levels of the quantitative factors used in the factorial design. The values of zero level (central point) are based on the suggestions of the literature [33].

Each experiment is conducted combining the two different levels that define to each parameter considering as a problem an image histogram example. Since the factors are five, a 2^{5-1} fractional factorial design is chosen, requiring sixteen optimization experiments plus one optimization trial for the central point. The results obtained from seventeen runs were analyzed according to [32, 33] using a general linear form of analysis of variance (ANOVA) [34] considering a 95 % of

Table 8.1 Levels of the factors used for the factorial design method

HSA parameters (factors)	(−) Level	Central point	(+) Level
HM	50	100	200
HMCR	0.5	0.75	0.9
PAR	0.1	0.5	0.9
BW	0.1	0.3	0.5
NI	200	250	300

Table 8.2 HSMA parameter values obtained by the factorial design method

HM	*HMCR*	*PAR*	*BW*	*NI*
100	0.75	0.5	0.5	300

confidence. After such analysis, it was found that the best possible configuration of HSA is shown in Table 8.2. These results were consistent considering six replications using different image histograms and the Otsu (Eq. 8.14) or Kapur (Eq. 8.20) functions, indistinctly. For more information on how to build fractional factorial designs, the reader is referred to [32, 33].

8.5 Experimental Results

The HSMA has been tested under a set of 11 benchmark images. Some of these images are widely used in the image processing literature to test different methods (Lena, Cameraman, Hunter, Baboon, etc.) [3, 9]. All the images have the same size (512 × 512 pixels) and they are in JPGE format. For the sake of representation, only five images which are presented in Fig. 8.1 have been used to show the visual results; however, the numerical outcomes are analyzed considering the complete set.

Since HSMA is stochastic, it is necessary to employ an appropriate statistical metrics to measure its efficiency. Hence, the results have been reported executing the algorithm 35 times for each image. In order to maintain compatibility with similar works reported in the literature [14, 15, 18, 19], the number of thresholds points used in the test are $th = 2, 3, 4, 5$. In the experiments, the stop criterion is the number of times in which the best fitness values remains with no change. Therefore, if the fitness value for the best harmony remains unspoiled in 10 % of the total number of iterations (*NI*), then the HSA is stopped.

To evaluate the stability and consistency, it has been computed the standard deviation (STD) from the results obtained in the 35 executions. Since the STD represents a measure about how the data are dispersed, the algorithm becomes more instable as the STD value increases [19]. Equation 8.23 shows the model used to calculate the STD value.

$$STD = \sqrt{\sum_{i=1}^{NI} \frac{(bf_i - av)}{Ru}}, \tag{8.23}$$

where bf_i is the best fitness of the i-th iteration, av is the average value of bf and Ru is the number of total executions ($Ru = 35$).

On the other hand, as an index of quality, the peak-to-signal ratio (PSNR) is used to assess the similarity of the segmented image against a reference image (original

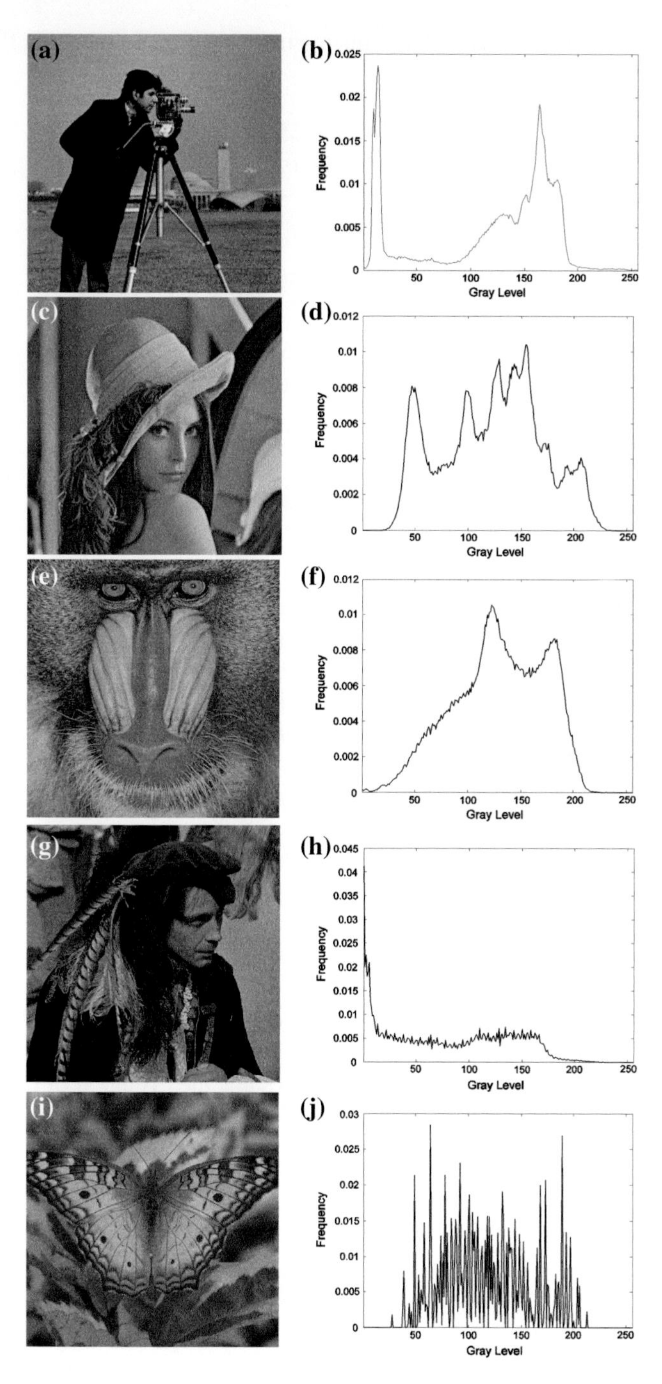

Fig. 8.1 **a** Camera man, **c** Lena, **e** Baboon, **g** Hunter and **i** Butterfly, the selected benchmark images. **b**, **d**, **f**, **h**, **j** histograms of the images

image) based on the produced mean square error (MSE) [18, 35]. Both PSNR and MSE are defined as:

$$
\begin{aligned}
PSNR &= 20\log_{10}\left(\frac{255}{RMSE}\right), \quad (\mathrm{dB}) \\
RMSE &= \sqrt{\frac{\sum_{i=1}^{ro}\sum_{j=1}^{co}\left(I_o^c(i,j) - I_{th}^c(i,j)\right)}{ro \times co}},
\end{aligned}
\tag{8.24}
$$

where I_o^c is the original image, I_{th}^c is the segmented image, $c = 1$ for gray scale and $c = 3$ for RGB images whereas ro, co are the total number of rows and columns of the image, respectively.

8.5.1 Otsu's Results

This section analyzes the results of HSMA after considering the variance among classes (Eq. 8.14) as the objective function, just as it has been proposed by Otsu [6]. The approach is applied over the complete set of benchmark images whereas the results are registered in Table 8.3. Such results present the best threshold values after testing the presented method with four different threshold points $th = 2, 3, 4, 5$. The table also features the *PSNR* and the *STD* values. It is evident that the *PSNR* and *STD* values increase their magnitude as the number of threshold points also increases.

For the sake of representation, it has been selected only five images of the set to show (graphically) the segmentation results. Figure 8.1 presents the images selected from the benchmark set and their respective histograms which possess irregular distributions (see Fig. 8.1j in particular). Under such circumstances, classical methods face great difficulties to find the best threshold values. The processing results for the selected original images are presented in five tables: Tables 8.4, 8.5, 8.6, 8.7 and 8.8. Such results show the segmented images considering four different threshold points $th = 2, 3, 4, 5$. The tables also show the evolution of the objective function during one execution.

8.5.2 Kapur's Results

This section analyzes the performance of HSMA after considering the entropy function (Eq. 8.20) as objective function, as it has been proposed by Kapur in [7]. Table 8.9 presents the experimental results after the application of HSMA over the

Table 8.3 Result after apply the HSMA using Otsu's function to the set of benchmark images

Image	*k*	Thresholds x^{c}_{best}	*PSNR*	*STD*
Camera man	2	70, 144	17.247	2.30 E−12
	3	59, 119, 156	20.211	1.55 E−02
	4	42, 95, 140, 170	21.533	2.76 E−12
	5	36, 82, 122, 149, 173	23.282	5.30 E−03
Lena	2	91, 150	15.401	9.22 E−13
	3	79, 125, 170	17.427	2.99 E−02
	4	73, 112, 144, 179	18.763	2.77 E−01
	5	71, 107, 134, 158, 186	19.443	3.04 E−01
Baboon	2	97, 149	15.422	6.92 E−13
	3	85, 125, 161	17.709	1.92 E−02
	4	71, 105, 136, 167	20.289	5.82 E−02
	5	66, 97, 123, 147, 173	21.713	4.40 E−01
Hunter	2	51, 116	17.875	2.30 E−12
	3	36, 86, 135	20.350	2.30 E−12
	4	27, 65, 104, 143	22.203	1.22 E−02
	5	22, 53,88, 112, 152	23.703	1.84 E−12
Airplane	2	113, 173	15.029	9.22 E−13
	3	92, 144, 190	18.854	4.83 E−01
	4	84, 129, 172, 203	20.735	7.24 E−01
	5	68, 106, 143, 180, 205	23.126	8.38 E−01
Peppers	2	72, 138	16.299	1.38 E−12
	3	65, 122, 169	18.359	4.61 E−13
	4	50, 88, 128, 171	20.737	4.61 E−13
	5	48, 85, 118, 150, 179	22.310	1.84 E−012
Living room	2	87, 145	15.999	1.15 E−12
	3	76, 123, 163	18.197	6.92 E−12
	4	56, 97, 132, 168	20.673	9.22 E−12
	5	49, 88, 120, 146, 178	22.225	2.86 E−02
Blonde	2	106, 155	14.609	6.92 E−13
	3	53, 112, 158	19.157	9.23 E−13
	4	50, 103, 139, 168	20.964	5.48 E−01
	5	49, 92, 121, 152, 172	22.409	6.50 E−01
Bridge	2	91, 56	13.943	4.61 E−13
	3	72, 120, 177	17.019	7.10 E−01
	4	63, 103, 145, 193	18.872	2.91 E−01
	5	56, 91, 124, 159, 201	20.539	3.57 E−01
Butterfly	2	99, 151	13.934	7.30 E−02
	3	82, 119, 160	16.932	6.17 E−01
	4	71, 102, 130, 163	19.259	3.07 E+00
	5	62, 77, 109, 137, 167	21.450	3.87 E+00

(continued)

Table 8.3 (continued)

Image	k	Thresholds x_{best}^{c}	*PSNR*	*STD*
Lake	2	85, 154	14.638	4.61 E−13
	3	78, 140, 194	15.860	1.84 E−12
	4	67, 110, 158, 198	17.629	2.68 E−01
	5	57, 88, 127, 166, 200	19.416	1.12 E−01

Table 8.4 Results after applying the HSMA by using Otsu over the cameraman image

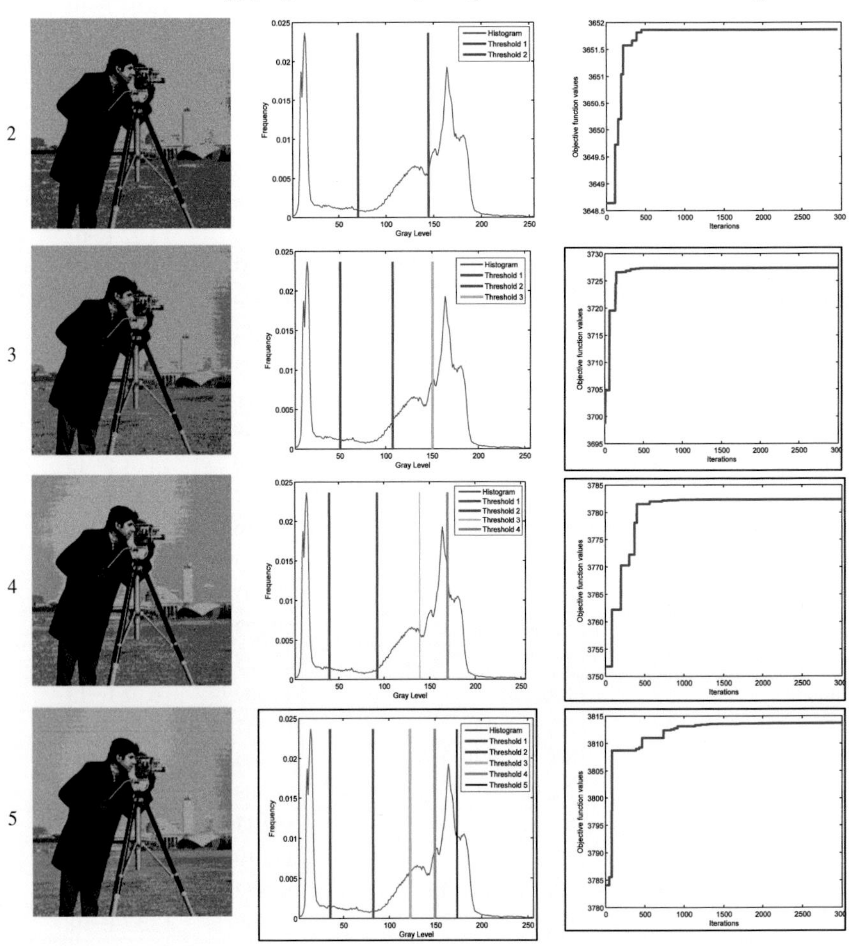

Table 8.5 Results after applying the HSMA by using Otsu over the lena image

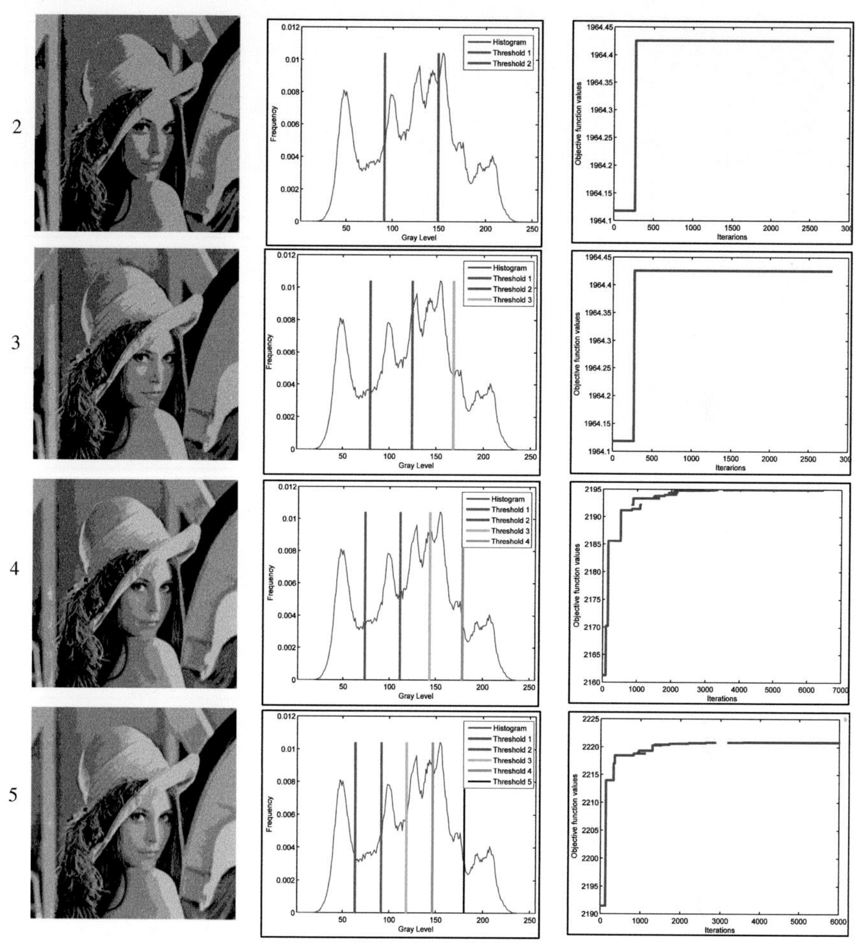

entire set of benchmark images. The values listed are: *PSNR*, *STD* and the best threshold values of the last population (x_t^B). The same test procedure that was previously applied to the Otsu's method (Sect. 8.5.1) is used with the Kapur's method, also considering the same stop criterion and a similar HSA parameter configuration.

Table 8.6 Results after applying the HSMA by using Otsu over the baboon image

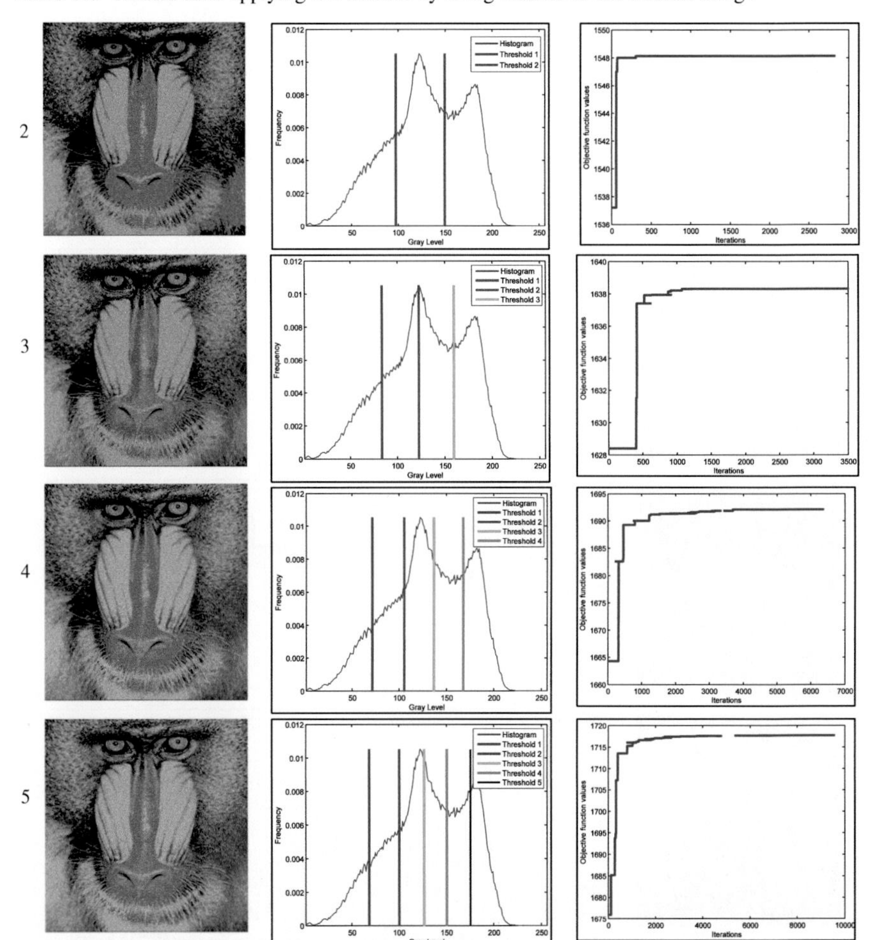

The results after apply the HSMA to the selected benchmark images are presented in Tables 8.10, 8.11, 8.12, 8.13 and 8.14. Four different threshold points have been employed: *th* = 2, 3, 4, 5. All tables exhibit the segmented image, the approximated histogram and the evolution of the fitness value during the execution of the HSA method.

From the results of both Otsu's and Kapur's methods, it is possible to appreciate that the HSMA converges (stabilizes) after a determined number of iterations depending on the *th* value. For experimental purposes HSMA continues running

Table 8.7 Results after applying the HSMA by using Otsu over the hunter image

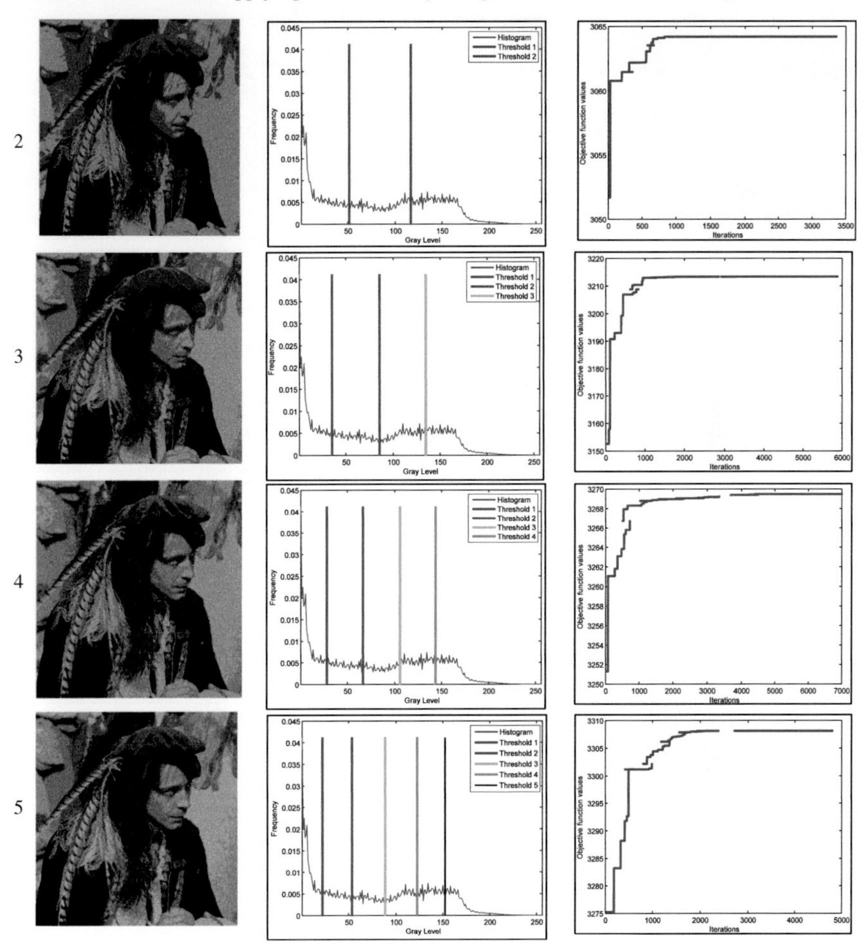

still further, even though the stop criterion is achieved. In this way, the graphics show that convergence is often reached in the first iterations of the optimization process. The segmented images provide evidence that the outcome is better with $th = 4$ and $th = 5$; however, if the segmentation task does not require to be extremely accurate then it is possible to select $th = 3$.

Table 8.8 Results after applying the HSMA by using Otsu over the butterfly image

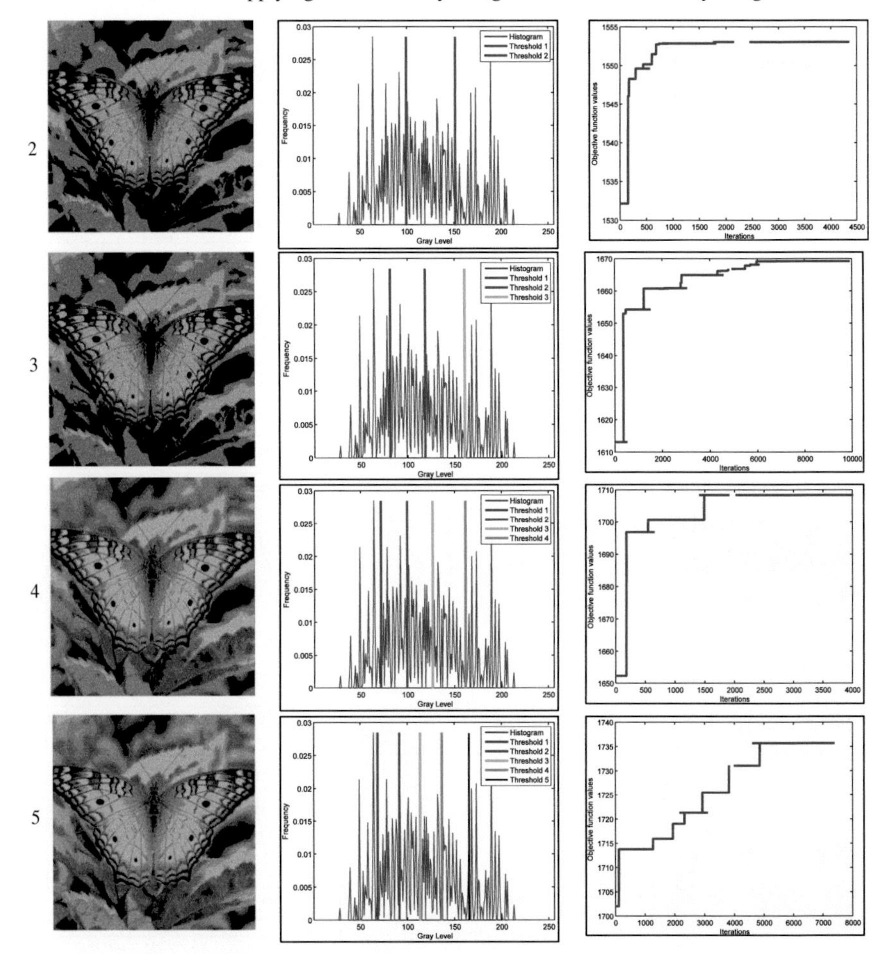

8.5.3 *Comparisons*

In order to analyze the results of HSMA, two different comparisons are executed. The first one involves the comparison between the two versions of the presented approach, one with the Otsu function and the other with the Kapur criterion. The second analyses the comparison between the HSMA and other state-of-the-art approaches.

Table 8.9 Results after apply the HSMA using Kapur's function to the set of benchmark images

Image	k	Thresholds x^{c}_{best}	*PSNR*	*STD*
Camera man	2	128, 196	13.626	3.60 E−15
	3	44, 103, 196	14.460	1.40 E−03
	4	44, 96, 146, 196	20.153	1.20 E−03
	5	24, 60, 98, 146, 196	20.661	2.75 E−02
Lena	2	96, 163	14.638	3.60 E−15
	3	23, 96, 163	16.218	7.66 E−02
	4	23, 80, 125, 173	19.287	1.44 E−14
	5	23, 71, 109, 144, 180	21.047	1.22 E−02
Baboon	2	79, 143	16.016	1.08 E−14
	3	79, 143, 231	16.016	7.19 E−02
	4	44, 98, 152, 231	18.485	8.47 E−02
	5	33, 74, 114, 159, 231	20.507	1.08 E−14
Hunter	2	92, 179	15.206	1.44 E−14
	3	59, 117, 179	18.500	4.16 E−04
	4	44, 89, 133, 179	21.065	4.31 E−04
	5	44, 89, 133, 179, 222	21.086	3.43 E−02
Airplane	2	70, 171	15.758	3.30 E−03
	3	68, 126, 182	18.810	1.08 E−14
	4	68, 126, 182, 232	18.810	1.82 E−01
	5	64, 104 143, 184, 232	20.321	1.80 E−01
Peppers	2	66, 143	16.265	7.21 E−15
	3	62, 112, 162	18.367	1.80 E−14
	4	62, 112, 162, 227	18.376	2.39 E−02
	5	48, 86, 127, 171, 227	18.827	4.17 E−04
Living room	2	89, 170	14.631	1.40 E−03
	3	47, 103, 175	17.146	2.70 E−03
	4	47, 98, 149, 197	19.144	1.34 E−02
	5	42, 85, 125, 162, 197	21.160	1.89 E−02
Blonde	2	125, 203	12.244	1.44 E−14
	3	65, 134, 203	16.878	1.00 E−03
	4	65, 113, 155, 203	20.107	5.50 E−03
	5	65, 113, 155, 203, 229	20.107	4.48 E−02
Bridge	2	94, 171	13.529	7.40 E−03
	3	65, 131, 195	16.806	1.44 E−02
	4	53, 102, 151, 199	18.902	2.47 E−02
	5	40, 85, 131, 171, 211	20.268	2.12 E−02
Butterfly	2	27, 213	8.1930	2.25 E−02
	3	27, 120, 213	13.415	8.60 E−04
	4	27, 96, 144, 213	16.725	3.80 E−03
	5	27, 83, 118, 152, 213	19.413	3.90 E−03

(continued)

Table 8.9 (continued)

Image	k	Thresholds x^{c}_{best}	*PSNR*	*STD*
Lake	2	91, 163	14.713	1.44 E−14
	3	73, 120, 170	16.441	3.05 E−04
	4	69, 112, 156, 195	17.455	4.53 E−02
	5	62, 96, 131, 166, 198	18.774	3.66 E−02

Table 8.10 Results after applying the HSMA by using Kapur over the cameraman image

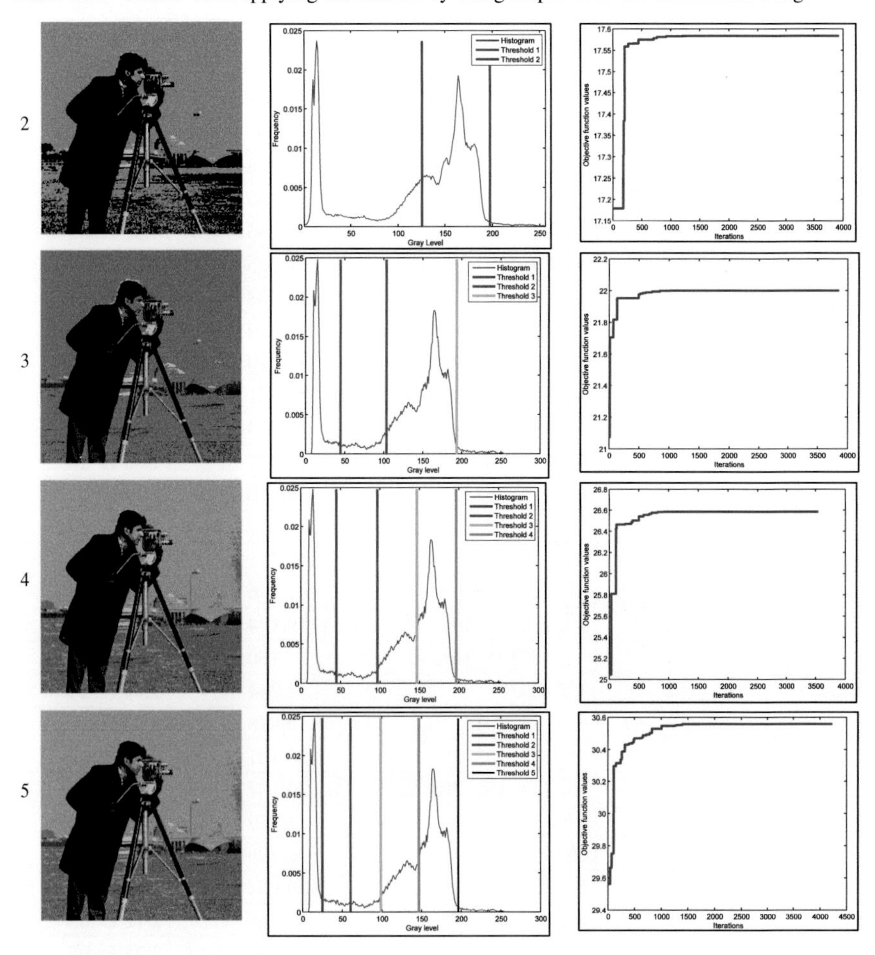

Table 8.11 Results after applying the HSMA by using Kapur over the lena image

2		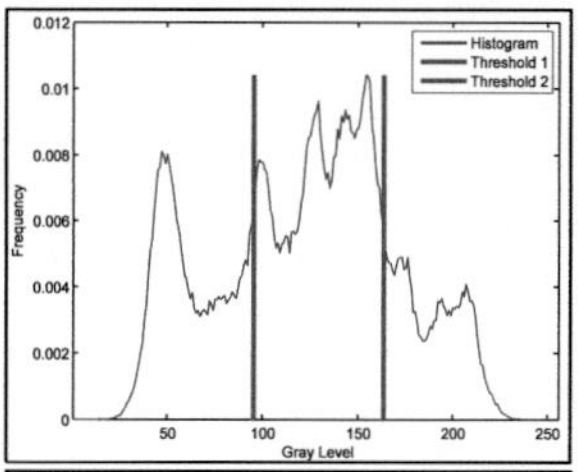	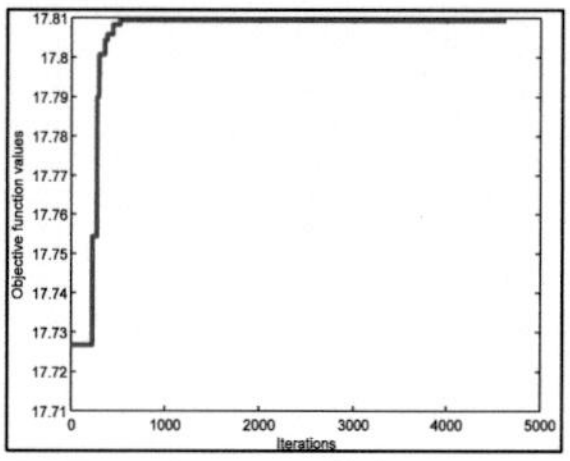
3		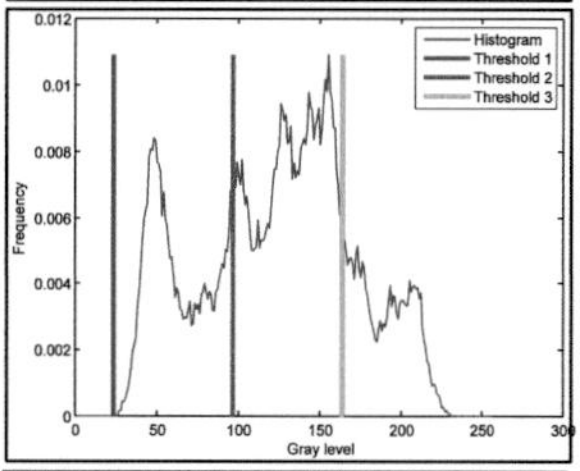	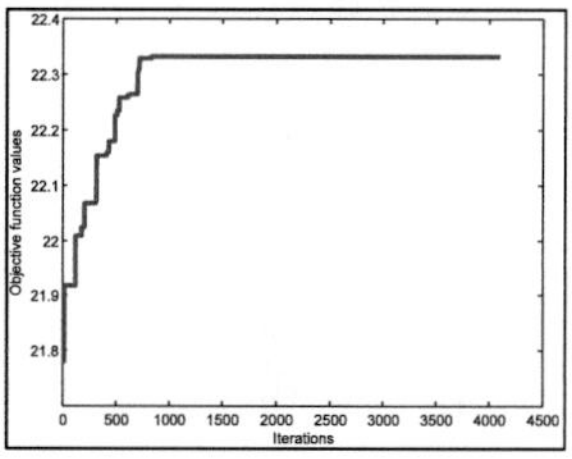
4		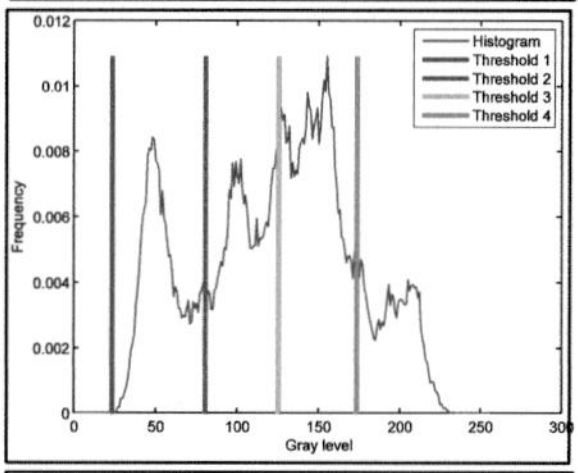	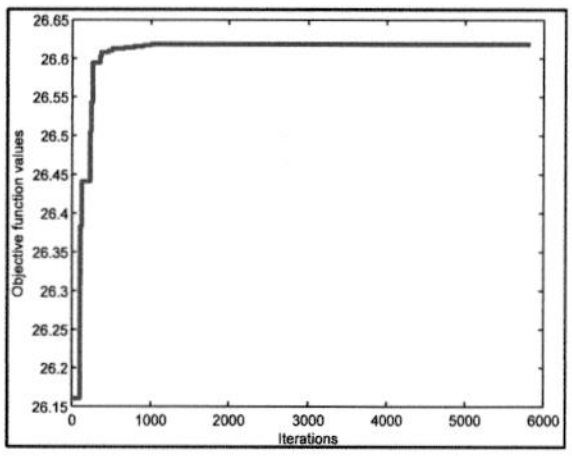
5		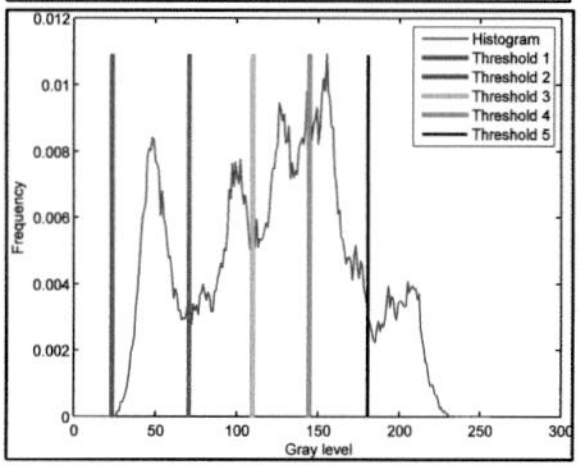	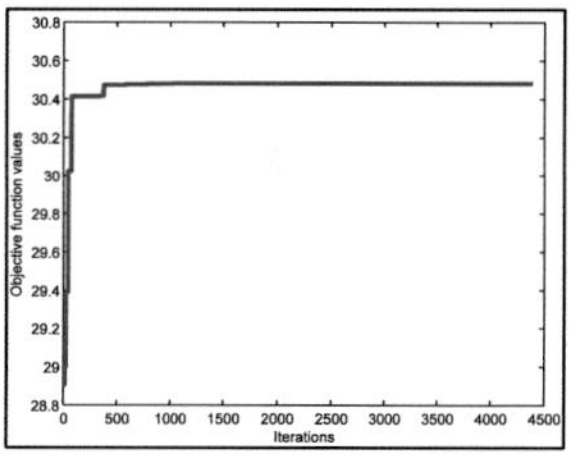

8.5.3.1 Comparison Between Otsu and Kapur HSMA

In order to statistically compare the results from Tables 8.3 and 8.9, a non-parametric significance proof known as the Wilcoxon's rank test [36, 37] for 35 independent samples has been conducted. Such proof allows assessing result differences among two related methods. The analysis is performed considering a 5 % significance level over the peak-to-signal ratio (PSNR) data corresponding to

Table 8.12 Results after applying the HSMA by using Kapur over the baboon image

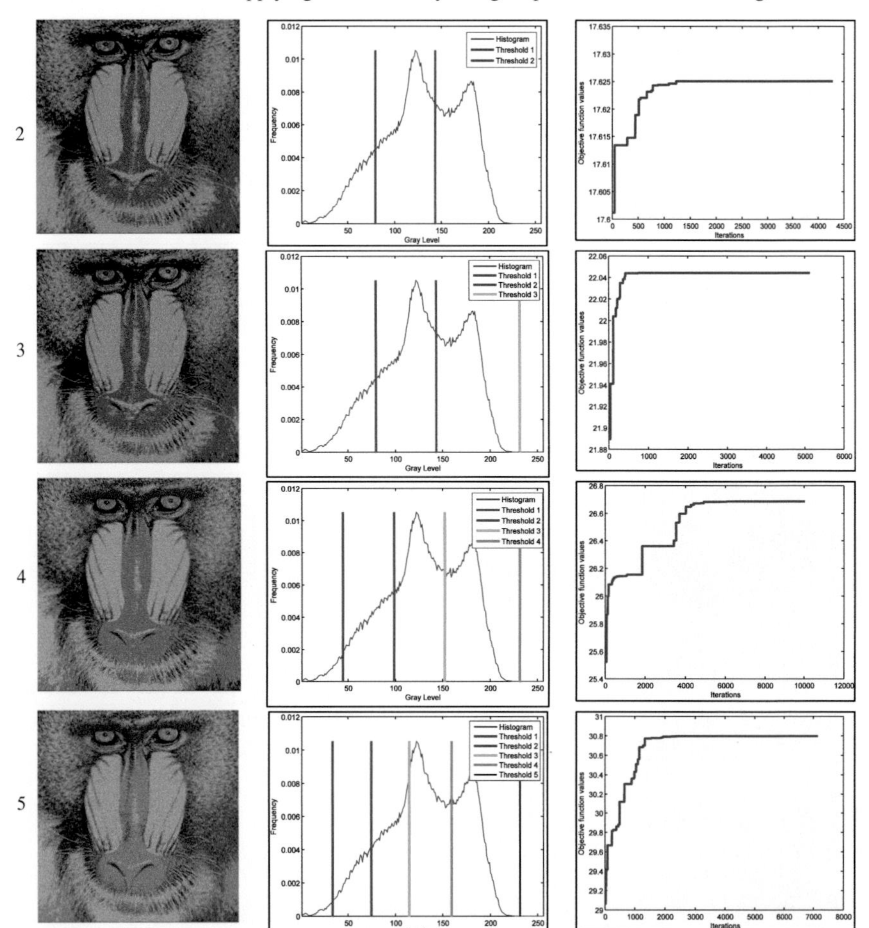

the five threshold points. Table 8.15 reports the p-values produced by Wilcoxon's test for a pair-wise comparison of the PSNR values between the Otsu and Kapur objective functions. As a null hypothesis, it is assumed that there is no difference between the values of the two objective functions. The alternative hypothesis considers an existent difference between the values of both approaches. All p-values reported in the Table 8.15 are less than 0.05 (5 % significance level) which is a strong evidence against the null hypothesis, indicating that the Otsu PSNR mean values for the performance are statistically better and it has not occurred by chance.

Table 8.13 Results after applying the HSMA by using Kapur over the hunter image

2		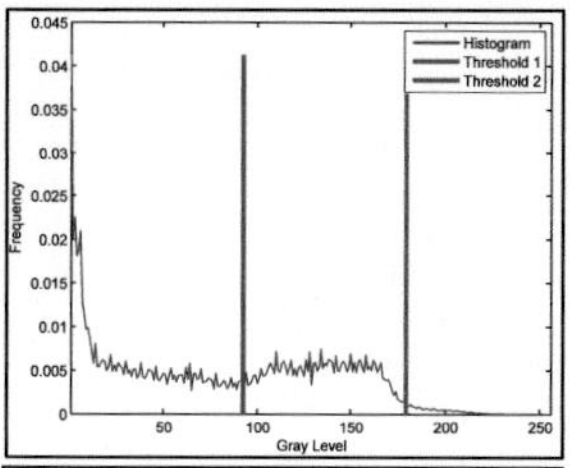	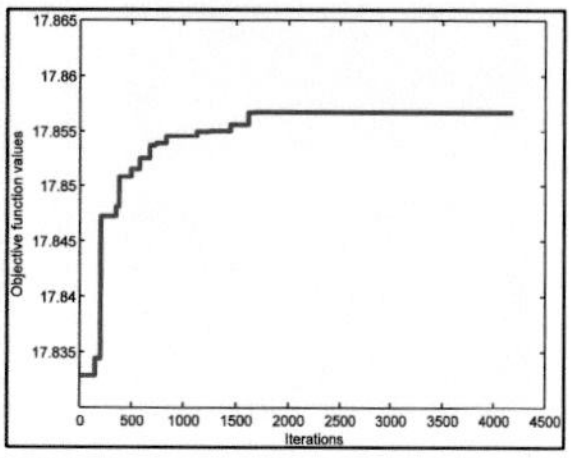
3		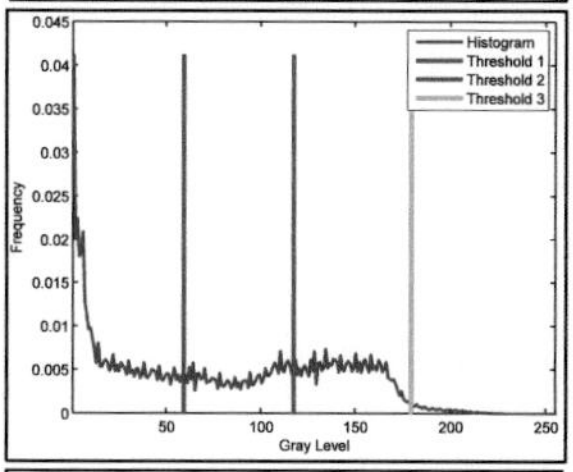	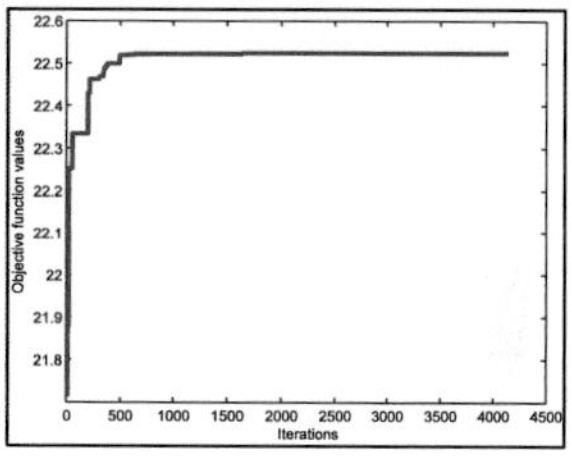
4		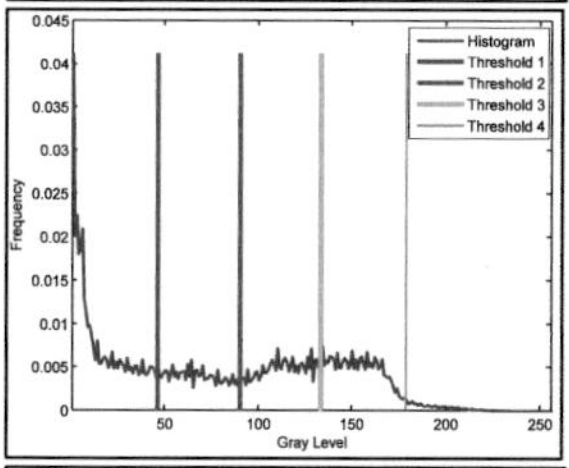	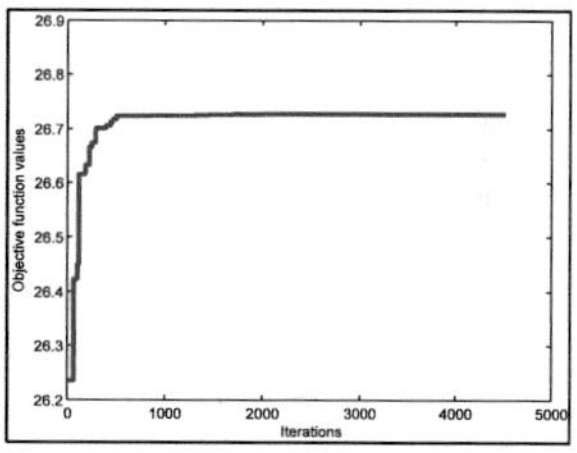
5		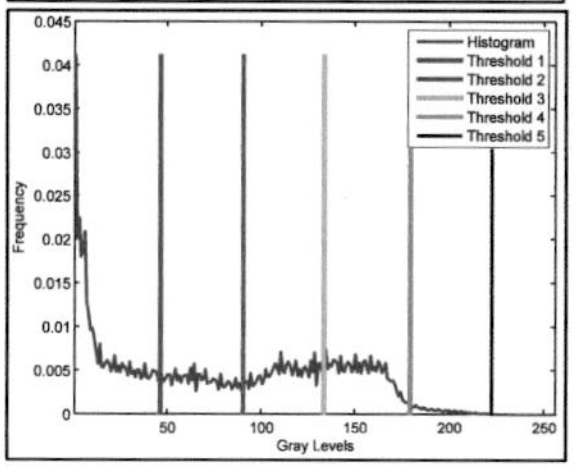	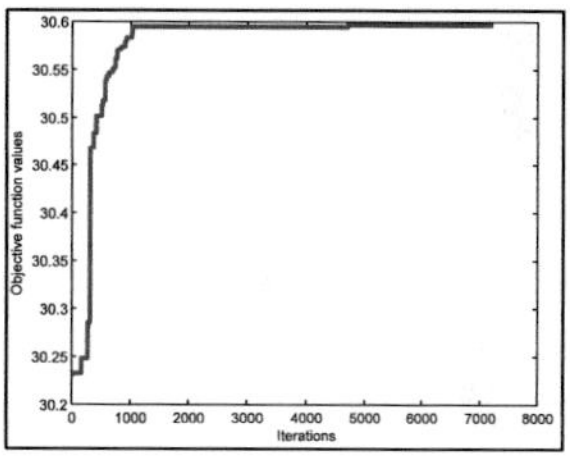

8.5.3.2 Comparison Among HSMA and Other MT Approaches

The results produced by HSMA have been compared with those generated by state-of-the-art thresholding methods such Genetic Algorithms (GA) [15], Particle Swarm Optimization (PSO) [18] and Bacterial Foraging (BF) [19].

Table 8.14 Results after applying the HSMA by using Kapur over the butterfly image

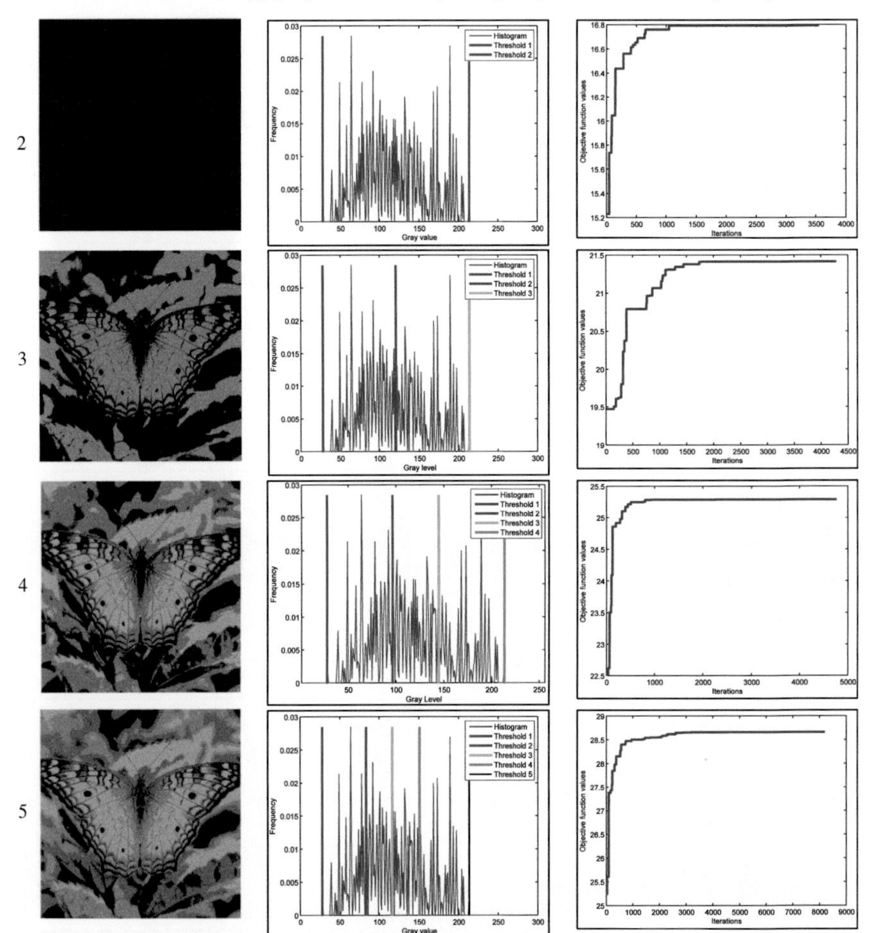

All the algorithms run 35 times over each selected image. The images at this test are the same as in Sects. 8.5.1 and 8.5.2 (Camera man, Lena, Baboon, Hunter and Butterfly). For each image, the *PSNR*, the *STD* and the mean of the objective function values are calculated. Moreover, the entire test is performed using both Otsu's and Kapur's objective functions.

Table 8.16 presents the computed values for a reduced benchmark test (five images). It is clear that the HSMA delivers better performance than the others. Such

Table 8.15 *p*-values produced by Wilcoxon's test comparing Otsu versus Kapur over the averaged PSNR from Tables 8.3 and 8.9

Image	k	p-value Otsu versus Kapur
Camera man	2	1.0425 E−16
	3	2.1435 E−15
	4	2.6067 E−16
	5	6.2260 E−16
Lena	2	1.0425 E−16
	3	9.4577 E−15
	4	9.7127 E−15
	5	1.2356 E−12
Baboon	2	1.0425 E−16
	3	1.7500 E−02
	4	5.3417 E−14
	5	1.4013 E−14
Hunter	2	1.0425 E−16
	3	2.6067 E−16
	4	6.6386 E−14
	5	6.4677 E−15
Airplane	2	1.0425 E−16
	3	4.6500 E−02
	4	1.7438 E−13
	5	6.2475 E−13
Peppers	2	1.0425 E−16
	3	1.0425 E−16
	4	6.0498 E−15
	5	5.3194 E−15
Living room	2	3.2405 E−15
	3	1.4223 E−15
	4	1.3175 E−14
	5	6.2220 E−14
Blonde	2	1.0425 E−16
	3	2.6067 E−16
	4	4.0480 E−13
	5	7.8167 E−04
Bridge	2	7.3588 E−06
	3	1.1300 E−04
	4	1.9400 E−02
	5	2.1900 E−02
Butterfly	2	1.1615 E−14
	3	2.5697 E−14
	4	3.7190 E−13
	5	1.7941 E−06

(continued)

Table 8.15 (continued)

Image	k	p-value Otsu versus Kapur
Lake	2	1.0425 E−16
	3	5.6120 E−16
	4	9.8174 E−14
	5	2.2292 E−14

values are computed using the Otsu's method as the objective function. On the other hand, the same experiment has been performed using the Kapur's method. Using the same criteria (as those described for the Otsu's method), the algorithm runs over 35 times at each image. The results of this experiment are presented in Table 8.17 and show that the presented HSMA algorithm is better in comparison with the GA, PSO and BF.

8.6 Conclusions

In this chapter, a MT-based method on the original Harmony Search Algorithm (HSA) is presented. The approach combines the good search capabilities of HSA algorithm and the use of some objective functions that have been proposed by the popular MT methods of Otsu and Kapur. In order to measure the performance of the presented approach, the peak signal-to-noise ratio (PSNR) is used to assess the segmentation quality by considering the coincidences between the segmented and the original images. In this work, a simple HSA implementation without any modification is considered in order to demonstrate that it can be applied to image processing tasks.

The study explores the comparison between two versions of HSMA: one employs the Otsu objective function while the other uses the Kapur criterion. Results show that the Otsu function delivers better results than the Kapur criterion. Such conclusion has been statistically proved considering the Wilcoxon test.

The presented approach has been compared to other techniques that implement different optimization algorithms like GA, PSO and BF. The efficiency of the algorithm has been evaluated in terms of the PSNR index and the STD value. Experimental results provide evidence on the outstanding performance, accuracy and convergence of the presented algorithm in comparison to other methods. Although the results offer evidence to demonstrate that the standard HSA method can yield good results on complicated images, the aim of our chapter is not to devise an MT algorithm that could beat all currently available methods, but to show that harmony search algorithms can be effectively considered as an attractive alternative for this purpose.

Table 8.16 Comparisons between HSMA, GA, PSO and BF, applied over the selected test images using Otsu's method

Image	k	HSMA			GA			PSO			BF		
		PSNR	*STD*	Mean	*PSNR*	*STD*	Mean	*PSNR*	*STD*	Mean	*PSNR*	*STD*	Mean
Camera man	2	17.247	2.30 E−12	3651.9	17.048	0.0232	3604.5	17.033	0.0341	3598.3	17.058	0.0345	3590.9
	3	20.211	1.55 E−02	3727.4	17.573	0.1455	3678.3	19.219	0.2345	3662.7	20.035	0.2459	3657.5
	4	21.533	2.76 E−12	3782.4	20.523	0.2232	3781.5	21.254	0.3142	3777.4	21.209	0.4560	3761.4
	5	23.282	5.30 E−03	3813.7	21.369	0.4589	3766.4	22.095	0.5089	3741.6	22.237	0.5089	3789.8
Lena	2	15.401	9.22 E−13	1964.4	15.040	0.0049	1960.9	15.077	0.0033	1961.4	15.031	2.99 E−04	1961.5
	3	17.427	2.99 E−02	2131.4	17.304	0.1100	2126.4	17.276	0.0390	2127.7	17.401	0.0061	2128.0
	4	18.763	2.77 E−01	2194.9	17.920	0.2594	2173.7	18.305	0.1810	2180.6	18.507	0.0081	2189.0
	5	19.443	3.04 E−01	2218.7	18.402	0.3048	2196.2	18.770	0.2181	2212.5	19.001	0.0502	2215.6
Baboon	2	15.422	6.92 E−13	1548.1	15.304	0.0031	1547.6	15.088	0.0077	1547.9	15.353	8.88 E−04	1548.0
	3	17.709	1.92 E−02	1638.3	17.505	0.1750	1633.5	17.603	0.0816	1635.3	17.074	0.0287	1637.0
	4	20.289	5.82 E−02	1692.1	18.708	0.2707	1677.7	19.233	0.0853	1684.3	19.654	0.0336	1690.7
	5	21.713	4.40 E−01	1717.5	20.203	0.3048	1712.9	20.526	0.1899	1712.9	21.160	0.1065	1716.7
Hunter	2	17.875	2.30 E−12	3054.2	17.088	0.0470	3064.1	17.932	0.2534	3064.1	17.508	0.0322	3064.1
	3	20.350	2.30 E−12	3213.4	20.045	0.1930	3212.9	19.940	0.9727	3212.4	20.350	0.9627	3213.4
	4	22.203	1.22 E−02	3269.5	20.836	0.6478	3268.4	21.128	2.2936	3266.3	21.089	2.2936	3266.3
	5	23.703	1.84 E−12	3308.1	21.284	1.6202	3305.6	22.026	4.1811	3276.3	22.804	3.6102	3291.1
Butterfly	2	13.934	7.30 E−02	1553.0	13.007	0.0426	1553.0	13.092	0.0846	1553.0	13.890	0.0643	1553.0
	3	16.932	6.17 E−01	1669.2	15.811	0.3586	1669.0	17.261	2.6268	1665.7	17.285	1.2113	1667.2
	4	19.259	3.07 E+00	1708.3	17.104	0.6253	1709.9	17.005	3.7976	1702.9	17.128	2.2120	1707.0
	5	21.450	3.87 E+00	1728.0	18.593	0.5968	1734.4	18.099	6.0747	1730.7	18.9061	3.5217	1733.0

Table 8.17 Comparison between HSMA GA, PSO and BF, applied over selected test images using Kapur's method

Image	k	HSMA			GA			PSO			BF		
		PSNR	*STD*	Mean	*PSNR*	*STD*	Mean	*PSNR*	*STD*	Mean	*PSNR*	*STD*	Mean
Camera man	2	13.626	3.60 E−15	17.584	11.941	0.1270	15.341	12.259	0.1001	16.071	12.264	0.0041	16.768
	3	14.460	1.40 E−03	22.007	14.827	0.2136	20.600	15.211	0.1107	21.125	15.250	0.0075	21.498
	4	20.153	1.20 E−03	26.586	17.166	0.2857	24.267	18.000	0.2005	25.050	18.406	0.0081	25.093
	5	20.661	2.75 E−02	30.553	19.795	0.3528	28.326	20.963	0.2734	28.365	21.211	0.0741	30.026
Lena	2	14.638	3.60 E−15	17.809	12.334	0.0049	16.122	12.345	0.0033	16.916	12.345	2.99 E−4	16.605
	3	16.218	7.66 E−02	22.306	14.995	0.1100	20.920	15,133	0.0390	20.468	15.133	0.0061	20.812
	4	19.287	1.44 E−14	26.619	17.089	0.2594	23.569	17.838	0.1810	24.449	17.089	0.0081	26.214
	5	21.047	1.22 E−02	30.485	19.549	0.3043	27.213	20.442	0.2181	27.526	19.549	0.0502	28.046
Baboon	2	16.016	1.08 E−14	17.625	12.184	0.0567	16.425	12.213	0.0077	16.811	12.216	8.88 E−4	16.889
	3	16.016	7.19 E−02	22.117	14.745	0.1580	21.069	15.008	0.0816	21.088	15.211	0.0287	21.630
	4	18.485	8.47 E−02	26.671	16.935	0.1765	25.489	17.574	0.0853	24.375	17.999	0.0336	25.446
	5	20.507	1.08 E−14	30.800	19.662	0.2775	29.601	20.224	0.1899	30.994	20.720	0.1065	30.887
Hunter	2	15.206	1.44 E−14	17.856	12.349	0.0148	16.150	12.370	0.0068	15.580	12.373	0.0033	16.795
	3	18.500	4.16 E−04	22.525	14.838	0.1741	21.026	15.128	0.0936	20.639	15.553	0.1155	21.860
	4	21.065	4.31 E−04	26.728	17.218	0.2192	25.509	18.040	0.1560	27.085	18.381	0.0055	26.230
	5	21.086	3.43 E−02	30.612	19.563	0.3466	29.042	20.533	0.2720	29.013	21.256	0.0028	28.856
Butterfly	2	8.1930	2.25 E−02	16.791	10.470	0.0872	15.481	10.474	0.0025	14.098	10.474	0.0014	15.784
	3	13.415	8.60 E−04	21.417	11.628	0.2021	20.042	12.313	0.1880	19.340	12.754	0.0118	21.308
	4	16.725	3.80 E−03	25.292	13.314	0.2596	23.980	14.231	0.2473	25.190	14.877	0.0166	25.963
	5	19.413	3.90 E−03	28.664	15.756	0.3977	27.411	16.337	0.2821	27.004	16.828	0.0877	27.980

References

1. Gonzalez, R.C., Woods, R.E.: Digital Image Processing. Addison Wesley, Reading, MA (1992)
2. Guo, R., Pandit, S.M.: Automatic threshold selection based on histogram modes and discriminant criterion. Mach. Vis. Appl. **10**, 331–338 (1998)
3. Pal, N.R., Pal, S.K.: A review on image segmentation techniques. Pattern Recognit. **26**, 1277–1294 (1993)
4. Shaoo, P.K., Soltani, S., Wong, A.K.C., Chen, Y.C.: Survey: a survey of thresholding techniques. Comput. Vis. Graph. Image Process. **41**, 233–260 (1988)
5. Snyder, W., Bilbro, G., Logenthiran, A., Rajala, S.: Optimal thresholding: a new approach. Pattern Recognit. Lett. **11**, 803–810 (1990)
6. Otsu, N.: A threshold selection method from gray-level histograms. IEEE Trans. Syst. Man Cybern. **SMC-9**, 62–66 (1979)
7. Kapur, J.N., Sahoo, P.K., Wong, A.K.C.: A new method for gray-level picture thresholding using the entropy of the histogram. Comput. Vis. Graph. Image Process. **2**, 273–285 (1985)
8. Kittler, J., Illingworth, J.: Minimum error thresholding. Pattern Recognit. **19**, 41–47 (1986)
9. Hammouche, K., Diaf, M., Siarry, P.: A comparative study of various meta-heuristic techniques applied to the multilevel thresholding problem. Eng. Appl. Artif. Intell. **23**, 676–688 (2010)
10. Storn, R., Price, K.: Differential evolution—A simple and efficient heuristic for global optimization over continuous spaces. J. Global Optim. **11**(4), 341–359 (1995)
11. Kirkpatrick, S., Gelatt, C.D., Vecchi, M.P.: Optimization by Simulated Annealing, Science, New Series, vol. 220, No. 4598, pp. 671–680 (1983)
12. Glover, F.: Tabu search—Part I. ORSA J. Comput. **1**(3), 190–206 (1989)
13. Holland, J.H.: Adaptation in Natural and Artificial Systems. University of Michigan Press, Ann Arbor, MI (1975)
14. Lai, C., Tseng, D.: A hybrid approach using gaussian smoothing and genetic algorithm for multilevel thresholding. Int. J. Hybrid Intell. Syst. **1**, 143–152 (2004)
15. Yin, P.-Y.: A fast scheme for optimal thresholding using genetic algorithms. Signal Process. **72**, 85–95 (1999)
16. Kennedy, J., Eberhart, R.: Particle swarm optimization. In: Proceedings of the 1995 IEEE International Conference on Neural Networks, vol. 4, pp. 1942–1948 (1995)
17. Karaboga, D.: An idea based on honey bee swarm for numerical optimization. Technical Report-TR06. Engineering Faculty, Computer Engineering Department, Erciyes University (2005)
18. Akay, B.: A study on particle swarm optimization and artificial bee colony algorithms for multilevel thresholding. Appl. Soft Comput. **13**(6), 3066–3091 (2012). doi:10.1016/j.asoc.2012.03.072
19. Sathya, P.D., Kayalvizhi, R.: Optimal multilevel thresholding using bacterial foraging algorithm. Expert Syst. Appl. **38**, 15549–15564 (2011)
20. Geem, Z.W., Kim, J.H., Loganathan, G.V.: A new heuristic optimization algorithm: harmony search. Simulations **76**, 60–68 (2001)
21. Mahdavi, M., Fesanghary, M., Damangir, E.: An improved harmony search algorithm for solving optimization problems. Appl. Math. Comput. **188**, 1567–1579 (2007)
22. Lee, K.S., Geem, Z.W.: A new meta-heuristic algorithm for continuous engineering optimization, harmony search theory and practice. Comput. Methods Appl. Mech. Eng. **194**, 3902–3933 (2005)
23. Lee, K.S., Geem, Z.W., Lee, S.H., Bae, K.-W.: The harmony search heuristic algorithm for discrete structural optimization. Eng. Optim. **37**, 663–684 (2005)
24. Kim, J.H., Geem, Z.W., Kim, E.S.: Parameter estimation of the nonlinear Muskingummodel using harmony search. J. Am. Water Resour. Assoc. **37**, 1131–1138 (2001)

25. Geem, Z.W.: Optimal cost design of water distribution networks using harmony search. Eng. Optim. **38**, 259–280 (2006)
26. Geem, Z.W., Lee, K.S., Park, Y.J.: Application of harmony search to vehicle routing. Am. J. Appl. Sci. **2**, 1552–1557 (2005)
27. Vasebi, A., Fesanghary, M., Bathaee, S.M.T.: Combined heat and power economic dispatch by harmony search algorithm. Electr. Power Energy Syst. **29**, 713–719 (2007)
28. Degertekin, S.O.: Optimum design of steel frames using harmony search algorithm. Struct. Multidiscipl. Optim. **36**(4), 393–401 (2008)
29. Cuevas, E., Ortega-Sánchez, N., Zaldivar, D., Pérez-Cisneros, M.: Circle detection by harmony search optimization. J. Intell. Robot. Syst. **66**(3), 359–376 (2012)
30. Alia, O., Mandava, R.: The variants of the harmony search algorithm: an overview. J. Artif. Intell. Rev. **36**(1), 49–68 (2011)
31. Lobo, F.G., Lima, C.F., Michalewicz, Z. (eds.): Parameter Setting in Evolutionary Algorithms, Studies in Computational Intelligence, vol. 54. Springer, Berlin (2007)
32. Costa, C.B.B., Maciel, M.R.W., MacielFilho, R.: Factorial design technique applied to genetic algorithm parameters in a batch cooling crystallization optimization. Comput. Chem. Eng. **29**, 2229–2241 (2005)
33. Khadwilard, A., Luangpaiboon, P., Pongcharoen, P.: Full factorial experimental design for parameters selection of harmony search algorithm. J. Ind. Technol. **8**(2), 1–10 (2012)
34. Box, G.E.P., Hunter, W.G., Hunter, J.S.: Statistic for Experimenters—An Introduction to Design Data Analysis and Model Building, pp. 306–342, 374–409. Wiley, New York. (1978)
35. Pal, S.K. Bhandari, D. Kundu, M.K.: Genetic algorithms, for optimal image enhancement. Pattern Recognit. Lett. **15**, 261–271 (1994)
36. Wilcoxon, F.: Individual comparisons by ranking methods. Biometrics **1**, 80–83 (1945)
37. Garcia, S., Molina, D., Lozano, M., Herrera, F.: A study on the use of non-parametric tests for analyzing the evolutionary algorithms' behaviour: a case study on the CEC'2005 Special session on real parameter optimization. J. Heurist. (2008). doi:10.1007/s10732-008-9080-4

Chapter 9
Leukocyte Detection by Using Electromagnetism-like Optimization

Abstract Automatic circle detection in digital images has been considered as an important and complex task for the computer vision community that has devoted important research efforts into optimal circle detectors. On the other hand, medical imaging is a relevant field of application of image processing algorithms. In particular, the analysis of white blood cell (WBC) images has engaged researchers from fields of medicine and computer vision alike. Since WBC's can be approximated by a quasi-circular form, a circular detector algorithm may be successfully applied. This chapter presents an algorithm for the automatic detection of white blood cells embedded into complicated and cluttered smear images that considers the complete process as a circle detection problem. The approach is based on a nature-inspired technique called the Electromagnetism-Like Optimization (EMO) which is a heuristic method that follows electromagnetism principles for solving complex optimization problems. The EMO algorithm is based on the electromagnetic attraction and repulsion among charged particles whose charge represents the fitness solution for each particle (a given solution). The algorithm uses the encoding of three non-collinear edge points as candidate circles over an edge map. A new objective function has been derived to measure the resemblance of a candidate circle to an actual WBC based on the information from the edge map and segmentation results. Guided by the values of such objective function, the set of encoded candidate circles (charged particles) are evolved by using the EMO algorithm so that they can fit into the actual blood cells contained in the edge-only map of the image. Experimental results from blood cell images with a varying range of complexity are included to validate the efficiency of the presented technique regarding detection, robustness and stability.

E. Cuevas et al., *Applications of Evolutionary Computation in Image Processing and Pattern Recognition*, Intelligent Systems Reference Library 100,
DOI 10.1007/978-3-319-26462-2_9

9.1 Introduction

Nature-inspired computing is a field of research that is concerned with both the use of biology as an inspiration for solving computational problems and the use of the natural physical phenomena to solve real world problems. Moreover, nature-inspired computing has proved to be useful in several application areas [1] with relevant contributions to optimization, pattern recognition, shape detection and machine learning. In particular, it has gained considerable research interest from the computer vision community as nature-based algorithms have successfully contributed to solve challenging computer vision problems.

On the other hand, White Blood Cells (WBC) also known as leukocytes play a significant role in the diagnosis of different diseases. Although digital image processing techniques have successfully contributed to generate new methods for cell analysis, which in turn, have lead into more accurate and reliable systems for disease diagnosis. However, high variability on cell shape, size, edge and localization, complicates the data extraction process. Moreover, the contrast between cell boundaries and the image's background may vary due to unstable lighting conditions during the capturing process.

Many works have been conducted in the area of blood cell detection. In [2] a method based on boundary support vectors is proposed to identify WBC. In such approach, the intensity of each pixel is used to construct feature vectors whereas a Support Vector Machine (SVM) is used for classification and segmentation. By using a different approach, in [3], Wu et al. developed an iterative Otsu method based on the circular histogram for leukocyte segmentation. According to such technique, the smear images are processed in the Hue-Saturation-Intensity (HSI) space by considering that the Hue component contains most of the WBC information. One of the latest advances in white blood cell detection research is the algorithm proposed by Wang et al. [4] that is based on the fuzzy cellular neural network (FCNN). Although such method has proved successful in detecting only one leukocyte in the image, it has not been tested over images containing several white cells. Moreover, its performance commonly decays when the iteration number is not properly defined, yielding a challenging problem itself with no clear clues on how to make the best choice.

Since blood cells can be approximated with a quasi-circular form, a circular detector algorithm may be handy. The problem of detecting circular features holds paramount importance for image analysis, in particular for medical image analysis [5]. The circle detection in digital images is commonly performed by the Circular Hough Transform [6]. A typical Hough-based approach employs an edge detector whose information guides the inference for circle locations and radius values. Peak detection is then performed by averaging, filtering and histogramming the transform space. However, such approach requires a large storage space given the required 3-D cells to cover all parameters (x, y, r). It also implies a high computational complexity yielding a low processing speed. The accuracy of the extracted parameters for the detected circle is poor, particularly in presence of noise [7]. For a digital image holding a significant width and height and a densely populated edge pixel map, the required processing time for Circular Hough Transform makes it prohibitive to be deployed in real time applications. In order to overcome such a problem, some other researchers have proposed new approaches based on the Hough transform, for instance the probabilistic Hough transform [8, 9], the Randomized Hough transform (RHT) [10] and the Fuzzy Hough Transform [11]. Alternative transformations have also been presented in the literature as the one proposed by Becker et al. in [12]. Although those new approaches demonstrated better processing speeds in comparison to the original Hough Transform, they are still very sensitive to noise.

As an alternative to Hough Transform-based techniques, the circle detection problem has also been handled through optimization methods. In general, they have demonstrated to deliver better results than those based on HT considering accuracy, speed and robustness [13]. Such approaches have produced several robust circle detectors using different optimization algorithms such as Genetic algorithms (GA) [13], Harmony Search (HSA) [14], Differential Evolution (DE) [15] and the Electromagnetism-like Optimization algorithm (EMO) [16].

Although detection algorithms based on optimization approaches present several advantages in comparison to those based on the Hough Transform, they have been scarcely applied to WBC detection. One exception is the work presented by Karkavitsas and Rangoussi [17] that solves the WBC detection problem through the use of GA. However, since the evaluation function, which assesses the quality of each solution, considers the number of pixels contained inside of a circle with fixed radius, the method is prone to produce misdetections particularly for images that contained overlapped or irregular WBC.

In this chapter, the WBC detection task is approached as an optimization problem and the EMO-based circle detector [16] is used to build the circular approximation. The EMO algorithm [18] is a stochastic evolutionary computation technique based on the electromagnetism theory. It considers each solution to be a charged particle. The charge of each particle is determined by an objective function. Thereby, EMO moves each particle according to its charge within an attraction or repulsion field among the population using the Coulomb's law and the superposition principle. This attraction–repulsion mechanism of the EMO algorithm corresponds to the reproduction, crossover and mutation in GA [19]. In general, the EMO algorithm can be considered as a fast and robust algorithm representing an alternative to solve complex, nonlinear, non-differentiable and non-convex optimization problems. The principal advantages of the EMO algorithm lies on several facts: it has no gradient operation, it can be used directly on a decimal system, it needs only few particles to converge and the convergence existence has been already verified [20].

The EMO-based circle detector uses the encoding of three edge points that represent candidate circles in the edge map of the scene. The quality of each individual is calculated by using an objective function which evaluates if such candidate circles are really present in the edge map of the image. The better a candidate circle approximates the actual edge-circle, the objective function value decreases. Therefore, the detection performance depends on the quality of the edge map as it is obtained from the original images. However, since smear images present different imaging conditions and staining intensities, they produce edge maps partially damaged by noisy pixels. Under such conditions, the use of the EMO-based circle detector cannot be directly applied to WBC detection.

This chapter presents an algorithm for the automatic detection of blood cell images based on the EMO algorithm. The presented method modifies the EMO-based circle detector by incorporating a new objective function. Such function allows to accurately measure the resemblance of a candidate circle with an actual WBC on the image which is based on the information not only from the edge map, but also from the segmentation results. Guided by the values of the new objective function, the set of encoded candidate circles are evolved using the EMO algorithm so that they can fit into actual WBC on the image. The approach generates a sub-pixel detector which can effectively identify leukocytes in real images. Experimental evidence shows the effectiveness of such method in detecting leukocytes despite complex conditions. Comparison to the state-of-the-art WBC detectors on multiple images demonstrates a better performance of the presented method.

The main contribution of this study is the proposal of a new WBC detector algorithm that efficiently recognize WBC under different complex conditions while considering the whole process as an circle detection problem. Although circle detectors based on optimization present several interesting properties, to the best of our knowledge, they have not yet been applied to any medical image processing up to date.

9.2 Electromagnetism-like Optimization Algorithm (EMO)

Initially designed for bound constrained optimization problems, the EMO method [18] utilizes N, n-dimensional points $x_{i,k}$, i = 1, 2, …, N, as a population for searching the feasible set $X = \{x \in R^n | l_i \leq x \leq u_i\}$, where the index k denotes the iteration (or generation) number of the algorithm while the lower and upper parameter limits are represented by l_i and u_i respectively. The initial population, $S_k = \{x_{1,k}, x_{2,k}, \ldots, x_{N,k}\}$ (being $k = 1$), is taken from uniformly distributed samples of the search region X. We also denote the population set at the kth iteration by S_k. It does contain the members of the set S_k that have changed with k. After the initialization of S_k, EMO continues its iterative process until a stopping condition (e.g. the maximum number of iterations) is met. An iteration of EMO consists of two steps. At first step, each point in S_k moves to a different location by using the attraction–repulsion mechanism of the electromagnetism theory [21]. In a second step, points that have been moved by the electromagnetism theory, are further moved locally by a local search and then become the members of S_{k+1} in the $(k + 1)$

th iteration. Both the attraction–repulsion mechanism and the local search in EMO are responsible for driving the members, $x_{i,k}$, of S_k to the close proximity of the global minimizer.

Similar to the electromagnetism theory for charged particles, each point $x_{i,k} \in S_k$ in the search space X is assumed as a charged particle where the charge of a point relates to its objective function value. Points holding better objective function values possess higher EMO's charges than other points.

The attraction–repulsion mechanism in EMO states that points holding more charge attract other points in S_k, while points showing less charge repel other. Finally, a total force vector, F_i^k, exerted over a point, e.g. the ith point $x_{i,k}$, is calculated by adding the resultant attraction–repulsion forces and each $x_{i,k} \in S_k$ is moved in the direction of its total force, denoting its location as $y_{i,k}$.

A local search is used to explore the neighborhood of each $y_{i,k}$. Considering a determined number of steps, known as I_l and a fixed neighbourhood search δ, the procedure iterates as follows: Point $y_{i,k}$ is assigned to a temporary point $z_{i,k}$ to store the initial information. Next, for a given coordinate i $(\in 1, \ldots, n)$, a random number is selected and combined with δ as a step length, which in turns moves the point $z_{i,k}$ along the direction d. If point $z_{i,k}$ observes a better performance within a set of I_l repetitions, point $y_{i,k}$ is replaced by $z_{i,k}$, otherwise $y_{i,k}$ is held. Therefore, the members, $x_{i,k+1} \in S_{k+1}$, of the $(k + 1)$th iteration are defined through:

$$x_{i,k+1} = \begin{cases} z_{i,k} & \text{if } f\left(z_{i,k}\right) < f\left(y_{i,k}\right) \\ y_{i,k} & \text{otherwise} \end{cases} \tag{9.1}$$

Algorithm 9.1 shows the general scheme of EMO. We also provide the description of each step as follows:

Algorithm 9.1 [EMO (N, $MAXITER$, I_l, δ)]

1. Input parameters: The maximum number of iterations $MAXITER$, the values for the local search parameters such as I_l and δ, and the size N of the population, are all defined.
2. Initialize: Set the iteration counter $k = 1$, initialize the members of S_k uniformly in X, and identify the best point in S_k.
3. while $k < MAXITER$ do
4. $F_i^k \leftarrow \text{CalcF}(S_k)$
5. $y_{i,k} = \text{Move}\left(x_{i,k}, F_i^k\right)$
6. $z_{i,k} = \text{Local}\left(I_l, \delta, y_{i,k}\right)$
7. $x_{i,k+1} = \text{Select}\left(S_{k+1}, y_{i,k}, z_{i,k}\right)$
8. End while

Input parameter values (Line 1): The EMO algorithm is run for $MAXITER$ iterations. In the local search phase, $n \times I_l$ is the maximum number of locations $z_{i,k}$, within a δ distance of $y_{i,k}$, for each i.

Initialize (Line 2): The points $x_{i,k}$, $k = 1$, are selected uniformly distributed in X, considering lower l_i and upper u_i parameter limits where $i = 1, 2, \ldots, n$. The objective function values $f(x_{i,k})$ are computed, and the best point

$$x_k^B = \arg \min_{x_{i,k} \in S_k} \{f(x_{i,k})\}, \tag{9.2}$$

is also identified.

Calculate force (Line 4): In this step, a charged-like value ($q_{i,k}$) is assigned to each point ($x_{i,k}$). The charge $q_{i,k}$ of $x_{i,k}$ is dependent on $f(x_{i,k})$, and points holding better objective function have more charge than others. The charges are computed as follows:

$$q_{i,k} = \exp\left\{ -n \frac{f(x_{i,k}) - f(x_k^B)}{\sum_{j=1}^{N} f(x_{j,k}) - f(x_k^B)} \right\} \tag{9.3}$$

where x_k^B represents the best particle in the population (see Eq. 9.2). Then, the force, $F_{i,j}^k$, between two points, $x_{i,k}$ and $x_{j,k}$, is calculated by using

$$F_{i,j}^k = \begin{cases} (x_{j,k} - x_{i,k}) \frac{q_{i,k} \cdot q_{j,k}}{\| x_{j,k} - x_{i,k} \|^2} & \text{if } f(x_{i,k}) > f(x_{j,k}) \\ (x_{i,k} - x_{j,k}) \frac{q_{i,k} \cdot q_{j,k}}{\| x_{j,k} - x_{i,k} \|^2} & \text{if } f(x_{i,k}) \le f(x_{j,k}) \end{cases} \tag{9.4}$$

The total force, F_i^k, corresponding to $x_{i,k}$ is now calculated as

$$F_i^k = \sum_{j=1, j \neq 1}^{N} F_{i,j}^k \tag{9.5}$$

Moving point $x_{i,k}$ along F_i^k (Line 5): In this step, each point $x_{i,k}$, except for x_k^B, is moved along the total force vector F_i^k by considering

$$x_{i,k} = x_{i,k} + \lambda \frac{F_i^k}{\| F_i^k \|} (RNG), \quad i = 1, 2, \ldots, N;\ i \neq B, \tag{9.6}$$

where λ is a random number between 0-1, and RNG denotes the allowed range of movement towards the lower l_i or upper u_i bound for the corresponding dimension. Local search (Line 6): For each $y_{i,k}$ a maximum of I_l points are generated at each coordinate direction in δ, the neighborhood of $y_{i,k}$. The process of generating local points is continued for each $y_{i,k}$ until either a better $z_{i,k}$ is found or the I_l trial is reached. Selection for the next iteration (Line 7): In this step, members $x_{i,k+1} \in S_{k+1}$ are selected from $y_{i,k}$ and $z_{i,k}$ using Eq. (9.1) and the best point is identified using Eq. (9.2).

9.3 Circle Detection Using EMO

9.3.1 Data Preprocessing

In order to detect circle shapes, candidate images must be preprocessed first by the well-known Canny algorithm which yields a single-pixel edge-only image. Then, the (x_i, y_i) coordinates for each edge pixel p_i are stored inside the edge vector $P = \{p_1, p_2, \ldots, p_{N_p}\}$, with N_p being the total number of edge pixels.

9.3.2 Particle Representation

In order to construct each particle C (circle candidate), the indexes e_1, e_2 and e_3, which represent three edge points previously stored inside the vector P, must be grouped assuming that the circle's contour connects them. Therefore, the circle $C = \{p_{e1}, p_{e2}, p_{e3}\}$ passing over such points may be considered as a potential solution for the detection problem. Considering the configuration of the edge points shown by Fig. 9.1, the circle center (x_0, y_0) and the radius r of C can be characterized as follows:

$$(x - x_0)^2 + (y - y_0)^2 = r^2 \tag{9.7}$$

where x_0 and y_0 are computed through the following equations:

$$\begin{aligned} x_0 &= \frac{\det(\mathbf{A})}{4((x_{e_2} - x_{e_1})(y_{e_3} - y_{e_1}) - (x_{e_3} - x_{e_1})(y_{e_2} - y_{e_1}))}, \\ y_0 &= \frac{\det(\mathbf{B})}{4((x_{e_2} - x_{e_1})(y_{e_3} - y_{e_1}) - (x_{e_3} - x_{e_1})(y_{e_2} - y_{e_1}))}, \end{aligned} \tag{9.8}$$

with $\det(\mathbf{A})$ and $\det(\mathbf{B})$ representing determinants of matrices $\mathbf{A}$ and $\mathbf{B}$ respectively, considering:

$$\begin{aligned} \mathbf{A} &= \begin{bmatrix} x_{e_2}^2 + y_{e_2}^2 - \left(x_{e_1}^2 + y_{e_1}^2\right) & 2 \cdot (y_{e_1} - y_{e_1}) \\ x_{e_3}^2 + y_{e_3}^2 - \left(x_{e_1}^2 + y_{e_1}^2\right) & 2 \cdot (y_{e_3} - y_{e_1}) \end{bmatrix} \\ \mathbf{B} &= \begin{bmatrix} 2 \cdot (x_{e_2} - x_{e_1}) & x_{e_2}^2 + y_{e_2}^2 - \left(x_{e_1}^2 + y_{e_1}^2\right) \\ 2 \cdot (x_{e_3} - x_{e_1}) & x_{e_3}^2 + y_{e_3}^2 - \left(x_{e_1}^2 + y_{e_1}^2\right) \end{bmatrix}, \end{aligned} \tag{9.9}$$

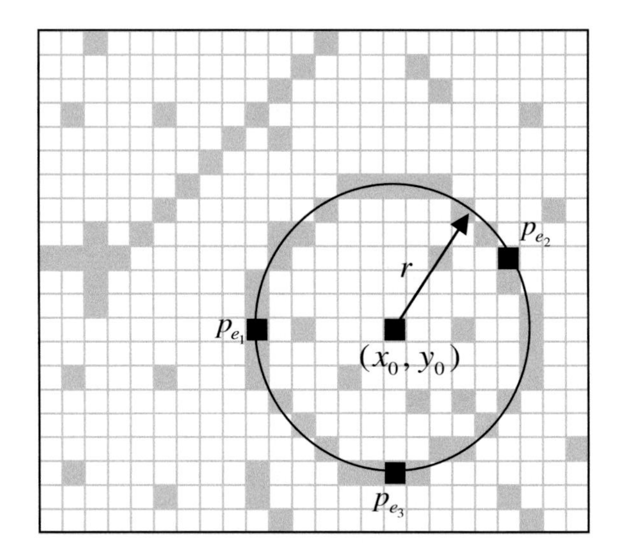

Fig. 9.1 Circle candidate (charged-particle) built from the combination of points p_{e_1}, p_{e_2} and p_{e_3}

the radius r can therefore be calculated using:

$$r = \sqrt{(x_0 - x_{e_d})^2 + (y_0 - y_{e_d})^2} \tag{9.10}$$

where $d \in \{1, 2, 3\}$, and (x_{e_d}, y_{e_d}) are the coordinates of any of the three selected points which define the particle C. Figure 9.1 illustrates main parameters defined by Eqs. (9.7)–(9.10). Therefore, the shaping parameters for the circle, $[x_0, y_0, r]$ can be represented as a transformation T of the edge vector indexes e_1, e_2 and e_3.

$$[x_0, y_0, r] = T(e_1, e_2, e_3) \tag{9.11}$$

By exploring each index as a particle, it is possible to sweep the continuous space while looking for shape parameters by means of the EMO algorithm. This approach reduces the search space by eliminating unfeasible solutions.

9.3.3 *Objective Function*

In order to model the fitness function, the circumference coordinates of the circle candidate C are calculated as a virtual shape which, in turn, must be validated, i.e. if it really exists in the edge image. The circumference coordinates are grouped within the test set $H = \{h_1, h_2, \ldots, h_{N_s}\}$, with N_s representing the number of points over which the existence of an edge point, that corresponds to C, must be verified.

The test H is generated by the midpoint circle algorithm (MCA) [22] which is a well-known algorithm to determine the required points for drawing a circle. MCA requires as inputs only the radius r and the center point (x_0, y_0) considering only the first octant over the circle equation $x^2 + y^2 = r^2$. It draws a curve starting at point (r, 0) and proceeds upwards-left by using integer additions and subtractions. The MCA aims to calculate the required points H in order to represent a circle candidate. Although the algorithm is considered as the quickest providing a sub-pixel precision, it is important to assure that points lying outside the image plane must not be considered in H.

The objective function $J(C)$ represents the matching error produced between the pixels H of the circle candidate C (particle) and the pixels that actually exist in the edge-only image, yielding:

$$J(C) = 1 - \frac{\sum_{v=1}^{N_s} E(h_v)}{N_s} \quad (9.12)$$

where $E(h_v)$ is a function that verifies the pixel existence in h_v, being $h_v \in H$ and p_{e_1} is the number of elements of H. Hence the function $E(h_v)$ is defined as:

$$E(h_v) = \begin{cases} 1 & \text{if the test pixel } h_v \text{ is an edge point} \\ 0 & \text{otherwise} \end{cases} \quad (9.13)$$

A value of $J(C)$ near to zero implies a better response from the "circularity" operator. Figure 9.2 shows the procedure to evaluate a candidate solution C with its representation as a virtual shape H. Figure 9.2a shows the original edge map, while Fig. 9.2b presents the virtual shape H representing the particle $C = \{p_{e_1}, p_{e_2}, p_{e_3}\}$. In Fig. 9.2c, the virtual shape H is compared to the original image, point by point, in order to find coincidences between virtual and edge points. The individual has been built from points p_{e_1}, p_{e_2} and p_{e_3} which are shown by Fig. 9.2a. The virtual shape H, obtained by MCA, gathers 56 points ($N_s = 56$) with only 18 of them existing in both images (shown as blue points plus red points in Fig. 9.2c) and yielding: $\sum_{h=1}^{Ns} E(h_v) = 18$, therefore $J(C) \approx 0.67$.

9.3.4 EMO Implementation

The implementation of the presented algorithm can be summarized into the following steps:

Step 1: The Canny filter is applied to find the edges and store them in the $P = \{p_1, p_2, \ldots, p_{N_p}\}$ vector. The index k is set to 1.

Step 2: m initial particles are generated $(C_{a,1}, a \in [1, m])$. Particles belonging to a seriously small or to a quite big radius are eliminated (collinear points are discarded).

Step 3: The objective function $J(C_{a,k})$ is evaluated to determine the best particle C^B(where $C^B \leftarrow \arg\min\{J(C_{a,k})\}$).

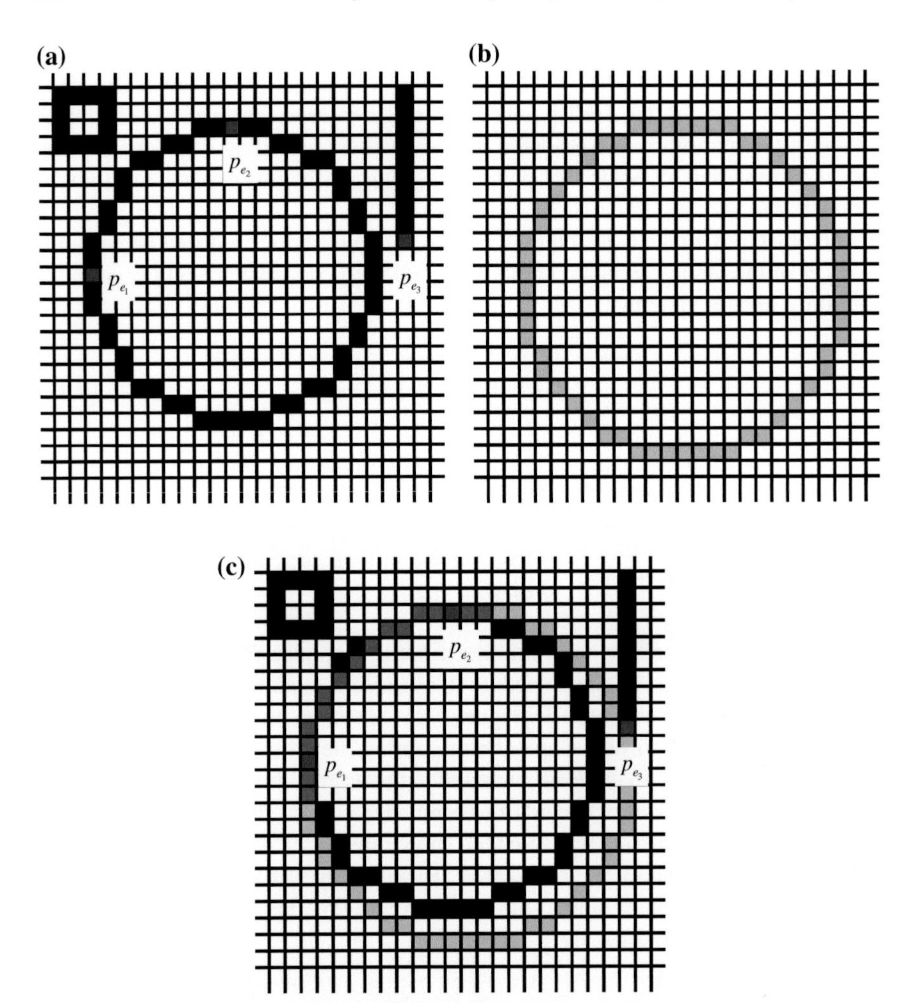

Fig. 9.2 Procedure to evaluate the objective function $J(C)$: The image shown by (**a**) presents the original edge image while (**b**) portraits the virtual shape H corresponding to C. The image in (**c**) shows coincidences between both images through *blue* or *red* pixels while the virtual shape is also depicted in *green*

Step 4: The charge between particles is calculated using expression (9.3), and its vector force is calculated through Eqs. (9.4) and (9.5). The particle with a better objective function holds a bigger charge and therefore a bigger attraction force.

Step 5: The particles are moved according to their force magnitude. The new particle's position C_a^y is calculated by expression (9.6). $C^B$is not moved because it has the biggest force and it attracts others particles to itself.

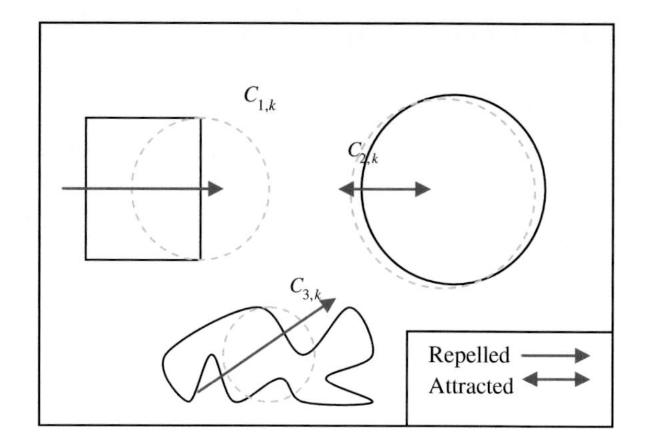

Fig. 9.3 An analogy to the Coulomb's law

Step 6: For each C_a^y a maximum of I_l points are generated at each coordinate direction in the δ neighborhood of C_a^y. The process of generating local points is continued for each C_a^y until either a better C_a^z is found or the $n \times I_l$ trial is reached.

Step 7: The new particles $C_{a,k+1}$ are selected from C_a^y and C_a^z using (1).

Step 8: The k index is increased. If $k = MAXITER$ or if $J(C_{a,k})$ value is as smaller as the pre-defined threshold value then the algorithm is stopped and the flow jumps to step 9. Otherwise, it jumps to step 3.

Step 9: The best C^B particle is selected from the last iteration.

Step 10: From the original edge map, the algorithm marks points corresponding to C^B. In case of multi-circle detection, it jumps to step 2.

Step 11: Finally the best particle C_{Nc}^B from each circle is used to draw (over the original image) the detected circles, considering Nc as the number of detected circles.

Figure 9.3 shows an analogy to the Coulomb's law. The original figures to be detected are represented by a solid black line while the shapes with discontinuous gray lines represent the candidate circles. Since the candidate circles $C_{1,\mathrm{k}}$ and $C_{3,\mathrm{k}}$ present a high value in the fitness function $J(C_{\mathrm{a,k}})$, they are repelled (blue lines), moving away the shapes. In contrast the circle candidate $C_{2,\mathrm{k}}$ that holds a small value of $J(C_{\mathrm{a,k}})$, is attracted (red line) to the circular shape contained in the image.

9.4 The White Blood Cell Detector

In order to detect WBC, the presented detector combines the EMO-based circle detector presented in Sect. 9.3 with a new objective function.

9.4.1 Image Preprocessing

To employ the presented detector, smear images must be preprocessed to obtain two new images: the segmented image and its corresponding edge map. The segmented image is produced by using a segmentation strategy whereas the edge map is generated by a border extractor algorithm. Both images are considered by the new objective function to measure the resemblance of a candidate circle with an actual WBC.

The goal of the segmentation strategy is to isolate the white blood cells (WBC's) from other structures such as red blood cells and background pixels. Information of color, brightness and gradients are commonly used within a thresholding scheme to generate the labels to classify each pixel. Although a simple histogram thresholding can be used to segment the WBC's, at this work the Diffused Expectation-Maximization (DEM) has been used to assure better results [23]. DEM is an Expectation-Maximization (EM) based algorithm which has been used to segment complex medical images [24]. In contrast to classical EM algorithms, DEM considers the spatial correlations among pixels as a part of the minimization criteria. Such adaptation allows to segment objects in spite of noisy and complex conditions. For the WBC's segmentation, the DEM has been configured considering three different classes ($K = 3$), $g(\nabla h_{ik}) = |\nabla h_{ik}|^{-9/5}$, $\lambda = 0.1$ and $m = 10$ iterations. These values have been found as the best configuration set according to [23]. As a final result of the DEM operation, three different thresholding points are obtained: the first corresponds to the WBC's, the second to the red blood cells whereas the third represents the pixels classified as background. Figure 9.4b presents the segmentation results obtained by the DEM approach employed at this work considering the Fig. 9.4a as the original image.

Once the segmented image has been produced, the edge map is computed. The purpose of the edge map is to obtain a simple image representation that preserves object structures. Optimization-based circle detectors [17–20] operate directly over the edge map in order to recognize circular shapes. Several algorithms can be used to extract the edge map; however, at this work, the morphological edge detection procedure [25] has been used to accomplish such a task. Morphological edge detection is a traditional method to extract borders from binary images in which original images (I_B) are eroded by a simple structure element (I_E). Then, the eroded image is inverted ($\bar{I}_E$) and compared with the original image ($\bar{I}_E \wedge I_B$) in order to detect pixels which are present in both images. Such pixels compose the computed edge map from I_B. Figure 9.4c shows the edge map obtained by using the morphological edge detection procedure.

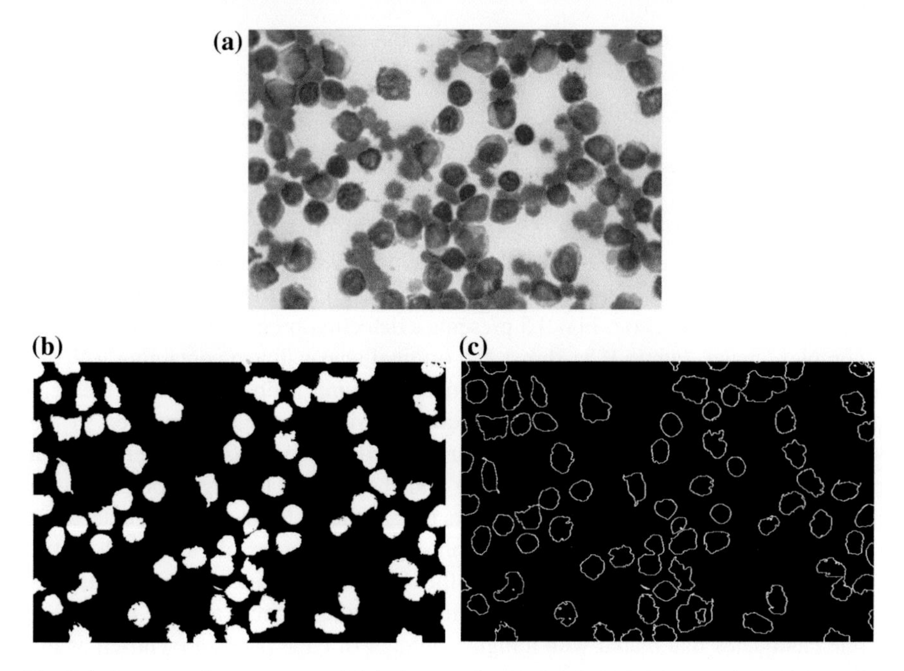

Fig. 9.4 Preprocessing process. **a** Original smear image, **b** segmented image obtained by DEM and **c** the edge map obtained by using the morphological edge detection procedure

9.4.2 The Modified EMO-Based Circle Detector

The circle detection approach uses the encoding of three edge points that represent candidate circles in the image. In the original EMO-based circle detector, the quality of each individual is calculated by using an objective function which evaluates the existence of a candidate circle considering only information from the edge map (shape structures). The better a candidate circle approximates the actual edge-circle, the objective function value decreases. Therefore, the detection performance depends on the quality of the edge map that is obtained from the original images. However, since smear images present different imaging conditions and staining intensities, they produce edge maps partially damaged by noisy pixels. Under such conditions, the use of the EMO-based circle detector cannot be directly applied to WBC detection. In order to use the EMO-based circle detector within the context of WBC detection, it is necessary to change the fitness function presented in Eq. 9.11. At this work, a new objective function has been derived to measure the resemblance of a candidate circle to an actual WBC based on the information from the edge map and the segmented image. Such new objective function takes into consideration not only the information provided by the edge map, but also the relationship among the pixels falling inside of the candidate circle which is contained in the segmented image, validating the existence of the WBC. This new function $J(\mathbf{C})$ is thus calculated as follows:

$$J_{New}(C) = 2 - \frac{\sum_{v=1}^{N_s} E(h_v)}{N_s} - \frac{Wp}{Bp}, \tag{9.13}$$

where h_v and N_s keep the same meaning than Eq. 9.11. *Wp*is the amount of white pixel falling inside the candidate circle represented by *C*. Likewise, *Bp* corresponds to the total number of black pixels falling inside *C*. To illustrate the functionality of the new objective function, Fig. 9.5 presents a detection procedure which considers a complex image. Figure 9.5a shows the original smear image containing a WBC and a stain produced by the coloring process. Figures 9.5b, c represent the segmented image and the edge map, respectively. Since the stain contained in the smear image (Fig. 9.5a) possesses similar properties than a WBC, it remains as a part of the segmented image (Fig. 9.5b) and the edge map (Fig. 9.5c). Such an inconsistency produces big detection errors in case the EMO-based circle detector is used without modification. Figure 9.5d presents detection results obtained by the original EMO-based circle detector. As the original objective function considers only the number of coincidences between the candidate circle and the edge map, circle candidates that match with a higher number of edge pixels are chosen as the best circle instances. In Fig. 9.5d, the detected circle presents a coincidence of 37 different pixels in the edge map. Such coincidence is considered as the best possible under the restrictions of the original objective function.

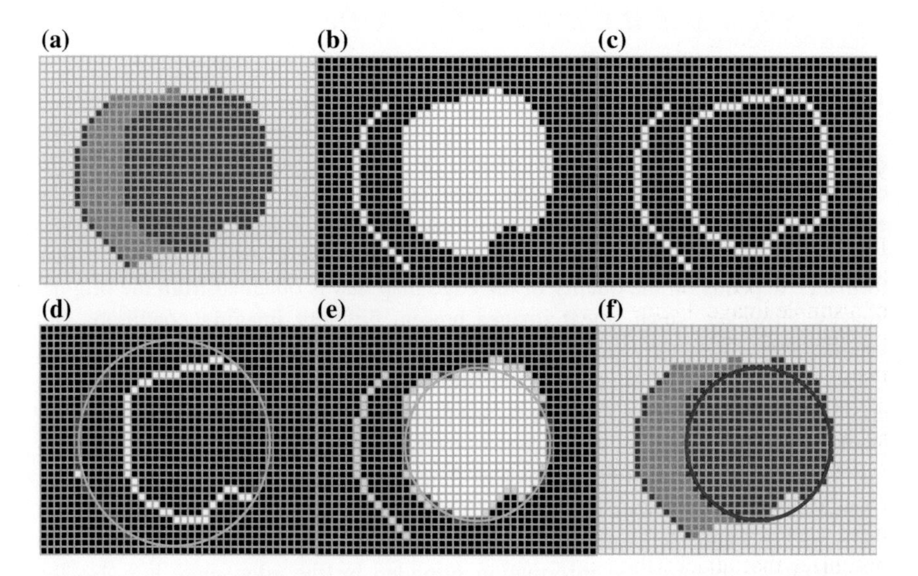

Fig. 9.5 WBC detection procedure. **a** Smear image. **b** Segmented image. **c** Edge map. **d** Detected circle by using the original objective function. *Red points* show the coincidences between the candidate circle and the edge map. **e** Detected circle by using the new objective function. *Yellow points* represent the edge pixels without coincidence. **f** Final result

Table 9.1 EMO parameters used for leukocites detection in medical images

m	n	$MAXITER$	δ	$LISTER$
50	3	5	4	4

On the other hand, when the modified objective function is used in the recognition procedure, the accuracy and the robustness of the detection are both significantly improved. By using the new objective function, information from the segmented image is employed to refine the solution that is provided by coincidences with the edge map. Figure 9.5e presents the detection result that has been produced by the modified EMO-based circle detector. In the Figure, the detected circle matches with only 32 pixels of the edge map. However, it is considered as the best instance due to the relationship of its internal pixels (the white pixels are much more than the black pixels). Finally, Fig. 9.5f shows final detection results over the original smear image. Table 9.1 presents the parameters for the EMO algorithm used in this work. They have been kept for all test images after being experimentally defined.

Under such assumptions, the complete process to detect WBC's is implemented as follows:

Step 1: Segment the WBC's using the DEM algorithm.
Step 2: Get the edge map from the segmented image by using the morphological edge detection method.
Step 3: Start the circle detector based in EMO over the edge map while saving best circles (Sect. 9.3.3).
Step 4: Define parameter values for each circle that identify the WBC's.

9.4.3 Numerical Example

In order to present the algorithm's step-by-step operation, a numerical example has been set by applying the presented method to detect a single leukocyte lying inside of a simple image. Figure 9.6a shows the image used in the example. After applying the threshold operation, the WBC is located besides few other pixels which are merely noise (see Fig. 9.6b). Then, the edge map is subsequently computed and stored pixel by pixel inside the vector P. Figure 9.6c shows the resulting image after such procedure.

The EMO-based circle detector is executed using information of the edge map and the segmented image (for the sake of easiness, it only considers a population of three particles). Like all evolutionary approaches, EMO is a population-based optimizer that attacks the starting point problem by sampling the search space at multiple, randomly chosen, initial particles. By taking three random pixels from vector P, three different particles are constructed. Figure 9.6d depicts the initial

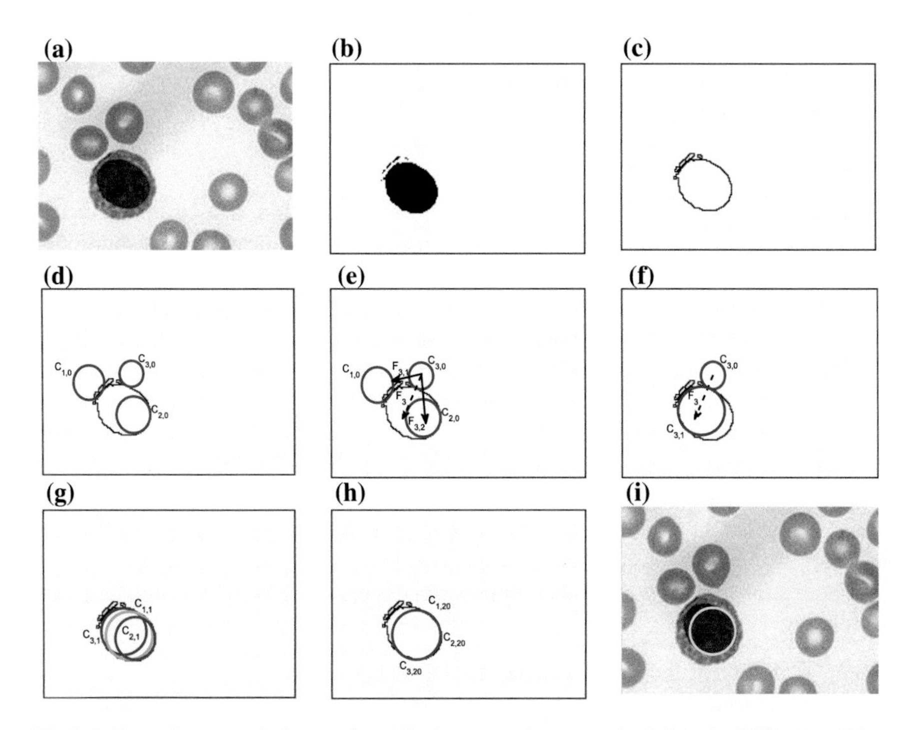

Fig. 9.6 Detection numerical example: **a** The image used as example. **b** Segmented image. **c** Edge map. **d** Initial particles. **e** Forces exerted over $C_{3,0}$. **F** New position of $C_{3,0}$. **g** Positions of all particles after the first generation. **h** Final particle configuration after 20 generations. **i** Final result overlapped the original image

particle distribution. Since the particle $C_{2,0}$ holds the best fitness value $J_{New}(C_{2,0})$ (it does possess a better coincidence with the edge map and a god pixel relationship), it is considered as the best particle C^B. Then, the charge of each particle is calculated through Eq. 9.3 and the forces exerted over each particle are computed. Figure 9.6e shows the forces exerted over the $C_{3,0}$ particle. Since the $C_{3,0}$ particle is the worst particle in terms of fitness value, it is attracted by particles $C_{1,0}$ and $C_{2,0}$. $F_{3,1}$ and $F_{3,2}$ represent the existent attracting forces of $C_{3,0}$ with respect to $C_{1,0}$ and $C_{2,0}$ whereas F_3 corresponds to the resultant force. Considering F_3 as the final force exerted over $C_{3,0}$, the position of $C_{3,0}$ is modified using Eq. 9.6. Figure 9.6f depicts the new position $C_{3,1}$ of particle $C_{3,0}$ (the second sub-index means the iteration number). If the same procedure is applied over all the particles (except for $C_{2,0}$ that is the best particle), it yields positions shown at Fig. 9.6g. Therefore, after 20 iterations, all particles converge to the same position presented in Fig. 9.6h whereas the Fig. 9.6i depicts the final result.

9.5 Experimental Results

Experimental tests have been developed in order to evaluate the performance of the WBC detector. It was tested over microscope images from blood-smears holding a 600×500 pixel resolution. They correspond to supporting images on the leukemia diagnosis. The images show several complex conditions such as deformed cells and overlapping with partial occlusions. The robustness of the algorithm has been tested under such demanding conditions. Figure 9.7a shows an example image employed in the test. It was used as input image for the WBC detector. Figure 9.7b presents the segmented WBC's obtained by the DEM algorithm. Figure 9.7c, d present the edge map and the white blood cells after detection, respectively. The results show that the presented algorithm can effectively detect and mark blood cells despite cell occlusion, deformation or overlapping. Other parameters may also be calculated through the algorithm: the total area covered by white blood cells and relationships between several cell sizes.

Other example is presented in Fig. 9.8. It represents a complex example with an image showing seriously deformed cells. Despite such imperfections, the presented approach can effectively detect the cells as it is shown in Fig. 9.8d.

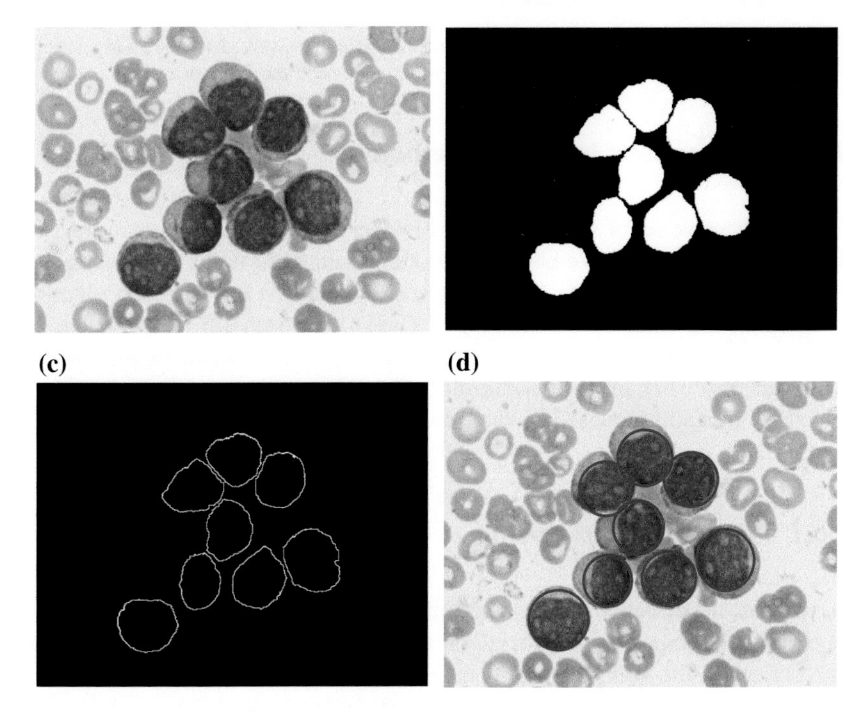

Fig. 9.7 Resulting images of the first test after applying the WBC detector: **a** Original image, **b** image segmented by the DEM algorithm, **c** edge map and **d** the white detected blood cells

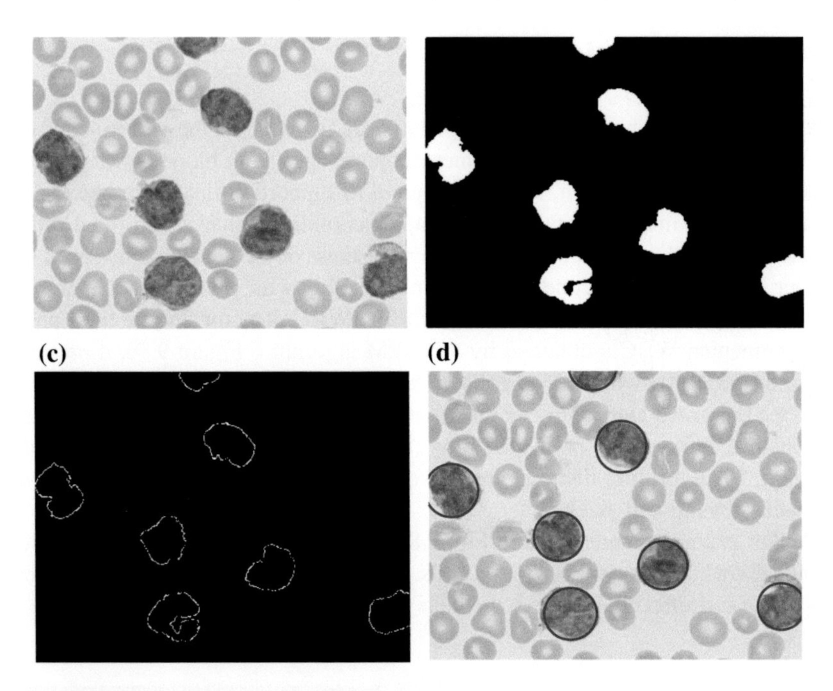

Fig. 9.8 Resulting images of the second test after applying the WBC detector: **a** Original image, **b** image segmented by the DEM algorithm, **c** edge map and **d** the white detected blood cells

9.6 Comparisons to Other Methods

A comprehensive set of smear-blood test images is used to test the performance of the presented approach. We have applied the presented EMO-based detector to test images in order to compare its performance to other WBC detection algorithms such as the Boundary Support Vectors (BSV) approach [2], the iterative Otsu (IO) method [3], the Wang algorithm [4] and the Genetic algorithm-based (GAB) detector [17]. In all cases, the algorithms are tuned according to the value set which is originally proposed by their own references.

9.6.1 Detection Comparison

To evaluate the detection performance of the presented detection method, Table 9.2 tabulates the comparative leukocyte detection performance of the BSV approach, the IO method, the Wang algorithm, the BGA detector and the presented method, in terms of detection rates and false alarms. The experimental data set includes 30 images which are collected from the Cellavision reference library (http://www.cellavision.com). Such images contain 426 leukocytes (222 bright leukocytes and

Table 9.2 Comparative leukocyte detection performance of the BSV approach, the IO method, the Wang algorithm, the BGA detector and the presented EMO method over the data set which contains 30 images and 426 leukocytes

Leukocyte type	Method	Leukocytes detected	Missing	False alarms	DR (%)	FAR (%)
Bright leukocytes (222)	BSV	104	118	67	46.85	30.18
	IO	175	47	55	78.83	24.77
	Wang	186	36	42	83.78	18.92
	BGA	177	45	22	79.73	9.91
	EMO	211	11	10	95.04	4.50
Dark leukocytes (204)	BSV	98	106	54	48.04	26.47
	IO	166	38	49	81.37	24.02
	Wang	181	23	38	88.72	18.63
	BGA	170	34	19	83.33	9.31
	EMO	200	4	6	98.04	2.94
Overall (426)	BSV	202	224	121	47.42	28.40
	IO	341	85	104	80.05	24.41
	Wang	367	59	80	86.15	18.78
	BGA	347	79	41	81.45	9.62
	EMO	411	15	16	96.48	3.75

204 dark leukocytes according to smear conditions) which have been detected and counted by a human expert. Such values act as ground truth for all the experiments. For the comparison, the detection rate (DR) is defined as the ratio between the number of leukocytes correctly detected and the number leukocytes determined by the expert. The false alarm rate (FAR) is defined as the ratio between the number of non-leukocyte objects that have been wrongly identified as leukocytes and the number leukocytes which have been actually determined by the expert.

Experimental results show that the presented EMO method, which achieves 96.48 % leukocyte detection accuracy with 3.75 % false alarm rate, is compared favorably against other WBC detection algorithms, such as the BSV approach, the IO method, the Wang algorithm and the BGA detector.

9.6.2 Robustness Comparison

Images of blood smear are often deteriorated by noise due to various sources of interference and other phenomena that affect the measurement processes in imaging and data acquisition systems. Therefore, the detection results depend on the algorithm's ability to cope with different kinds of noises. In order to demonstrate the robustness in the WBC detection, the presented EMO approach is compared to the BSV approach, the IO method, the Wang algorithm and the BGA detector under noisy environments. In the test, two different experiments have been studied. The

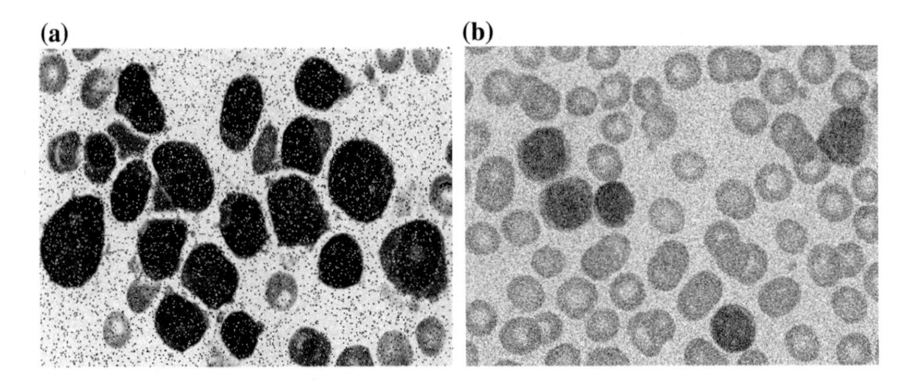

Fig. 9.9 Examples of images included in the experimental set for robustness comparison. **a** Image contaminated with 10 % of Salt and Pepper noise and **b** image polluted with $\sigma = 10$ of Gaussian noise

first inquest explores the performance of each algorithm when the detection task is accomplished over images corrupted by Salt and Pepper noise. The second experiment considers images polluted by Gaussian noise. Salt and Pepper and Gaussian noise are selected for the robustness analysis because they represent the most compatible noise types commonly found in images of blood smear [25]. The comparison considers the complete set of 30 images presented in Sect. 5.1 containing 426 leukocytes which have been detected and counted by a human expert. The added noise is produced by MatLab©, considering two noise levels of 5 and 10 % for Salt and Pepper noise whereas $\sigma = 5$ and $\sigma = 10$ are used for the case of Gaussian noise. Figure 9.9 shows only two images with different noise type as example. The outcomes in terms of the detection rate (DR) and the false alarm rate (FAR) are reported for each noise type in Tables 9.3 and 9.4. The results show that the presented EMO algorithm presents the best detection performance, achieving in the worst case a DR of 87.79 and 89.20 %, under contaminated conditions of Salt and Pepper and Gaussian noise, respectively. On the other hand, the EMO detector possesses the least degradation performance presenting a FAR value of 8.21 and 7.51 %.

9.6.3 *Stability Comparison*

In order to compare the stability performance of the presented method, its results are compared to those reported by Wang et al. in [4] which is considered as an accurate technique for the detection of WBC.

The Wang algorithm is an energy-minimizing method which is guided by internal constraint elements and influenced by external image forces, producing the segmentation of WBC's at a closed contour. As external forces, the Wang approach uses edge information which is usually represented by the gradient magnitude of the image. Therefore, the contour is attracted to pixels with large image gradients, i.e. strong

Table 9.3 Comparative WBC detection among methods that considers the complete data set of 30 images corrupted by different levels of salt and pepper noise

Noise level	Method	Leukocytes detected	Missing	False alarms	DR (%)	FAR (%)
5 % Salt and pepper noise 426 leukocytes	BSV	148	278	114	34.74	26.76
	IO	270	156	106	63.38	24.88
	Wang	250	176	118	58.68	27.70
	BGA	306	120	103	71.83	24.18
	EMO	390	36	30	91.55	7.04
10 % Salt and pepper noise 426 leukocytes	BSV	101	325	120	23.71	28.17
	IO	240	186	78	56.34	18.31
	Wang	184	242	123	43.19	28.87
	BGA	294	132	83	69.01	19.48
	EMO	374	52	35	87.79	8.21

Table 9.4 Comparative WBC detection among methods that considers the complete data set of 30 images corrupted by different levels of Gaussian noise

Noise level	Method	Leukocytes detected	Missing	False alarms	DR (%)	FAR (%)
$\sigma = 5$ Gaussian noise 426 leukocytes	BSV	172	254	77	40.37	18.07
	IO	309	117	71	72.53	16.67
	Wang	301	125	65	70.66	15.26
	BGA	345	81	61	80.98	14.32
	EMO	397	29	21	93.19	4.93
$\sigma = 10$ Gaussian noise 426 leukocytes	BSV	143	283	106	33.57	24.88
	IO	281	145	89	65.96	20.89
	Wang	264	162	102	61.97	23.94
	BGA	308	118	85	72.30	19.95
	EMO	380	46	32	89.20	7.51

edges. At each iteration, the Wang method finds a new contour configuration which minimizes the energy that corresponds to external forces and constraint elements.

In the comparison, the net structure and its operational parameters, corresponding to the Wang algorithm, follow the configuration suggested in [4] while the parameters for the EMO algorithm are taken from Table 9.1.

Figure 9.10 shows the performance of both methods considering a test image with only two white blood cells. Since the Wang method uses gradient information in order to appropriately find a new contour configuration, it needs to be executed iteratively in order to detect each structure (WBC). Figure 9.10b shows the results after the Wang approach has been applied considering only 200 iterations. Furthermore, Fig. 9.10c shows results after applying the EMO method which has been presented in this chapter.

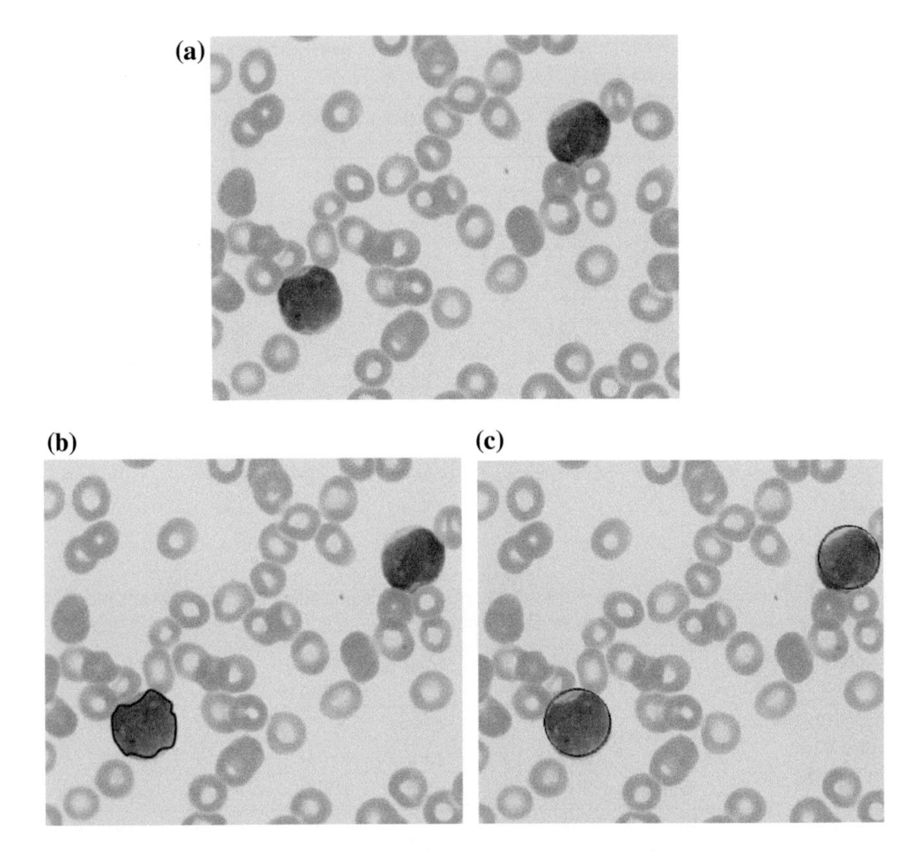

Fig. 9.10 Comparison of the EMO and the Wang's method for white blood cell detection in medical images. **a** Original image. **b** Detection using the Wang's method, **c** Detection after applying the EMO method

The Wang algorithm uses the fuzzy cellular neural network (FCNN) as optimization approach. It employs gradient information and internal states in order to find a better contour configuration. In each iteration, the FCNN tries, as contour points, different new pixel positions which must be located nearby the original contour position. Such fact might cause the contour solution to remain trapped into a local minimum. In order to avoid such a problem, the Wang method applies a considerable number of iterations so that a near optimal contour configuration can be found. However, when the number of iterations increases the possibility to cover other structures increases too. Thus, if the image has a complex background (as smear images), the method gets confused so that finding the correct contour configuration from the gradient magnitude is not easy. Therefore, a drawback of Wang's method is related to its optimal iteration number (instability). Such number must be determined experimentally as it depends on the image context and its complexity. Figure 9.11a shows the result of applying 400 cycles of the Wang's algorithm while Fig. 9.11b presents the detection of the same cell shapes after 1000

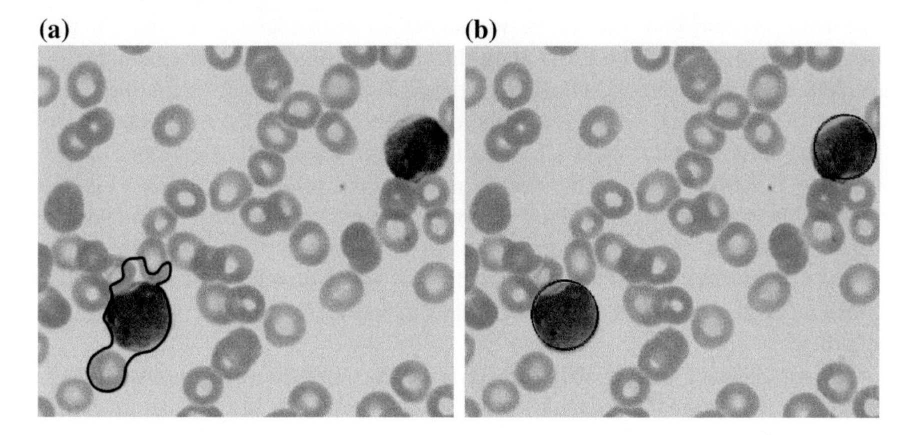

Fig. 9.11 Result comparison for the white blood cells detection showing **a** Wang's algorithm after 400 cycles and **b** EMO detector method considering 1000 cycles

Table 9.5 Error in cell's size estimation after applying the EMO algorithm and the Wang's method to detect one leukocite embedded into a blood-smear image

Algorithm	Iterations	Error (%)
Wang	60	70
	200	1
	400	121
EMO presented	60	8.22
	200	10.1
	400	10.8

The error is averaged over twenty experiments

iterations using the presented algorithm. From Fig. 9.11a, it can be seen that the contour produced by Wang's algorithm degenerates as the iteration process continues, wrongly covering other shapes lying nearby.

In order to compare the accuracy of both methods, the estimated WBC area which has been approximated by both approaches, is compared to the actual WBC size considering different degrees of evolution i.e. the cycle number for each algorithm. The comparison acknowledges only one WBC because it is the only detected shape in the Wang's method. Table 9.5 shows the averaged results over twenty repetitions for each experiment.

9.7 Conclusions

This chapter presented an algorithm for the automatic detection of white blood cells that are embedded into complicated and cluttered smear images by considering the complete process as a circle detection problem. The approach is based on a nature-inspired technique called the Electromagnetism-Like Optimization (EMO) which is a heuristic method that follows electromagnetism principles for solving complex optimization problems. The EMO algorithm is based on electromagnetic attraction and repulsion forces among charged particles whose charge represents the fitness solution for each particle (a given solution). The algorithm uses the encoding of three non-collinear edge points as candidate circles over an edge map. A new objective function has been derived to measure the resemblance of a candidate circle to an actual WBC based on the information from the edge map and segmentation results. Guided by the values of such objective function, the set of encoded candidate circles (charged particles) are evolved by using the EMO algorithm so that they can fit into the actual blood cells that are contained in the edge map.

The performance of the EMO-method has been compared to other existing WBC detectors (the Boundary Support Vectors (BSV) approach [2], the iterative Otsu (IO) method [3], the Wang algorithm [4] and the Genetic algorithm-based (GAB) detector [19]) considering several images which exhibit different complexity levels. Experimental results demonstrate the high performance of the presented method in terms of detection accuracy, robustness and stability.

References

1. Liu, J., Tsui, K.: Toward nature-inspired computing. ACM Commun. **49**, 59–64 (2006)
2. Wang, M., Chu, R.: A novel white blood cell detection method based on boundary support vectors. In: Proceedings of the 2009 IEEE International Conference on Systems, Man, and Cybernetics San Antonio, TX, USA, October 2009
3. Wu, J., Zeng, P., Zhou, Y., Oliver, C.: A novel color image segmentation method and its application to white blood cell image analysis. In: 8th International Conference on Signal Processing (2006)
4. Wang, S., Korris, F.L., Fu, D.: Applying the improved fuzzy cellular neural network IFCNN to white blood cell detection". Neurocomputing **70**, 1348–1359 (2007)
5. Karkavitsas, G., Rangoussi, M.: Object localization in medical images using genetic algorithms, World Academy of Science. Eng. and Tec. **2**, 6–9 (2005)
6. Muammar, H., Nixon, M.: Approaches to extending the hough transform. In: Proceedings of International Conference on Acoustics, Speech and Signal Processing ICASSP-89, vol. 3, pp. 1556–1559 (1989)
7. Atherton, T., Kerbyson, D.: Using phase to represent radius in the coherent circle Hough transform. In: IEEE Colloquium on the Hough Transform, pp. 1–4. IEEE (1993)
8. Fischer, M., Bolles, R.: Random sample consensus: a paradigm to model fitting with applications to image analysis and automated cartography. CACM **24**(6), 381–395 (1981)
9. Shaked, D., Yaron, O., Kiryati, N.: Deriving stopping rules for the probabilistic Hough transform by sequential analysis. Comput. Vis. Image. Und. **63**, 512–526 (1996)
10. Xu, L., Oja, E., Kultanen, P.: A new curve detection method: randomized hough transform (RHT). Pattern Recogn. Lett. **11**(5), 331–338 (1990)
11. Han, J., Koczy, L.: Fuzzy hough transform. In: Proceedings of 2nd International Conference on Fuzzy Systems, vol. 2, pp. 803–808 (1993)
12. Becker, J., Grousson, S., Coltuc, D.: From Hough transforms to integral transforms. In: International Geoscience and Remote Sensing Symposium, 2002 IGARSS-02, vol. 3, pp. 1444–1446 (2002)

13. Ayala-Ramirez, V., Garcia-Capulin, C., Perez-Garcia, A., Sanchez-Yanez, R.: Circle detection on images using genetic algorithms. Pattern Recogn. Lett. **27**, 652–657 (2006)
14. Cuevas, E., Ortega-Sánchez, N., Zaldivar, D., Pérez-Cisneros, M.: Circle detection by harmony search optimization. J. Intell. Rob. Syst. **66**(3), 359–376 (2012)
15. Cuevas, E., Zaldivar, D., Pérez-Cisneros, M., Ramírez-Ortegón, M.: Circle detection using discrete differential evolution optimization. Pattern Anal. Appl. **14**(1), 93–107 (2011)
16. Cuevas, E., Oliva, D., Zaldivar, D., Pérez-Cisneros, M., Sossa, H.: Circle detection using electro-magnetism optimization. Inf. Sci. **182**(1), 40–55 (2012)
17. Karkavitsas, G., Rangoussi, M.: Object Localization in Medical Images Using Genetic Algorithms. International Journal of Information and Communication Engineering **1**(4), 204–207 (2005)
18. Birbil, S., Fang, C.: An electromagnetism-like mechanism for global optimization. J. Global Optim. **25**, 263–282 (2003)
19. Rocha, A., Fernandes, E.: Hybridizing the electromagnetism-like algorithm with descent search for solving engineering design problems. Int. J. Comput. Math. **86**(10–11), 1932–1946 (2009)
20. Birbil, S., Fang, C., Sheu, R.: On the convergence of a population-based global optimization algorithm. J. Glob. Optim. **30**(2), 301–318 (2004)
21. Cowan, E.W.: Basic Electromagnetism. Academic Press, New York (1968)
22. Bresenham, J.: A linear algorithm for incremental digital display of circular arcs. Commun. ACM **20**, 100–106 (1987)
23. Boccignone, G., Ferraro, M., Napoletano, P.: Diffused expectation maximisation for image segmentation. Electron. Lett. **40**, 1107–1108 (2004)
24. Boccignonea, G., Napoletano, P., Caggiano, V., Ferraro, M.: A multi-resolution diffused expectation–maximization algorithm for medical image segmentation. Comput. Biol. Med. **37**, 83–96 (2007)
25. Gonzalez, R.C., Woods, R.E.: Digital Image Processing. Addison Wesley, Reading, MA (1992)

Chapter 10
Automatic Segmentation by Using an Algorithm Based on the Behavior of Locust Swarms

Abstract As an alternative to classical techniques, the problem of image segmentation has also been handled through evolutionary methods. Recently, several algorithms based on evolutionary principles have been successfully applied to image segmentation with interesting performances. However, most of them maintain two important limitations: (1) they frequently obtain sub-optimal results (misclassifications) as a consequence of an inappropriate balance between exploration and exploitation in their search strategies; (2) the number of classes is fixed and known in advance. This chapter presents an algorithm for the automatic selection of pixel classes for image segmentation. The presented method combines a recent evolutionary method with the definition of a new objective function that appropriately evaluates the segmentation quality with respect to the number of classes. The evolutionary algorithm, called Locust Search (LS), is based on the behavior of swarms of locusts. Different to the most of existent evolutionary algorithms, it explicitly avoids the concentration of individuals in the best positions, avoiding critical flaws such as the premature convergence to sub-optimal solutions and the limited exploration-exploitation balance. Experimental tests over several benchmark functions and images validate the efficiency of the presented technique with regard to accuracy and robustness.

10.1 Introduction

Image segmentation [1] consists in grouping image pixels based on some criteria such as intensity, color, texture, etc., and still represents a challenging problem within the field of image processing. Edge detection [2], region-based segmentation [3] and thresholding methods [4] are the most popular solutions for image segmentation problems.

Among such algorithms, thresholding is the simplest method. It works by considering threshold (points) values to adequately separate distinct pixels regions within the image being processed. In general, thresholding methods are divided into

E. Cuevas et al., *Applications of Evolutionary Computation in Image Processing and Pattern Recognition*, Intelligent Systems Reference Library 100,
DOI 10.1007/978-3-319-26462-2_10

two types depending on the number of threshold values namely, bi-level and multilevel. In bi-level thresholding, only a threshold value is required to separate the two objects of an image (e.g. foreground and background). On the other hand, multilevel thresholding divides pixels into more than two homogeneous classes that require several threshold values.

The thresholding methods use a parametric or nonparametric approach [5]. In parametric approaches [6, 7], it is necessary to estimate the parameters of a probability density function that is capable of modelling each class. A nonparametric technique [8–11] employs a given criteria such as the between-class variance or the entropy and error rate, in order to determine optimal threshold values.

A common method to accomplish parametric thresholding is the modeling of the image histogram through a Gaussian mixture model [12] whose parameters define a set of pixel classes (threshold points). Therefore, each pixel that belongs to a determined class is labeled according to its corresponding threshold points with several pixel groups gathering those pixels that share a homogeneous gray-scale level.

The problem of estimating the parameters of a Gaussian mixture that better model an image histogram has been commonly solved through the Expectation Maximization (EM) algorithm [13, 14] or Gradient-based methods such as Levenberg-Marquardt, LM [15]. Unfortunately, EM algorithms are very sensitive to the choice of the initial values [16], meanwhile Gradient-based methods are computationally expensive and may easily get stuck within local minima [17].

As an alternative to classical techniques, the problem of Gaussian mixture identification has also been handled through evolutionary methods. In general, they have demonstrated to deliver better results than those based on classical approaches in terms of accuracy and robustness [18]. Under these methods, an individual is represented by a candidate Gaussian mixture model. Just as the evolution process unfolds, a set of evolutionary operators are applied in order to produce better individuals. The quality of each candidate solution is evaluated through an objective function whose final result represents the similarity between the mixture model and the histogram. Some examples of these approaches involve optimization methods such as Artificial Bee Colony (ABC) [19], Artificial Immune Systems (AIS) [20], Differential Evolution (DE) [21], Electromagnetism optimization (EO) [22], Harmony Search (HS) [23] and Learning Automata (LA) [24]. Although these algorithms own interesting results, they present two important limitations. (1) They frequently obtain sub-optimal approximations as a consequence of a limited balance between exploration and exploitation in their search strategies. (2) They are based on the assumption that the number of Gaussians (classes) in the mixture is pre-known and fixed, otherwise they cannot work. The cause of the first limitation is associated to their evolutionary operators employed to modify the individual positions. In such algorithms, during their evolution, the position of each agent for the next iteration is updated yielding an attraction towards the position of the best particle seen so-far or towards other promising individuals. Therefore, as the algorithm evolves, these behaviors cause that the entire population rapidly concentrates around the best particles, favoring the premature convergence and damaging the appropriate exploration of the search space [25, 26]. The second limitation is produced as a consequence of the objective function that evaluates the similarity between the mixture model and the histogram. Under such an objective function, the number of Gaussians functions in the mixture is fixed. Since the number of threshold values (Gaussian functions) used for image segmentation varies depending on the image, the best threshold number and values are obtained by an exhaustive trial and error procedure.

On the other hand, bio-inspired algorithms represent a field of research that is concerned with the use of biology as a metaphor for producing optimization algorithms. Such approaches use our scientific understanding of biological systems as an inspiration that, at some level of abstraction, can be represented as optimization processes.

In the last decade, several optimization algorithms have been developed by a combination of deterministic rules and randomness, mimicking the behavior of natural phenomena. Such methods include the social behavior of bird flocking and fish schooling such as the Particle Swarm Optimization (PSO) algorithm [27] and the emulation of the differential evolution in species such as the Differential Evolution (DE) [28]. Although PSO and DE are the most popular algorithms for solving complex optimization problems, they present serious flaws such as premature convergence and difficulty to overcome local minima [29, 30]. The cause for such problems is associated to the operators that modify individual positions. In such algorithms, during the evolution, the position of each agent for the next iteration is updated yielding an attraction towards the position of the best particle seen so-far (in case of PSO) or towards other promising individuals (in case of DE). As the algorithm evolves, these behaviors cause that the entire population rapidly concentrates around the best particles, favoring the premature convergence and damaging the appropriate exploration of the search space [31, 32].

Recently, the collective intelligent behavior of insect or animal groups in nature has attracted the attention of researchers. The intelligent behavior observed in these groups provides survival advantages, where insect aggregations of relatively simple and "unintelligent" individuals can accomplish very complex tasks using only limited local information and simple rules of behavior [33]. Locusts (*Schistocerca gregaria*) are a representative example of such collaborative insects [34]. Locust is a kind of grasshopper that can change reversibly between a solitary and a social phase, with clear behavioral differences among both phases [35]. The two phases show many differences regarding the overall level of activity and the degree to which locusts are attracted or repulsed among them [36]. In the solitary phase, locusts avoid contact to each other (locust concentrations). As consequence, they distribute throughout the space, exploring sufficiently over the plantation [36]. On other hand, in the social phase, locusts frantically concentrate around those elements that have already found good food-sources [37]. Under such a behavior, locust attempt to efficiently find better nutrients by devastating promising areas within the plantation.

This chapter presents an algorithm for the automatic selection of pixel classes for image segmentation. The presented method combines a novel evolutionary method with the definition of a new objective function that appropriately evaluates the segmentation quality with regard to the number of classes. The new evolutionary algorithm, called Locust Search (LS), is based on the behavior presented in swarms of locusts. In the presented algorithm, individuals emulate a group of locusts which interact to each other based on the biological laws of the cooperative swarm. The algorithm considers two different behaviors: solitary and social. Depending on the behavior, each individual is conducted by a set of evolutionary operators which mimics different cooperative conducts that are typically found in the swarm. Different to most of existent evolutionary algorithms, the behavioral model in the presented approach explicitly avoids the concentration of individuals in the current best positions. Such fact allows to avoid critical flaws such as the premature convergence to sub-optimal solutions and the incorrect exploration-exploitation balance. In order to automatically define the optimal number of pixel classes (Gaussian functions in the mixture), a new objective function has been also incorporated. The new objective function is divided in two parts. The first part evaluates the quality of each candidate solution in terms of its similarity with regard to the image histogram. The second part penalizes the overlapped area among Gaussian functions (classes). Under these circumstances, Gaussian functions that do not "positively" participate in the histogram approximation could be easily eliminated in the final Gaussian mixture model.

In order to illustrate the proficiency and robustness of the presented approach, several numerical experiments have been conducted. Such experiments are divided into two sections. In the first part, the presented LS method is compared to other well-known evolutionary techniques over a set of benchmark functions. In the second part, the performance of the presented segmentation algorithm is compared to other segmentation methods which are also based on evolutionary principles. The results in both cases validate the efficiency of the presented technique with regard to accuracy and robustness.

This chapter is organized as follows: in Sect. 10.2 basic biological issues of the algorithm analogy are introduced and explained. Section 10.3 describes the novel LS algorithm and its characteristics. A numerical study on different benchmark function is presented in Sect. 10.4 while Sect. 10.5 presents the modelling of an image histogram through a Gaussian mixture. Section 10.6 exposes the LS segmentation algorithm and Sect. 10.7 the performance of the presented segmentation algorithm. Finally, Sect. 10.8 draws some conclusions.

10.2 Biological Fundamentals and Mathematical Models

Social insect societies are complex cooperative systems that self-organize within a set of constraints. Cooperative groups are good at manipulating and exploiting their environment, defending resources and breeding, yet allowing task specialization among group members [38, 39]. A social insect colony functions as an integrated unit that not only possesses the ability to operate at a distributed manner, but also undertakes a huge construction of global projects [40]. It is important to acknowledge that global order for insects can arise as a result of internal interactions among members.

Locusts are a kind of grasshoppers that exhibit two opposite behavioral phases: solitary and social (gregarious). Individuals in the solitary phase avoid contact to each other (locust concentrations). As consequence, they distribute throughout the space while sufficiently exploring the plantation [36]. In contrast, locusts in the gregarious phase gather into several concentrations. Such congregations may contain up to 10^{10} members, cover cross-sectional areas of up to 10 km^2, and a travelling capacity up to 10 km per day for a period of days or weeks as they feed causing a devastating crop loss [41]. The mechanism to switch from the solitary phase to the gregarious phase is complex and has been a subject of significant biological inquiry. Recently, a set of factors has been implicated to include geometry of the vegetation landscape and the olfactory stimulus [42].

Only few works [36, 37] that mathematically model the locust behavior have been published. Both approaches develop two different minimal models with the goal of reproducing the macroscopic structure and motion of a group of locusts. Considering that the method in [36] focuses on modelling the behavior for each locust in the group, its fundamentals have been employed to develop the algorithm that is presented in this chapter.

10.2.1 Solitary Phase

This sections describes how each locust's position is modified as a result of its behavior under the solitary phase. Considering that $\mathbf{x}_i^k$ represents the current position of the ith locust in a group of N different elements, the new position $\mathbf{x}_i^{k+1}$ is calculated by using the following model:

$$\mathbf{x}_i^{k+1} = \mathbf{x}_i^k + \Delta\mathbf{x}_i, \tag{10.1}$$

with $\Delta\mathbf{x}_i$ corresponding to the change of position that is experimented by $\mathbf{x}_i^k$ as a consequence of its social interaction with all other elements in the group.

Two locusts in the solitary phase exert forces on each other according to basic biological principles of attraction and repulsion (see, e.g., [43]). Repulsion operates quite strongly over a short length scale in order to avoid concentrations. Attraction is weaker, and operates over a longer length scale, providing the social force that is required to maintain the group's cohesion. Therefore, the strength of such social forces can be modeled by the following function:

$$s(r) = F \cdot e^{-r/L} - e^{-r} \tag{10.2}$$

here, r is a distance, F describes the strength of attraction, and L is the typical attractive length scale. We have scaled the time and the space coordinates so that the repulsive strength and length scale are both represented by the unity. We assume that $F < 1$ and $L > 1$ so that repulsion is stronger and features in a shorter-scale, while attraction is applied in a weaker and longer-scale; both facts are typical for social organisms [21]. The social force exerted by locust j over locust i is:

$$\mathbf{s}_{ij} = s(r_{ij}) \cdot \mathbf{d}_{ij}, \tag{10.3}$$

where $r_{ij} = \|\mathbf{x}_j - \mathbf{x}_i\|$ is the distance between the two locusts and $\mathbf{d}_{ij} = (\mathbf{x}_j - \mathbf{x}_i)/r_{ij}$ is the unit vector pointing from $\mathbf{x}_i$ to $\mathbf{x}_j$. The total social force on each locust can be modeled as the superposition of all of the pairwise interactions:

$$\mathbf{S}_i = \sum_{\substack{j=1 \\ j \neq 1}}^{N} \mathbf{s}_{ij}. \tag{10.4}$$

The change of position $\Delta\mathbf{x}_i$ is modeled as the total social force experimented by $\mathbf{x}_i^k$ as the superposition of all of the pairwise interactions. Therefore, $\Delta\mathbf{x}_i$ is defined as follows:

$$\Delta\mathbf{x}_i = \mathbf{S}_i. \tag{10.5}$$

In order to illustrate the behavioral model under the solitary phase, Fig. 10.1 presents an example, assuming a population of three different members ($N = 3$) which adopt a determined configuration in the current iteration k. As a consequence of the social forces, each element suffers an attraction or repulsion to other elements depending on the distance among them. Such forces are represented by $\mathbf{s}_{12}$, $\mathbf{s}_{13}$, $\mathbf{s}_{21}$, $\mathbf{s}_{23}$, $\mathbf{s}_{31}$, $\mathbf{s}_{32}$. Since $\mathbf{x}_1$ and $\mathbf{x}_2$ are too close, the social forces $\mathbf{s}_{12}$ and $\mathbf{s}_{13}$ present a repulsive nature. On the other hand, as the distances $\|\mathbf{x}_1 - \mathbf{x}_3\|$ and $\|\mathbf{x}_2 - \mathbf{x}_3\|$ are

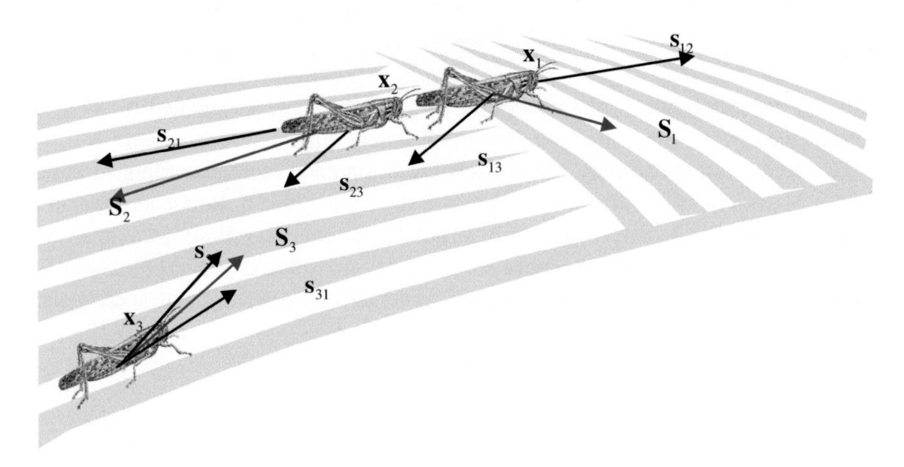

Fig. 10.1 Behavioral model under the solitary phase

quite long, the social forces $\mathbf{s}_{13}$, $\mathbf{s}_{23}$, $\mathbf{s}_{31}$ and $\mathbf{s}_{32}$ between $\mathbf{x}_1 \leftrightarrow \mathbf{x}_3$ and $\mathbf{x}_2 \leftrightarrow \mathbf{x}_3$, all belong to the attractive nature. Therefore, the change of position $\Delta\mathbf{x}_1$ is computed as the vector resultant between $\mathbf{s}_{12}$ and $\mathbf{s}_{13}$ ($\Delta\mathbf{x}_1 = \mathbf{s}_{12} + \mathbf{s}_{13}$) is $\mathbf{S}_1$. The values $\Delta\mathbf{x}_2$ and $\Delta\mathbf{x}_3$ are also calculated accordingly.

In addition to the presented model [36], some studies [44–46] suggest that the social force $\mathbf{s}_{ij}$ is also affected by the dominance of the involved individuals $\mathbf{x}_i$ and $\mathbf{x}_j$ in the pairwise process. Dominance is a property that relatively qualifies the capacity of an individual to survive, in relation to other elements in a group. The locust's dominance is determined by several characteristics such as size, chemical emissions, location with regard to food sources, etc. Under such circumstances, the social force is magnified or weakened depending on the most dominant individual that is involved in the repulsion-attraction process.

10.2.2 Social Phase

In this phase, locusts frantically concentrate around the elements that have already found good food sources. They attempt to efficiently find better nutrients by devastating promising areas within the plantation. In order to simulate the social phase, the food quality index Fq_i is assigned to each locust $\mathbf{x}_i$ of the group as such index reflects the quality of the food source where $\mathbf{x}_i$ is located.

Under this behavioral model, each of the N elements of the group are ranked according to its corresponding food quality index. Afterwards, the b elements featuring the best food quality indexes are selected ($b \ll N$). Considering a concentration radius R_c that is created around each selected element, a set of c new locusts is randomly generated inside R_c. As a result, most of the locusts will be concentrated around the best b elements. Figure 10.2 shows a simple example of the behavioral model under the social phase. In the example, the configuration includes eight locust ($N = 8$), just as it is illustrated by Fig. 10.2a that also presents the food quality index for each locust. A food quality index near to one indicates a better food source. Therefore, Fig. 10.2b presents the final configuration after the social phase, assuming $b = 2$.

10.3 The Locust Search (LS) Algorithm

In this paper, some behavioral principles drawn from a swarm of locusts have been used as guidelines for developing a new swarm optimization algorithm. The LS assumes that entire search space is a plantation, where all the locusts interact to each other. In the proposed approach, each solution within the search space represents a locust position inside the plantation. Every locust receives a food quality index according to the fitness value of the solution that is symbolized by the locust's position. As it has been previously discussed, the algorithm implements two

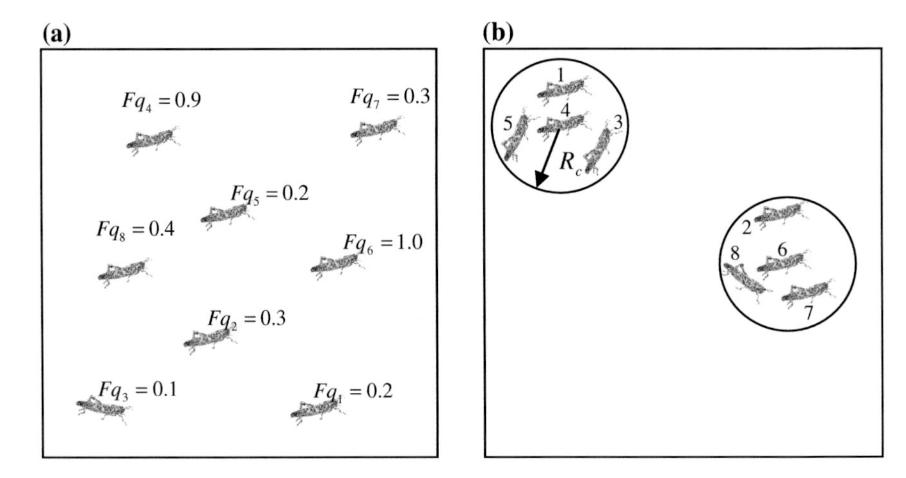

Fig. 10.2 Behavioral model under the social phase. **a** Initial configuration and food quality indexes, **b** final configuration after the operation of the social phase

different behavioral schemes: solitary and social. Depending on each behavioral phase, each individual is conducted by a set of evolutionary operators which mimics the different cooperative operations that are typically found in the swarm. The proposed method formulates the optimization problem in the following form:

$$\begin{array}{lll} \text{maximize/minimize} & f(\mathbf{l}), & \mathbf{l} = (l_1, \ldots, l_n) \in \mathbf{R}^n \\ \text{subject to} & & \mathbf{l} \in \mathbf{S} \end{array} \tag{10.6}$$

where f: $\mathbf{R}^n \rightarrow \mathbf{R}$ is a nonlinear function whereas $\mathbf{S} = \{\mathbf{l} \in \mathbf{R}^n | lb_d \leq l_d \leq ub_d,\ d = 1, \ldots, n\}$ is a bounded feasible search space, which is constrained by the lower (lb_d) and upper (ub_d) limits.

In order to solve the problem formulated in Eq. 10.6, a population $\mathbf{L}^k(\{\mathbf{l}_1^k, \mathbf{l}_2^k, \ldots, \mathbf{l}_N^k\})$ of N locusts (individuals) is evolved inside the LS operation from the initial point (k = 0) to a total *gen* number of iterations (k = *gen*). Each locust $\mathbf{l}_i^k$ ($i \in [1, \ldots, N]$) represents an n-dimensional vector $\{l_{i,1}^k, l_{i,2}^k, \ldots, l_{i,n}^k\}$ where each dimension corresponds to a decision variable of the optimization problem to be solved. The set of decision variables constitutes the feasible search space $\mathbf{S} = \{\mathbf{l}_i^k \in \mathbb{R}^n | lb_d \leq l_{i,d}^k \leq ub_d\}$, where lb_d and ub_d corresponds to the lower and upper bounds for the dimension d, respectively. The food quality index that is associated to each locust $\mathbf{l}_i^k$ (candidate solution) is evaluated through an objective function $f(\mathbf{l}_i^k)$ whose final result represents the fitness value of $\mathbf{l}_i^k$. In the LS algorithm, each iteration of the evolution process consists of two operators: (A) solitary and (B) social. Beginning by the solitary stage, the set of locusts is operated in order to sufficiently explore the search space. On the other hand, the social operation refines existent solutions within a determined neighborhood (exploitation) subspace.

10.3.1 Solitary Operation (A)

One of the most interesting features of the proposed method is the use of the solitary operator to modify the current locust positions. Under this approach, locusts are displaced as a consequence of the social forces produced by the positional relations among the elements of the swarm. Therefore, near individuals tend to repel each other, avoiding the concentration of elements in regions. On the other hand, distant individuals tend to attract to each other, maintaining the cohesion of the swarm. A clear difference to the original model in [20] considers that social forces are also magnified or weakened depending on the most dominant (best fitness values) individuals that are involved in the repulsion-attraction process.

In the solitary phase, a new position $\mathbf{p}_i$ ($i \in [1,\ldots,N]$) is produced by perturbing the current locust position $\mathbf{l}_i^k$ with a change of position $\Delta\mathbf{l}_i$ ($\mathbf{p}_i = l_i^k + \Delta\mathbf{l}_i$). The change of position $\Delta\mathbf{l}_i$ is the result of the social interactions experimented by $\mathbf{l}_i^k$ as a consequence of its repulsion-attraction behavioral model. Such social interactions are pairwise computed among $\mathbf{l}_i^k$ and the other $N - 1$ individuals in the swarm. In the original model, social forces are calculated by using Eq. 10.3. However, in the proposed method, it is modified to include the best fitness value (the most dominant) of the individuals involved in the repulsion-attraction process. Therefore, the social force, that is exerted between $\mathbf{l}_j^k$ and $\mathbf{l}_i^k$, is calculated by using the following new model:

$$\mathbf{s}_{ij}^{m} = \rho(\mathbf{l}_i^k, \mathbf{l}_j^k) \cdot s(r_{ij}) \cdot \mathbf{d}_{ij} + rand(1,-1), \tag{10.7}$$

where $s(r_{ij})$ is the social force strength defined in Eq. 10.2 and $\mathbf{d}_{ij} = (\mathbf{l}_j^k - \mathbf{l}_i^k)/r_{ij}$ is the unit vector pointing from $\mathbf{l}_i^k$ to $\mathbf{l}_j^k$. Besides, $rand(1,-1)$ is a randomly generated number between 1 and -1. Likewise, $\rho(\mathbf{l}_i^k, \mathbf{l}_j^k)$ is the dominance function that calculates the dominance value of the most dominant individual from $\mathbf{l}_j^k$ and $\mathbf{l}_i^k$. In order to operate $\rho(\mathbf{l}_i^k, \mathbf{l}_j^k)$, all the individuals from $\mathbf{L}^k$ ($\{\mathbf{l}_1^k, \mathbf{l}_2^k, \ldots, \mathbf{l}_N^k\}$) are ranked according to their fitness values. The ranks are assigned so that the best individual receives the rank 0 (zero) whereas the worst individual obtains the rank $N - 1$. Therefore, the function $\rho(\mathbf{l}_i^k, \mathbf{l}_j^k)$ is defined as follows:

$$\rho(\mathbf{l}_i^k, \mathbf{l}_j^k) = \begin{cases} e^{-\left(5\cdot\mathrm{rank}(\mathbf{l}_i^k)/N\right)} & \text{if } \mathrm{rank}(\mathbf{l}_i^k) < \mathrm{rank}(\mathbf{l}_j^k) \\ e^{-\left(5\cdot\mathrm{rank}(\mathbf{l}_j^k)/N\right)} & \text{if } \mathrm{rank}(\mathbf{l}_i^k) > \mathrm{rank}(\mathbf{l}_j^k) \end{cases}, \tag{10.8}$$

where the function rank(α) delivers the rank of the α-individual. According to Eq. 10.8, $\rho(\mathbf{l}_i^k, \mathbf{l}_j^k)$ yields a value within the interval (1,0). Its maximum value of one in $\rho(\mathbf{l}_i^k, \mathbf{l}_j^k)$ is reached when either individual $\mathbf{l}_j^k$ or $\mathbf{l}_i^k$ is the best element of the population $\mathbf{L}^k$ regarding their fitness values. On the other hand, a value close to zero is obtained when both individuals $\mathbf{l}_j^k$ and $\mathbf{l}_i^k$ possess quite bad fitness values.

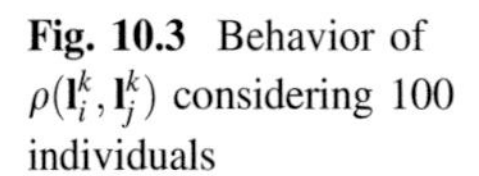

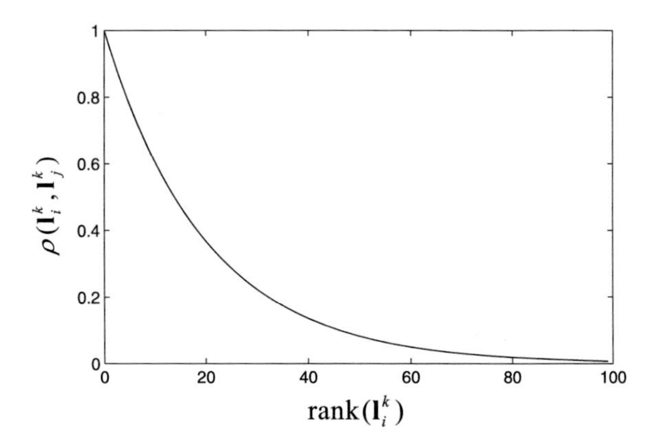

Fig. 10.3 Behavior of $\rho(\mathbf{l}_i^k, \mathbf{l}_j^k)$ considering 100 individuals

Figure 10.3 shows the behavior of $\rho(\mathbf{l}_i^k, \mathbf{l}_j^k)$ considering 100 individuals. In the Figure, it is assumed that $\mathbf{l}_i^k$ represents one of the 99 individuals with ranks between 0 and 98 whereas $\mathbf{l}_j^k$ is fixed to the element with the worst fitness value (rank 99).

Under the incorporation of $\rho(\mathbf{l}_i^k, \mathbf{l}_j^k)$ in Eq. 10.7, social forces are magnified or weakened depending on the best fitness value (the most dominant) of the individuals involved in the repulsion-attraction process.

Finally, the total social force on each individual $\mathbf{l}_i^k$ is modeled as the superposition of all of the pairwise interactions exerted over it:

$$\mathbf{S}_i^m = \sum_{\substack{j=1 \\ j \neq i}}^{N} \mathbf{s}_{ij}^m, \tag{10.9}$$

Therefore, the change of position $\Delta\mathbf{l}_i$ is considered as the total social force experimented by $\mathbf{l}_i^k$ as the superposition of all of the pairwise interactions. Thus, $\Delta\mathbf{l}_i$ is defined as follows:

$$\Delta\mathbf{l}_i = \mathbf{S}_i^m, \tag{10.10}$$

After calculating the new positions $\mathbf{P}$ ($\{\mathbf{p}_1, \mathbf{p}_2, \ldots, \mathbf{p}_N\}$) of the population $\mathbf{L}^k$ ($\{\mathbf{l}_1^k, \mathbf{l}_2^k, \ldots, \mathbf{l}_N^k\}$), the final positions $\mathbf{F}$ ($\{\mathbf{f}_1, \mathbf{f}_2, \ldots, \mathbf{f}_N\}$) must be calculated. The idea is to admit only the changes that guarantee an improvement in the search strategy. If the fitness value of $\mathbf{p}_i$ ($f(\mathbf{p}_i)$) is better than $\mathbf{l}_i^k$ ($f(\mathbf{l}_i^k)$), then $\mathbf{p}_i$ is accepted as the final solution. Otherwise, $\mathbf{l}_i^k$ is retained. This procedure can be resumed by the following statement (considering a minimization problem):

$$\mathbf{f}_i = \begin{cases} \mathbf{p}_i & \text{if } f(\mathbf{p}_i) < f(\mathbf{l}_i^k) \\ \mathbf{l}_i^k & \text{otherwise} \end{cases} \tag{10.11}$$

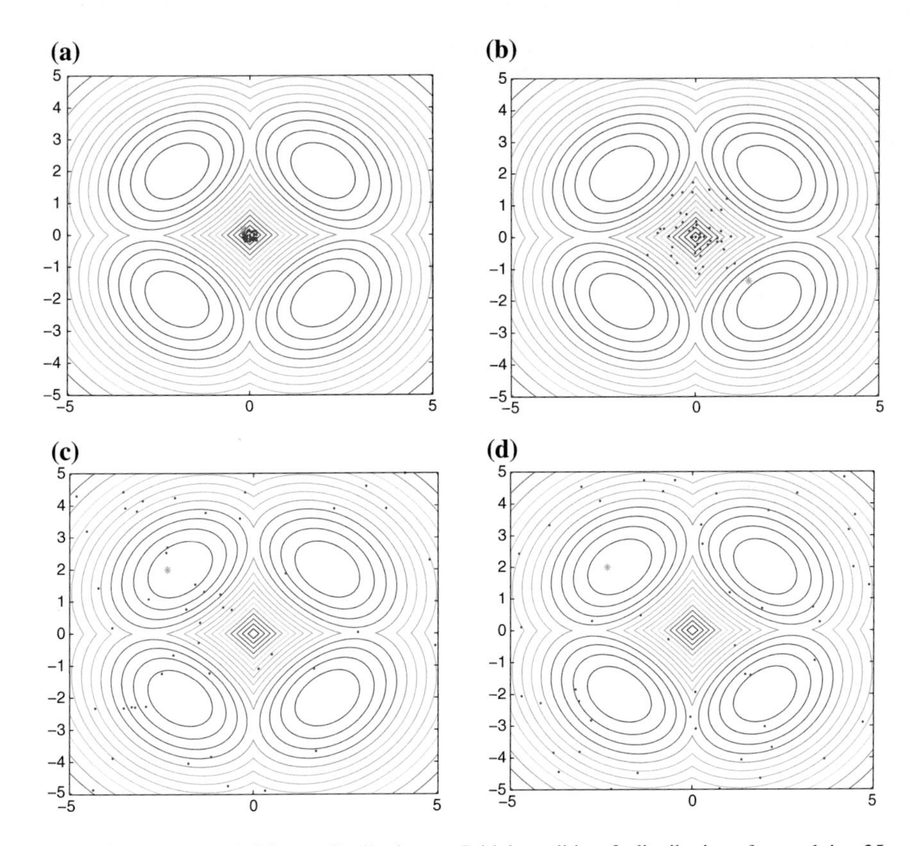

Fig. 10.4 Examples of different distributions. **a** Initial condition, **b** distribution after applying 25, **c** 50 and **d** 100 operations. The *green asterisk* represents the minimum value so-far

In order to illustrate the performance of the solitary operator, Fig. 10.4 presents a simple example with the solitary operator being iteratively applied. It is assumed a population of 50 different members (N = 50) which adopt a concentrated configuration as initial condition (Fig. 10.4a). As a consequence of the social forces, the set of elements tends to distribute throughout the search space. Examples of different distributions are shown in Fig. 10.4b–d after applying 25, 50 and 100 different solitary operations, respectively.

10.3.2 Social Operation (B)

The social procedure represents the exploitation phase of the LS algorithm. Exploitation is the process of refining existent individuals within a small neighborhood in order to improve their solution quality.

The social procedure is a selective operation which is applied only to a subset **E** of the final positions **F** (where $\mathbf{E} \subseteq \mathbf{F}$). The operation starts by sorting **F** with respect to fitness values, storing the elements in a temporary population $\mathbf{B} = \{\mathbf{b}_1, \mathbf{b}_2, \ldots, \mathbf{b}_N\}$. The elements in **B** are sorted so that the best individual receives the position $\mathbf{b}_1$ whereas the worst individual obtains the location $\mathbf{b}_N$. Therefore, the subset **E** is integrated by only the first g locations of **B** (promising solutions). Under this operation, a subspace C_j is created around each selected particle $\mathbf{f}_j \in \mathbf{E}$.

The size of C_j depends on the distance e_d which is defined as follows:

$$e_d = \frac{\sum_{q=1}^{n} \left(ub_q - lb_q \right)}{n} \cdot \beta \tag{10.12}$$

where ub_q and lb_q are the upper and lower bounds in the qth dimension, n is the number of dimensions of the optimization problem, whereas $\beta \in [0, 1]$ is a tuning factor. Therefore, the limits of C_j can be modeled as follows:

$$\begin{aligned} uss_j^q &= b_{j,q} + e_d \\ lss_j^q &= b_{j,q} - e_d \end{aligned} \tag{10.13}$$

where uss_j^q and lss_j^q are the upper and lower bounds of the qth dimension for the subspace C_j, respectively.

Considering the subspace C_j around each element $\mathbf{f}_j \in \mathbf{E}$, a set of h new particles ($\mathbf{M}_j^h = \{\mathbf{m}_j^1, \mathbf{m}_j^2, \ldots, \mathbf{m}_j^h\}$) are randomly generated inside bounds fixed by Eq. 10.13. Once the h samples are generated, the individual $\mathbf{l}_j^{k+1}$ of the next population $\mathbf{L}^{k+1}$ must be created. In order to calculate $\mathbf{l}_j^{k+1}$, the best particle $\mathbf{m}_j^{best}$, in terms of its fitness value from the h samples (where $\mathbf{m}_j^{best} \in [\mathbf{m}_j^1, \mathbf{m}_j^2, \ldots, \mathbf{m}_j^h]$), is compared to $\mathbf{f}_j$. If $\mathbf{m}_j^{best}$ is better than $\mathbf{f}_j$ according to their fitness values, $\mathbf{l}_j^{k+1}$ is updated with $\mathbf{m}_j^{best}$, otherwise $\mathbf{f}_j$ is selected. The elements of **F** that have not been processed by the procedure ($\mathbf{f}_w \notin \mathbf{E}$) transfer their corresponding values to $\mathbf{L}^{k+1}$ with no change. The social operation is used to exploit only prominent solutions. According to the proposed method, inside each subspace C_j, h random samples are selected. Since the number of selected samples at each subspace is very small (typically $h < 4$), the use of this operator substantially reduces the number of fitness function evaluations.

In order to demonstrate the social operation, a numerical example has been set by applying the proposed process to a simple function. Such function considers the interval of $-3 \leq d_1, d_2 \leq 3$ whereas the function possesses one global maxima of value 8.1 at (0, 1.6). Notice that d_1 and d_2 correspond to the axis coordinates (commonly x and y). For this example, it is assumed a final position population **F** of six 2-dimensional members ($N = 6$). Figure 10.5 shows the initial configuration of the proposed example, with the black points representing half of the particles with the best fitness values (the first three element of **B**, $g = 3$) whereas the grey points

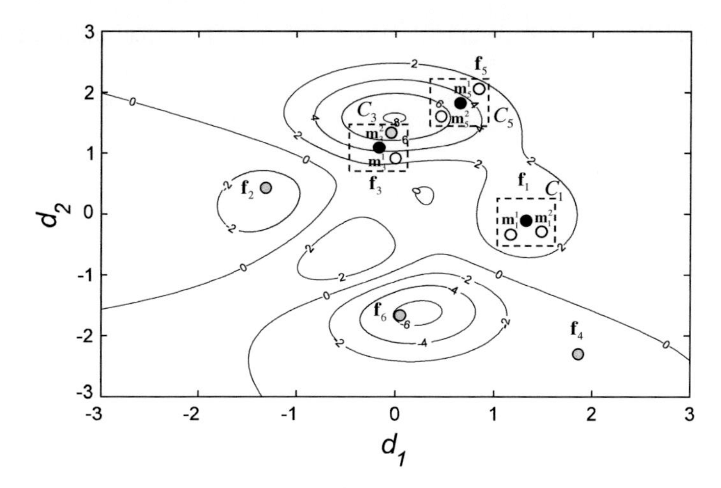

Fig. 10.5 Operation of the social procedure

$(\mathbf{f}_2, \mathbf{f}_4, \mathbf{f}_6 \notin \mathbf{E})$ correspond to the remaining individuals. From Fig. 10.5, it can be seen that the social procedure is applied to all black particles ($\mathbf{f}_5 = \mathbf{b}_1$, $\mathbf{f}_3 = \mathbf{b}_2$ and $\mathbf{f}_1 = \mathbf{b}_3$, $\mathbf{f}_5, \mathbf{f}_3, \mathbf{f}_1 \in \mathbf{E}$) yielding two new random particles (h = 2), which are characterized by white points $\mathbf{m}_1^1$, $\mathbf{m}_1^2$, $\mathbf{m}_3^1$, $\mathbf{m}_3^2$, $\mathbf{m}_5^1$ and $\mathbf{m}_5^2$, for each black point inside of their corresponding subspaces C_1, C_3 and C_5. Considering the particle $\mathbf{f}_3$ in Fig. 10.5, the particle $\mathbf{m}_3^2$ corresponds to the best particle ($\mathbf{m}_3^{best}$) from the two randomly generated particles (according to their fitness values) within C_3. Therefore, the particle $\mathbf{m}_3^{best}$ will substitute $\mathbf{f}_3$ in the individual $\mathbf{l}_3^{k+1}$ for the next generation, since it holds a better fitness value than $\mathbf{f}_3 (f(\mathbf{f}_3) < f(\mathbf{m}_3^{best}))$.

The LS optimization procedure is defined over a bounded search space **S**. Search points that do not belong to such area are considered to be infeasible. However, during the evolution process, some candidate solutions could fall outside the search space. In the proposed approach, such infeasible solutions are arbitrarily placed with a random position inside the search space **S**.

10.3.3 Complete LS Algorithm

LS is a simple algorithm with only seven adjustable parameters: the strength of attraction F, the attractive length L, number of promising solutions g, the population size N, the tuning factor β, the number of random samples h and the number of generations *gen*. The operation of LS is divided into three parts: Initialization of the solitary and social operations. In the initialization ($k = 0$), the first population $\mathbf{L}^0$ $(\{\mathbf{l}_1^0, \mathbf{l}_2^0, \ldots, \mathbf{l}_N^0\})$ is produced. The values $\{l_{i,1}^0, l_{i,2}^0, \ldots, l_{i,n}^0\}$ of each individual $\mathbf{l}_i^k$ and each dimension d are randomly and uniformly distributed between the pre-specified lower initial parameter bound lb_d and the upper initial parameter bound ub_d.

$$l_{i,j}^{0} = lb_d + \text{rand} \cdot (ub_d - lb_d); \quad i = 1,2,\ldots,N; \; d = 1,2,\ldots,n. \tag{10.14}$$

In the evolution process, the solitary (A) and social (B) operations are iteratively applied until the number of iterations $k = gen$ has been reached. The complete LS procedure is illustrated in the Algorithm 10.1.

Algorithm 10.1 Locust Search (LS) algorithm

1: **Input:** F, L, g, N and gen
2: Initialize $\mathbf{L}^0$ (k=0)
3: **until** (k=gen)
5: $\mathbf{F}\leftarrow$SolitaryOperation($\mathbf{L}^k$)
6: $\mathbf{L}^{k+1}\leftarrow$SocialOperation($\mathbf{L}^k$, $\mathbf{F}$)
8: k=k+1
7: **end until**

10.3.4 Discussion About the LS Algorithm

Evolutionary algorithms (EA) have been widely employed for solving complex optimization problems. These methods are found to be more powerful than conventional methods that are based on formal logics or mathematical programming [47]. In the EA algorithm, search agents have to decide whether to explore unknown search positions or to exploit already tested positions in order to improve their solution quality. Pure exploration degrades the precision of the evolutionary process but increases its capacity to find new potentially solutions. On the other hand, pure exploitation allows refining existent solutions but adversely drives the process to local optimal solutions. Therefore, the ability of an EA to find a global optimal solution depends on its capacity to find a good balance between the exploitation of found-so-far elements and the exploration of the search space [48].

Most of swarm algorithms and other evolutionary algorithms tend to exclusively concentrate the individuals in the current best positions. Under such circumstances, such algorithms seriously limit their exploration-exploitation capacities.

Different to most of existent evolutionary algorithms, in the presented approach, the modeled behavior explicitly avoids the concentration of individuals in the current best positions. Such fact allows not only to emulate the cooperative behavior of the locust colony in a good realistic way, but also to incorporate a computational mechanism to avoid critical flaws that are commonly present in the popular evolutionary algorithms, such as the premature convergence and the incorrect exploration-exploitation balance.

It is important to emphasize that the presented approach conducts two operators (solitary and social) within a single iteration. Such operators are similar to those that are used by other evolutionary methods such as ABC (employed bees, onlooker bees and scout bees), AIS (clonal proliferation operator, affinity maturation operator and clonal selection operator) and DE (mutation, crossover and selection), which are all executed in a single iteration.

10.4 Numerical Experiments over Benchmark Functions

A comprehensive set of 13 functions, all collected from Refs. [49–51], has been used to test the performance of the LS approach as an optimization method. Tables B.1 and B.2 in the Appendix B present the benchmark functions used in our experimental study. Such functions are classified into two different categories: unimodal test functions (Table B.1) and multimodal test functions (Table B.2). In these tables, n is the function dimension, f_{opt} is the minimum value of the function, with $\mathbf{S}$ being a subset of R^n. The optimum location ($\mathbf{x}_{opt}$) for functions in Tables B.1 and B.2, are in $[0]^n$, except for f_5, f_{12}, f_{13} with $\mathbf{x}_{opt}$ in $[1]^n$ and f_8 in $[420.96]^n$. A detailed description of optimum locations is given in Tables B.1 and B.2 of Appendix B.

We have applied the LS algorithm to 13 functions whose results have been compared to those produced by the Particle Swarm Optimization (PSO) method [27] and the Differential Evolution (DE) algorithm [28], both considered as the most popular algorithms for many optimization applications. In all comparisons, the population has been set to 40 ($N = 40$) individuals. The maximum iteration number for all functions has been set to 1000. Such stop criterion has been selected to maintain compatibility to similar works reported in the literature [49, 50].

The parameter setting for each of the algorithms in the comparison is described as follows:

1. PSO: In the algorithm, $c_1 = c_2 = 2$ while the inertia factor (ω) is decreased linearly from 0.9 to 0.2.
2. DE: The DE/Rand/1 scheme is employed. The parameter settings follow the instructions in [28, 52]. The crossover probability is $CR = 0.9$ and the weighting factor is $F = 0.8$.
3. In LS, F and L are set to 0.6 and L, respectively. Besides, g is fixed to 20 ($N/2$), $h = 2$, $\beta = 0.6$ whereas gen and N are configured to 1000 and 40, respectively. Once such parameters have been experimentally determined, they are kept for all experiments in this section.

In the comparison, three indexes are considered: the average best-so-far solution (ABS), the standard deviation (SD) and the number of function evaluations (NFE). The first two indexes assess the accuracy of the solution whereas the last one measures the computational cost. The average best-so-far solution (ABS) expresses the average value of the best function evaluations that have been obtained from 30 independent executions. The standard deviation (SD) indicates the dispersion of the ABS values. Evolutionary methods are, in general, complex pieces of software with several operators and stochastic branches. Therefore, it is difficult to conduct a complexity analysis from a deterministic perspective. Under such circumstances, it is more appropriate to use the number of function evaluations (NFE), just as it is used in the literature [53, 54], to evaluate and assess the computational effort (time) and the complexity among optimizers. It represents how many times an algorithm uses the objective function to evaluate the objective (fitness) function until the best solution of a determined execution has been found. Since the experiments require 30 different executions, the NFE index corresponds to the averaged value obtained from these executions. A small NFE value indicates that less time is needed to reach the global optimum.

10.4.1 Uni-modal Test Functions

Functions f_1–f_7 are unimodal functions. The results for unimodal functions over 30 runs are reported in Table 10.1 considering the average best-so-far solution (ABS), the standard deviation (SD) and the number of function evaluations (NFE). According to this table, LS provides better results than PSO and DE for all functions in terms of ABS and SD. In particular, the test yields the largest performance difference in functions $f_4 - f_7$. Such functions maintain a narrow curving valley that is hard to optimize, in case the search space cannot be explored properly and the direction changes cannot be kept up with [55]. For this reason, the performance differences are directly related to a better trade-off between exploration and exploitation that is produced by LS operators. In the practice, a main goal of an optimization algorithm is to find a solution as good as possible within a small time window. The computational cost for the optimizer is represented by its NFE values. According to Table 10.1, the NFE values that are obtained by the presented method are smaller than its counterparts. Lower NFE values are more desirable since they correspond to less computational overload and, therefore, faster results. In the results perspective, it is clear that PSO and DE need more than 1000 generations in order to produce better solutions. However, this number of generations is considered in the experiments aiming for producing a visible contrast among the approaches. If the number of generations has been set to an exaggerated value, then all methods would converge to the best solution with no significant troubles.

A non-parametric statistical significance proof known as the Wilcoxon's rank sum test for independent samples [56, 57] has been conducted with an 5 % significance level, over the "average best-so-far" data of Table 10.1. Table 10.2 reports the p-values produced by Wilcoxon's test for the pair-wise comparison of the "average best so-far" of two groups. Such groups are formed by LS versus PSO and LS versus DE. As a null hypothesis, it is assumed that there is no significant difference between mean values of the two algorithms. The alternative hypothesis considers a significant difference between the "average best-so-far" values of both

Table 10.1 Minimization results from the benchmark functions test in Table B.1 with $n = 30$

		PSO	DE	LS
f_1	ABS	1.66×10^{-1}	6.27×10^{-3}	4.55×10^{-4}
	SD	3.79×10^{-1}	1.68×10^{-1}	6.98×10^{-4}
	NFE	28,610	20,534	16,780
f_2	ABS	4.83×10^{-1}	2.02×10^{-1}	5.41×10^{-3}
	SD	1.59×10^{-1}	0.66	1.45×10^{-2}
	NFE	28,745	21,112	16,324
f_3	ABS	2.75	5.72×10^{-1}	1.61×10^{-3}
	SD	1.01	0.15	1.32×10^{-3}
	NFE	38,320	36,894	20,462
f_4	ABS	1.84	0.11	1.05×10^{-2}
	SD	0.87	0.05	6.63×10^{-3}
	NFE	37,028	36,450	21,158
f_5	ABS	3.07	2.39	4.11×10^{-2}
	SD	0.42	0.36	2.74×10^{-3}
	NFE	39,432	37,264	21,678
f_6	ABS	6.36	6.51	5.88×10^{-2}
	SD	0.74	0.87	1.67×10^{-2}
	NFE	38,490	36,564	22,238
f_7	ABS	6.14	0.12	2.71×10^{-2}
	SD	0.73	0.02	1.18×10^{-2}
	NFE	37,274	35,486	21,842

Maximum number of iterations = 1000

Table 10.2 p-values produced by Wilcoxon's test that compares LS versus PSO and DE over the "average best-so-far" values from Table 10.1

LS	PSO	DE
f_1	1.83×10^{-4}	1.73×10^{-2}
f_2	3.85×10^{-3}	1.83×10^{-4}
f_3	1.73×10^{-4}	6.23×10^{-3}
f_4	2.57×10^{-4}	5.21×10^{-3}
f_5	4.73×10^{-4}	1.83×10^{-3}
f_6	6.39×10^{-5}	2.15×10^{-3}
f_7	1.83×10^{-4}	2.21×10^{-3}

approaches. All p-values reported in the table are less than 0.05 (5 % significance level) which is a strong evidence against the null hypothesis, indicating that the LS results are statistically significant and that it has not occurred by coincidence (i.e. due to the normal noise contained in the process).

10.4.2 Multimodal Test Functions

Multimodal functions possess many local minima which make the optimization a difficult task to be accomplished. In multimodal functions, the results reflect the algorithm's ability to escape from local optima. We have applied the algorithms over functions f_8–f_{13} where the number of local minima increases exponentially as the dimension of the function increases. The dimension of such functions is set to 30. The results are averaged over 30 runs, with performance indexes being reported in Table 10.3 as follows: the average best-so-far solution (ABS), the standard deviation (SD) and the number of function evaluations (NFE). Likewise, p-values of the Wilcoxon signed-rank test of 30 independent runs are listed in Table 10.4. From the results, it is clear that LS yields better solutions than others algorithms for functions f_9, f_{10}, f_{11} and f_{12}, in terms of the indexes ABS and SD. However, for functions f_8 and f_{13}, LS produces similar results to DE. The Wilcoxon rank test results, that are presented in Table 10.6, confirm that LS performed better than PSO and DE considering four problems f_9–f_{12}, whereas, from a statistical viewpoint, there is not difference between results from LS and DE for f_8 and f_{13}. According to Table 10.3, the NFE values obtained by the presented method are smaller than those produced by other optimizers. The reason of this remarkable performance is associated with its two operators: (i) the solitary operator allows a better particle distribution in the search space, increasing the algorithm's ability to find the global

Table 10.3 Minimization results from the benchmark functions test in Table B.2 with n = 30

		PSO	DE	LS
f_8	ABS	-6.7×10^3	-1.26×10^4	-1.26×10^4
	SD	6.3×10^2	3.7×10^2	1.1×10^2
	NFE	38,452	35,240	21,846
f_9	ABS	14.8	4.01×10^{-1}	2.49×10^{-3}
	SD	1.39	5.1×10^{-2}	4.8×10^{-4}
	NFE	37,672	34,576	20,784
f_{10}	ABS	14.7	4.66×10^{-2}	2.15×10^{-3}
	SD	1.44	1.27×10^{-2}	3.18×10^{-4}
	NFE	39,475	37,080	21,235
f_{11}	ABS	12.01	1.15	1.47×10^{-4}
	SD	3.12	0.06	1.48×10^{-5}
	NFE	38,542	34,875	22,126
f_{12}	ABS	6.87×10^{-1}	3.74×10^{-1}	5.58×10^{-3}
	SD	7.07×10^{-1}	1.55×10^{-1}	4.18×10^{-4}
	NFE	35,248	30,540	16,984
f_{13}	ABS	1.87×10^{-1}	1.81×10^{-2}	1.78×10^{-2}
	SD	5.74×10^{-1}	1.66×10^{-2}	1.64×10^{-3}
	NFE	36,022	31,968	18,802

Maximum number of iterations = 1000

Table 10.4 p-values produced by Wilcoxon's test comparing LS versus PSO and DE over the "average best-so-far" values from Table 10.3

LS	PSO	DE
f_8	1.83×10^{-4}	0.061
f_9	1.17×10^{-4}	2.41×10^{-4}
f_{10}	1.43×10^{-4}	3.12×10^{-3}
f_{11}	6.25×10^{-4}	1.14×10^{-3}
f_{12}	2.34×10^{-5}	7.15×10^{-4}
f_{13}	4.73×10^{-4}	0.071

optima; and (ii) the use of the social operation provides of a simple exploitation operator that intensifies the capacity of finding better solutions during the evolution process.

10.5 Gaussian Mixture Modelling

In this section, the modeling of image histograms through Gaussian mixture models is presented. Let consider an image holding L gray levels $[0,\ldots,L-1]$ whose distribution is defined by a histogram $h(g)$ represented by the following formulation:

$$h(g) = \frac{n_g}{N_p}, \quad h(g) > 0,$$
$$Np = \sum_{g=0}^{L-1} n_g, \quad \text{and} \quad \sum_{g=0}^{L-1} h(g) = 1, \tag{10.15}$$

where n_g denotes the number of pixels with gray level g and Np the total number of pixels in the image. Under such circumstances, $h(g)$ can be modeled by using a mixture $p(x)$ of Gaussian functions of the form:

$$p(x) = \sum_{i=1}^{K} \frac{P_i}{\sqrt{2\pi}\sigma_i} \exp\left[\frac{-(x-\mu_i)}{2\sigma_i^2}\right] \tag{10.16}$$

where K symbolizes the number of Gaussian functions of the model whereas P_i is the apriori probability of function i. μ_i and σ_i represent the mean and standard deviation of the ith Gaussian function, respectively. Furthermore, the constraint $\sum_{i=1}^{K} P_i = 1$ must be satisfied by the model. In order to evaluate the similarity between the image histogram and a candidate mixture model, the mean square error can be used as follows:

$$J = \frac{1}{n}\sum_{j=1}^{L} \left[p(x_j) - h(x_j)\right]^2 + \omega \cdot \left| \left(\sum_{i=1}^{K} P_i \right) - 1 \right| \tag{10.17}$$

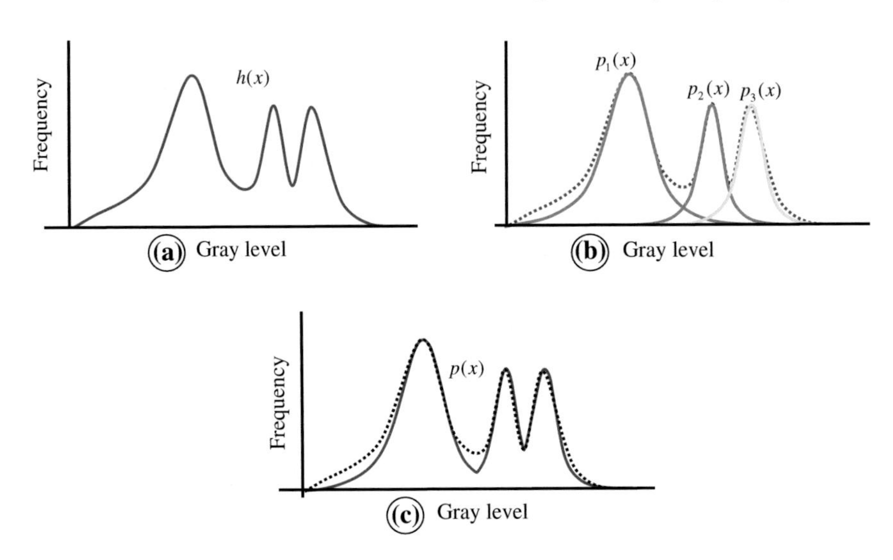

Fig. 10.6 Histogram approximation through a Gaussian mixture. **a** Original histogram, **b** configuration of the Gaussian components $p_1(x)$, $p_2(x)$ and $p_3(x)$, **c** final Gaussian mixture $p(x)$

where ω represents the penalty associated with the constrain $\sum_{i=1}^{K} P_i = 1$. Therefore, J is considered as the objective function which must be minimized in the estimation problem. In order to illustrate the histogram modeling through a Gaussian mixture, Fig. 10.6 presents an example, assuming three classes, i.e. $K = 3$. Considering the Fig. 10.6a as the image histogram $h(x)$, the Gaussian mixture $p(x)$, that is shown in Fig. 10.6c, is produced by adding the Gaussian functions $p_1(x)$, $p_2(x)$ and $p_3(x)$ in the configuration presented in Fig. 10.6b. Once the model parameters that better model the image histogram have been determined, the final step is to define the threshold values T_i $(i \in [1, \ldots, K])$ which can be calculated by simple standard methods, just as it is presented in [19–21].

10.6 Segmentation Algorithm Based on LS

In the presented method, the segmentation process is approached as an optimization problem. Computational optimization generally involves two distinct elements: (1) a search strategy to produce candidate solutions (individuals, particles, insects, locust, etc.) and (2) an objective function that evaluates the quality of each selected candidate solution. Several computational algorithms are available to perform the first element. The second element, where the objective function is designed, is unquestionably the most critical. In the objective function, it is expected to embody the full complexity of the performance, biases, and restrictions of the problem to be solved. In the segmentation problem, candidate solutions represent Gaussian mixtures. The objective function J (Eq. 10.17) is used as a fitness value to evaluate the similarity presented between the Gaussian mixture and the image histogram. Therefore, guided by the fitness values (J values), the set of encoded candidate solutions are evolved using the evolutionary operators until the best possible resemblance can be found.

Over the last decade, several algorithms based on evolutionary and swarm principles [19–22] have been presented to solve the problem of segmentation through a Gaussian mixture model. Although these algorithms own certain good performance indexes, they present two important limitations. (1) They frequently obtain sub-optimal approximations as a consequence of an inappropriate balance between exploration and exploitation in their search strategies. (2) They are based on the assumption that the number of Gaussians (classes) in the mixture is pre-known and fixed, otherwise they cannot work.

In order to eliminate such flaws, the presented approach includes (A) a new search strategy and (B) the definition of a new objective function. For the search strategy, it is adopted the LS method (Sect. 10.4). Under LS, the concentration of individuals in the current best positions is explicitly avoided. Such fact allows reducing critical problems such as the premature convergence to sub-optimal solutions and the incorrect exploration-exploitation balance.

10.6.1 New Objective Function $\mathbf{J}^{\mathbf{new}}$

Previous segmentation algorithms based on evolutionary and swarm methods use Eq. 10.17 as objective function. Under these circumstances, each candidate solution (Gaussian mixture) is only evaluated in terms of its approximation with the image histogram.

Since the presented approach aims to automatically select the number of Gaussian functions K in the final mixture $p(x)$, the objective function must be modified. The new objective function J^{new} is defined as follows:

$$J^{new} = J + \lambda \cdot Q, \tag{10.18}$$

where λ is a scaling constant. The new objective function is divided in two parts. The first part J evaluates the quality of each candidate solution in terms of its similarity with regard to the image histogram (Eq. 10.17). The second part Q penalizes the overlapped area among Gaussian functions (classes), with Q being defined as follows:

$$Q = \sum_{i=1}^{K} \sum_{\substack{j=1 \\ j \neq i}}^{K} \sum_{l=1}^{L} \min\left(P_i \cdot p_i(l), P_j \cdot p_j(l)\right), \tag{10.19}$$

where K and L represents the number of classes and the gray levels, respectively. The parameters $p_i(l)$ and $p_j(l)$ symbolize the Gaussian functions i and j respectively, that are to be evaluated on the point l (gray level) whereas the elements P_i and P_j represent the apriori probabilities of the Gaussian functions i and j, respectively. Under such circumstances, mixtures with Gaussian functions that do not "positively" participate in the histogram approximation are severely penalized.

Figure 10.7 illustrates the effect of the new objective function J^{new} in the evaluation of Gaussian mixtures (candidate solutions). From the image histogram (Fig. 10.7a), it is evident that two Gaussian functions are enough to accurately approximate the original histogram. However, if the Gaussian mixture is modeled by using a greater number of functions (for example four as it is shown in Fig. 10.7b), the original objective function J is unable to obtain a reasonable approximation. Under the new objective function J^{new}, it is penalized the overlapped area among Gaussian functions (classes). Such areas, in Fig. 10.7c, correspond to Q_{12}, Q_{23}, Q_{34}, where Q_{12} represents the penalization value produced between the Gaussian function $p_1(x)$ and $p_2(x)$. Therefore, due to the penalization, the Gaussian mixture shown in Fig. 10.7b, c provides a solution of low quality. On the other hand, the Gaussian mixture presented in Fig. 10.7d maintains a low penalty, thus, it represents a solution of high quality. From Fig. 10.7d, it is easy to see that functions $p_1(x)$ and $p_4(x)$ can be removed from the final mixture. This elimination could be performed by a simple comparison with a threshold value θ, since $p_1(x)$ and $p_4(x)$ have a reduced amplitude ($p_1(x) \approx p_2(x) \approx 0$). Therefore, under J^{new}, it is possible to find the reduced Gaussian mixture model starting from a considerable number of functions. Since the presented segmentation method is

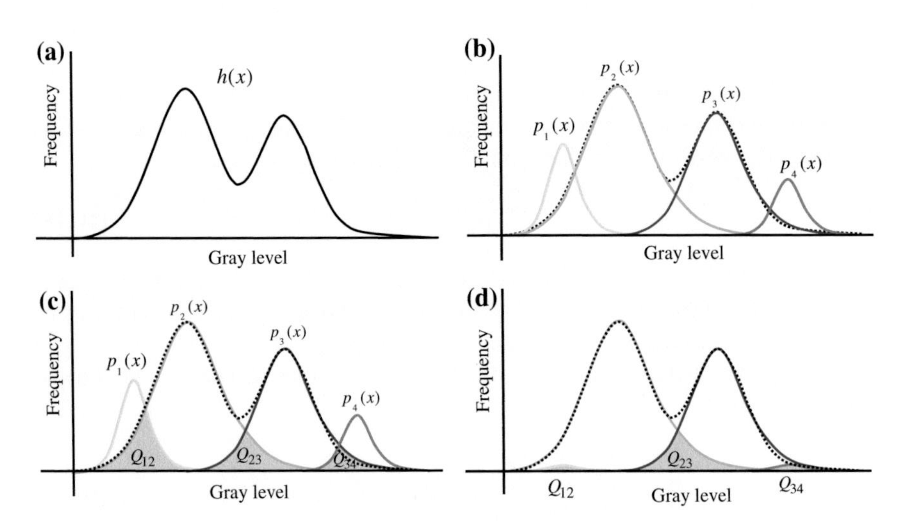

Fig. 10.7 Effect of the new objective function J^{new} in the evaluation of Gaussian mixtures (candidate solutions). **a** Original histogram, **b** Gaussian mixture considering four classes, **c** penalization areas and **d** Gaussian mixture of better quality solution

conceived as an optimization problem, the overall operation can be reduced to solve the formulation of Eq. 10.20 by using the LS algorithm.

$$\begin{aligned}
\text{minimize} \quad & J^{new}(\mathbf{x}) = J(\mathbf{x}) + \lambda \cdot Q(\mathbf{x}), \\
& \mathbf{x} = (P_1, \mu_1, \sigma_1, \ldots, P_K, \mu_K, \sigma_K) \in \mathbf{R}^{3 \cdot K}, \\
& 0 \le P_d \le 1, d \in (1, \ldots, K) \\
\text{subject to} \quad & 0 \le \mu_d \le 255 \\
& 0 \le \sigma_d \le 25
\end{aligned} \tag{10.20}$$

where P_d, μ_d and σ_d represent the probability, mean and standard deviation of the class d. It is important to remark that the new objective function J^{new} allows the evaluation of a candidate solution in terms of the necessary number of Gaussian functions and its approximation quality. Under such circumstances, it can be used in combination with any other evolutionary method and not only with the presented LS algorithm.

10.6.2 Complete Segmentation Algorithm

Once the new search strategy (LS) and objective function (J^{new}) have been defined, the presented segmentation algorithm can be summarized by the Algorithm 10.2. The new algorithm combines operators defined by LS and operations for calculating the threshold values.

Algorithm 10.2 Segmentation LS algorithm

1:	**Input:** F, L, g, gen, N, K, λ and θ.	
2:	Initialize $\mathbf{L}^0$ (k=0)	
3:	**until** (k=gen)	
5:	$\mathbf{F} \leftarrow$ SolitaryOperation($\mathbf{L}^k$)	Solitary operator (Section 3.1)
6:	$\mathbf{L}^{k+1} \leftarrow$ SocialOperation($\mathbf{L}^k$,$\mathbf{F}$)	Social operator (Section 3.2)
7:	k=k+1	
8:	**end until**	
9:	Obtain $\mathbf{l}_{best}^{gen}$	
10:	Reduce $\mathbf{l}_{best}^{gen}$	
11:	Calculate the threshold values T_i from the reduced model	
12:	Use T_i to segment the image	

(Line 1) The algorithm sets the operative parameters F, L, g, gen, N, K, λ and θ. They rule the behavior of the segmentation algorithm. (Line 2) Afterwards, the population $\mathbf{L}^0$ is initialized considering N different random Gaussian mixtures of K functions. The idea is to generate an N-random Gaussian mixture subjected to the constraints formulated in Eq. 10.20. The parameter K must hold a high value in order to correctly segment all images (recall that the algorithm is able to reduce the Gaussian mixture to its minimum expression). (Line 3) Then, the Gaussian mixtures are evolved by using the LS operators and the new objective function J^{new}. This process is repeated during *gen* cycles. (Line 8) After this procedure, the best Gaussian mixture $\mathbf{I}_{best}^{gen}$ is selected according to its objective function J^{new}. (Line 9) Then, the Gaussian mixture $\mathbf{I}_{best}^{gen}$ is reduced by eliminating those functions whose amplitudes are lower than θ $(p_i(x) \leq \theta)$. (Line 10) Then, it is calculated the threshold values T_i from the reduced model. (Line 11) Finally, the calculated T_i values are employed to segment the image. Figure 10.8 shows a flow chart of the complete segmentation procedure. The presented segmentation algorithm permits to

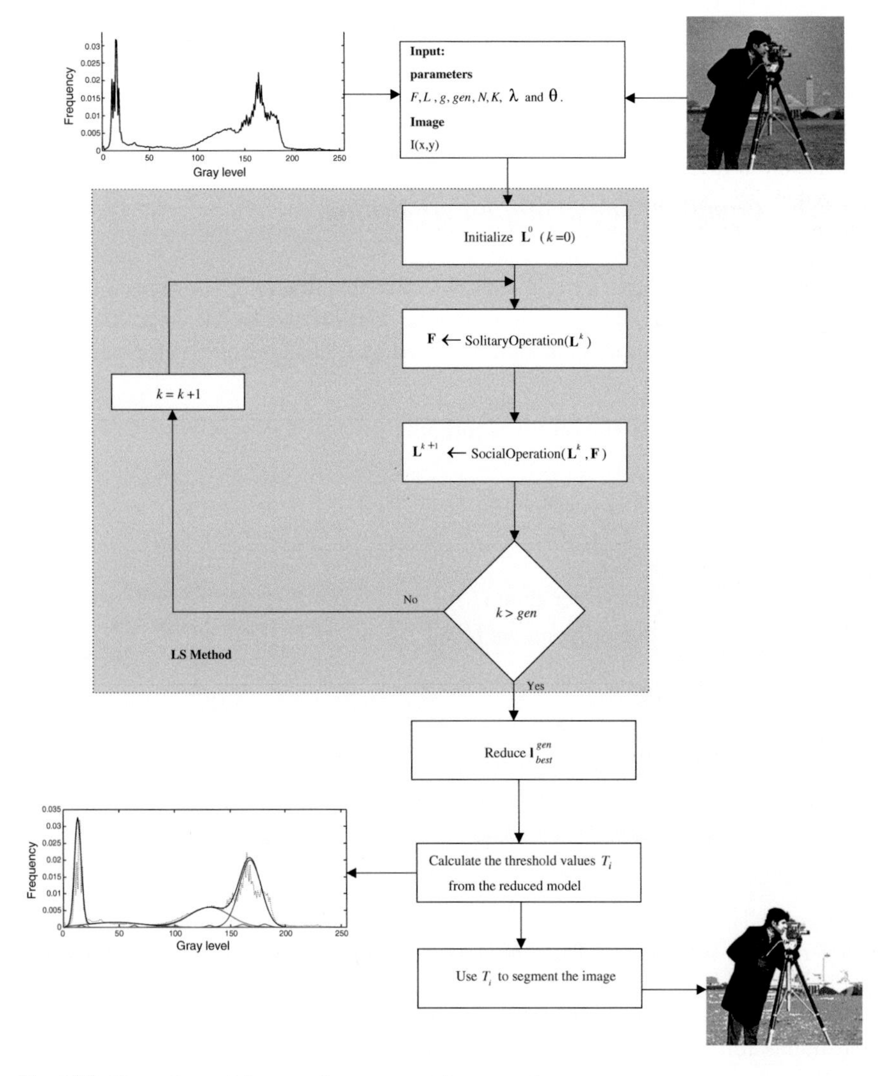

Fig. 10.8 Flow chart of the complete segmentation procedure

automatically detect the number of segmentation partitions (classes). Furthermore, due to its remarkable search capacities, the LS method maintains a better accuracy than previous algorithms based on evolutionary principles. However, the presented method presents two disadvantages: first, it is related to its implementation which in general is more complex than most of the other segmentators based on evolutionary basics. The second refers to the segmentation procedure of the presented approach which does not consider any spatial pixel characteristics. As a consequence, pixels that may belong to a determined region due to its position are labeled as a part of other region due to its gray-level intensity. Such a fact adversely affects the segmentation performance of the method.

10.7 Segmentation Results

This section analyses the performance of the presented segmentation algorithm. The discussion is divided into three parts: the first one shows the performance of the presented LS segmentation algorithm while the second presents a comparison between the presented approach and others segmentation algorithms that are based on evolutionary and swam methods. The comparison mainly considers the capacities of each algorithm to accurately and robustly approximate the image histogram. Finally, the third part presents an objective evaluation of segmentation results produced by all algorithms that have been employed in the comparisons.

10.7.1 Performance of LS Algorithm in Image Segmentation

This section presents two experiments that analyze the performance of the presented approach. Table 10.5 presents the algorithm's parameters that have been experimentally determined and kept for all the test images through all experiments.

10.7.1.1 First Image

The first test considers the histogram shown by Fig. 10.9b while Fig. 10.9a presents the original image. After applying the presented algorithm, just as it has been configured according to parameters in Table 10.5, a minimum model of four classes emerges after the start from Gaussian mixtures of 10 classes. Considering 30

Table 10.5 Parameter setup for the LS segmentation algorithm

F	L	g	gen	N	K	λ	θ
0.6	0.2	20	1000	40	10	0.01	0.0001

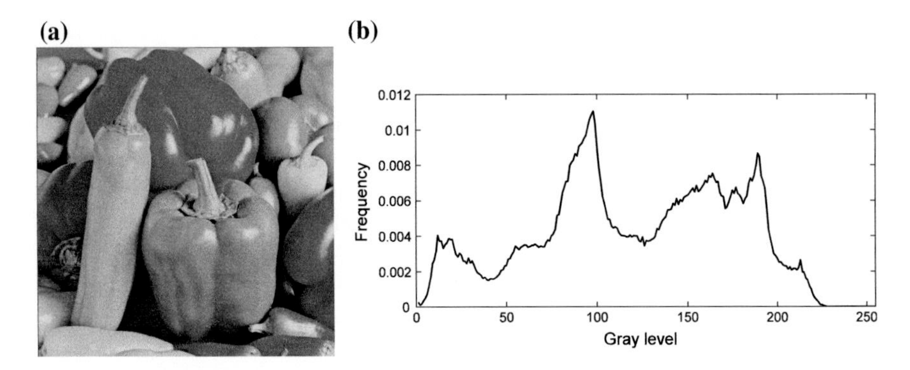

Fig. 10.9 **a** Original first image used on the first experiment, and **b** its histogram

Table 10.6 Results of the reduced Gaussian mixture for the first and the second image

Parameters	First image	Second image
$\overline{P_1}$	0.004	0.032
$\overline{\mu_1}$	18.1	12.1
$\overline{\sigma_1}$	8.2	2.9
$\overline{P_2}$	0.0035	0.0015
$\overline{\mu_2}$	69.9	45.1
$\overline{\sigma_2}$	18.4	24.9
$\overline{P_3}$	0.01	0.006
$\overline{\mu_3}$	94.9	130.1
$\overline{\sigma_3}$	8.9	17.9
$\overline{P_4}$	0.007	0.02
$\overline{\mu_4}$	163.1	167.2
$\overline{\sigma_4}$	29.9	10.1

independent executions, the averaged parameters of the resultant Gaussian mixture are presented in Table 10.6. One final Gaussian mixture (ten classes), which has been obtained by LS, is presented in Fig. 10.10. Furthermore, the approximation of the reduced Gaussian mixture is also visually compared with the original histogram in Fig. 10.10. On the other hand, Fig. 10.11 presents the segmented image after calculating the threshold points.

10.7.1.2 Second Image

For the second experiment, the image in Fig. 10.12 is tested. The method aims to segment the image by using a reduced Gaussian mixture model obtained by the LS approach. After executing the algorithm according to parameters from Table 10.5, the resulting averaged parameters of the resultant Gaussian mixture are presented in Table 10.6. In order to assure consistency, the results are averaged considering 30

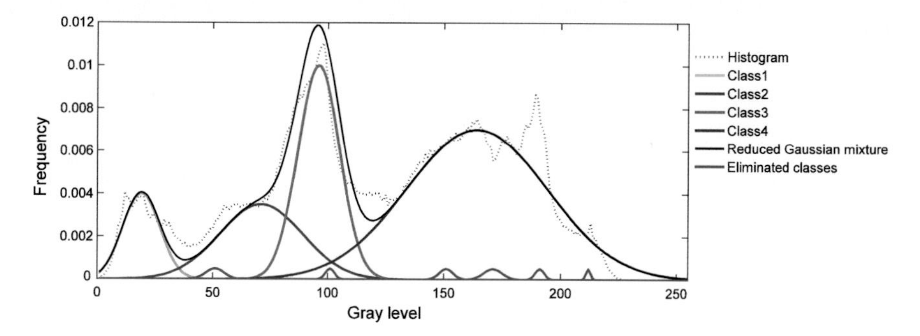

Fig. 10.10 Gaussian mixture obtained by LS for the first image

Fig. 10.11 Image segmented with the reduced Gaussian mixture

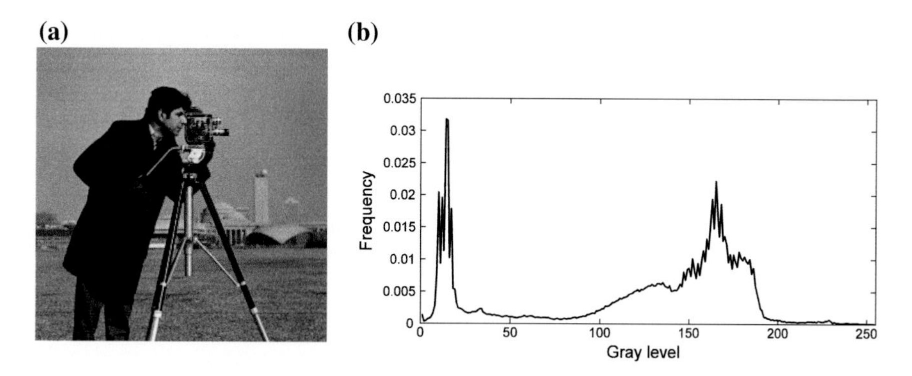

Fig. 10.12 **a** Original second image used on the first experiment, and **b** its histogram

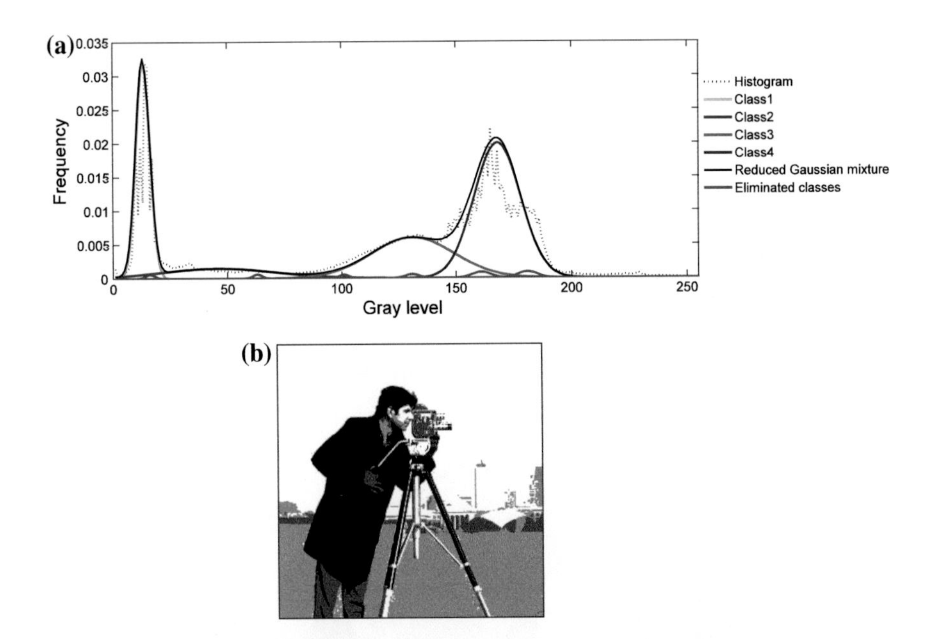

Fig. 10.13 **a** Segmentation result obtained by LS for the first image and **b** the segmented image

independent executions. Figure 10.13 shows the approximation quality that is obtained by the reduced Gaussian mixture model in (a) and the segmented image in (b).

10.7.2 Histogram Approximation Comparisons

This section discusses the comparison between the presented segmentation algorithm and other evolutionary-segmentation methods that have been proposed in the literature. The set of methods used in the experiments includes the J + ABC [19], J + AIS [20] and J + DE [21]. These algorithms consider the combination between the original objective function J (Eq. 10.17) and an evolutionary technique such as Artificial Bee Colony (ABC), the Artificial Immune Systems (AIS) and the Differential Evolution (DE) [21], respectively. Since the proposed segmentation approach considers the combination of the new objective function J^{new} (Eq. 10.18) and the LS algorithm, it will be referred as J^{new} + LS. The comparison focuses mainly on the capacities of each algorithm to accurately and robustly approximate the image histogram.

In the experiments, the populations have been set to 40 ($N = 40$) individuals. The maximum iteration number for all functions has been set to 1000. Such stop criterion has been considered to maintain compatibility to similar experiments that are reported in the literature [18]. The parameter setting for each of the segmentation algorithms in the comparison is described as follows:

1. $J + \mathrm{J} + \mathrm{ABC}$ [19]: In the algorithm, its parameters are configured as follows: the abandonment limit = 100, $\alpha = 0.05$ and $limit = 30$.
2. J + AIS [20]: It presents the following parameters, $h = 120$, $N_c = 80$, $\rho = 10$, $P_r = 20$, $L = 22$ and $T_e = 0.01$.
3. J + DE [21]: The DE/Rand/1 scheme is employed. The parameter settings follow the instructions in [21]. The crossover probability is $CR = 0.9$ and the weighting factor is $F = 0.8$.
4. In J^{new} + LS, the method is set according to the values described in Table 10.5.

In order to conduct the experiments, a synthetic image is designed to be used as a reference in the comparisons. The main idea is to know in advance the exact number of classes (and their parameters) that are contained in the image so that the histogram can be considered as a ground truth. The synthetic image is divided into four sections. Each section corresponds to a different class which is produced by setting each gray level pixel Pv^i to a value that is determined by the following model:

$$Pv_i = e^{-\left(\frac{(x-\mu_1)2}{2\sigma_i^2}\right)}, \tag{10.21}$$

where i represents the section, whereas μ_i and σ_i are the mean and the dispersion of the gray level pixels, respectively. The comparative study employs the image of 512 × 512 that is shown in Fig. 10.14a and the algorithm's parameters that have been presented in Table 10.7. Figure 10.14b illustrates the resultant histogram.

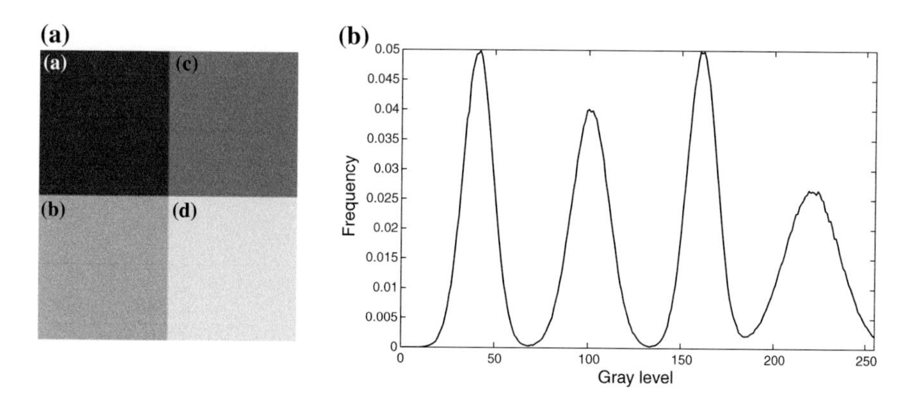

Fig. 10.14 **a** Synthetic image used in the comparison, and **b** its histogram

Table 10.7 Employed parameters for the design of the reference image

	P_i	μ_i	σ_i
(1)	0.05	40	8
(2)	0.04	100	10
(3)	0.05	160	8
(4)	0.027	220	15

In the comparison, the discussion focuses on the following issues: first of all, accuracy; second, convergence; and third, computational cost.

10.7.2.1 Convergence

This section analyzes the approximation convergence when the number of classes that are used by the algorithm during the evolution process is different to the actual number of classes in the image. Recall that the presented approach automatically finds the reduced Gaussian mixture which better adapts to the image histogram.

In the experiment, the methods: J + ABC, J + AIS and J + DE are executed considering Gaussian mixtures composed of 8 functions. Under such circumstances, the number of classes to be detected is higher than the actual number of classes in the image. On the other hand, the presented algorithm maintains the same configuration of Table 10.5. Therefore, it can detect and calculate up to ten classes ($K = 10$).

As a result, the techniques J + ABC, J + AIS and J + DE tend to overestimate the image histogram. This effect can be seen in Fig. 10.15a, where the resulting Gaussian functions are concentrated within actual classes. Such a behavior is a

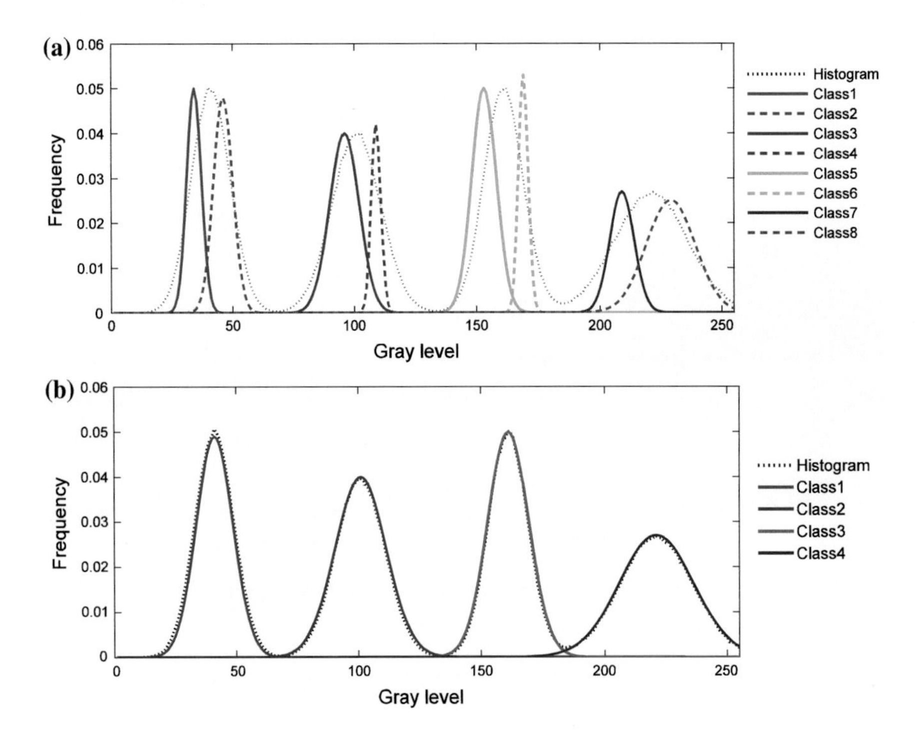

Fig. 10.15 Convergence results. **a** Convergence of the following methods: J + ABC, J + AIS and J + DE considering Gaussian mixtures of 8 classes, **b** convergence of the presented method (reduced Gaussian mixture)

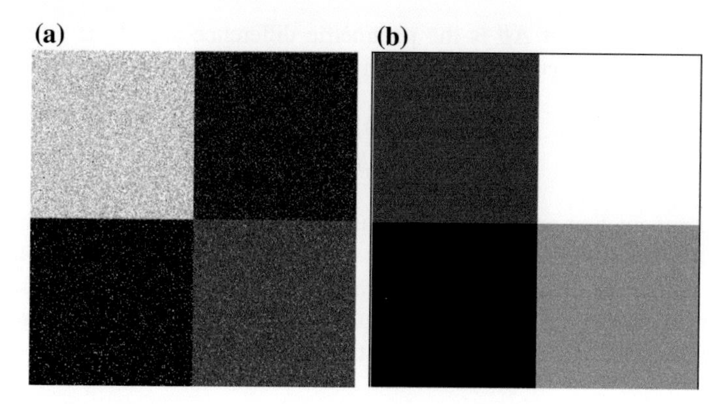

Fig. 10.16 Segmentation results obtained by **a** several methods including J + ABC, J + AIS and J + DE considering Gaussian mixtures of 8 classes; and **b** the presented method (reduced Gaussian mixture)

consequence of the evaluation that is considered by the original objective function J, which privileges only the approximation between the Gaussian mixture and the image histogram. This effect can be graphically illustrated by Fig. 10.16a that shows the pixel misclassification produced by the wrong segmentation of the Fig. 10.14a. On the other hand, the presented approach obtains a reduced Gaussian mixture model which allows the detection of each class from the actual histogram (see Fig. 10.15b). As a consequence, the segmentation is significantly improved by eliminating the pixel misclassification, as it is shown by Fig. 10.16b.

It is evident from Fig. 10.15 that the techniques, J + ABC, J + AIS and J + DE, all need an a priori knowledge of the number of classes that are contained in the actual histogram in order to obtain a satisfactory result. On the other hand, the presented algorithm is able to find a reduced Gaussian mixture whose classes coincide with the actual number of classes that are contained in the image histogram.

10.7.2.2 Accuracy

In this section, the comparison among the algorithms in terms of accuracy is reported. Most of the reported comparisons [19–26] are concerned about comparing the parameters of the resultant Gaussian mixtures by using real images. Under such circumstances, it is difficult to consider a clear reference in order to define a meaningful error. Therefore, the image defined in Fig. 10.14 has been used in the experiments because its construction parameters are clearly defined in Table 10.7.

Since the parameter values from Table 10.7 act as ground truth, a simple averaged difference between them and the values that are computed by each algorithm could be used as comparison error. However, as each parameter maintains different active intervals, it is necessary to express the differences in terms of percentage. Therefore, if $\Delta\beta$ is the parametric difference and β the ground truth parameter, the percentage error $\Delta\beta\,\%$ can be defined as follows:

$$\Delta\beta\,\% = \frac{\Delta\beta}{\beta} \cdot 100\,\% \tag{10.22}$$

In the segmentation problem, each Gaussian mixture represents a K-dimensional model where each dimension corresponds to a Gaussian function of the optimization problem to be solved. Since each Gaussian function possesses three parameters P_i, μ_i and σ_i, the complete number of parameters is $3 \cdot K$ dimensions. Therefore, the final error E produced by the final Gaussian mixture is:

$$E = \frac{1}{K \cdot 3} \sum_{v=1}^{K \cdot 3} \Delta\beta_v\,\%, \tag{10.23}$$

where $\beta_v \in (P_i, \mu_i, \sigma_i)$.

In order to compare accuracy, the algorithms, J + ABC, J + AIS, J + DE and the presented approach are all executed over the image shown by Fig. 10.14a. The experiment aims to estimate the Gaussian mixture that better approximates the actual image histogram. Methods J + ABC, J + AIS and J + DE consider Gaussian mixtures composed of 4 functions (K = 4). In case of the J^{new} + LS method, although the algorithm finds a reduced Gaussian mixture of four functions, it is initially set with ten functions (K = 10). Table 10.8 presents the final Gaussian mixture parameters and the final error E. The final Gaussian mixture parameters

Table 10.8 Results of the reduced Gaussian mixture in terms of accuracy

Algorithm	Gaussian function	$\overline{P_i}$	$\overline{\mu_i}$	$\overline{\sigma_i}$	E (%)
J + ABC	(1)	0.052	44.5	6.4	11.79
	(2)	0.084	98.12	12.87	
	(3)	0.058	163.50	8.94	
	(4)	0.025	218.84	17.5	
J + AIS	(1)	0.072	31.01	6.14	22.01
	(2)	0.054	88.52	12.21	
	(3)	0.039	149.21	9.14	
	(4)	0.034	248.41	13.84	
J + DE	(1)	0.041	35.74	7.86	13.57
	(2)	0.036	90.57	11.97	
	(3)	0.059	148.47	9.01	
	(4)	0.020	201.34	13.02	
J^{new} + LS	(1)	0.049	40.12	7.5	3.98
	(2)	0.041	102.04	10.4	
	(3)	0.052	168.66	8.3	
	(4)	0.025	110.92	15.2	

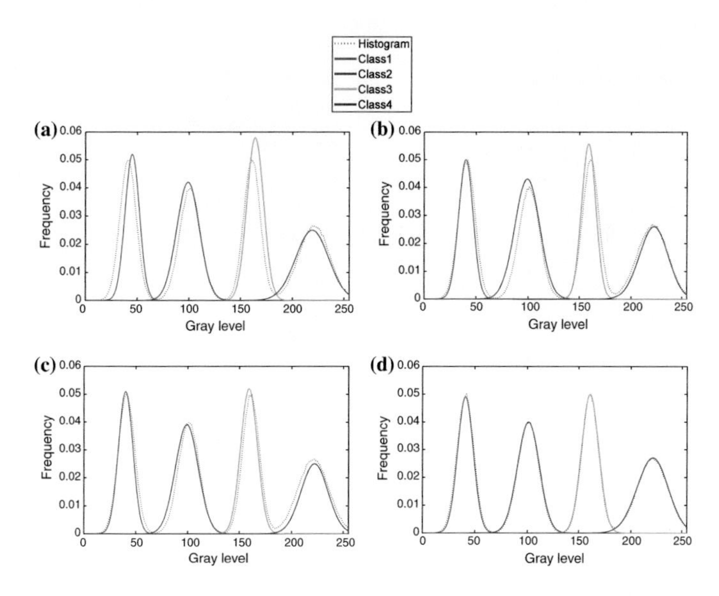

Fig. 10.17 Approximation results in terms of accuracy. **a** J + ABC, **b** J + AIS, **c** J + DE and **d** the presented J^{new} + LS approach

have been averaged over 30 independent executions in order to assure consistency. A close inspection of Table 10.8 reveals that the presented method is able to achieve the smallest error (E) in comparison to the other algorithms. Figure 10.16 presents the histogram approximations that are produced by each algorithm whereas Fig. 10.17 shows their correspondent segmented images. Both illustrations present the median case obtained throughout 30 runs. Figure 10.18 exhibits that J + ABC, J + AIS, and J + DE present different levels of misclassifications which are nearly absent in the presented approach case.

10.7.2.3 Computational Cost

The experiment aims to measure the complexity and the computing time spent by the J + ABC, the J + AIS, the J + DE and the Jnew + LS algorithm while calculating the parameters of the Gaussian mixture in benchmark images (see Figs. 10.9a–d. J + ABC, J + AIS and J + DE consider Gaussian mixtures that are composed of 4 functions (K = 4). In case of the J^{new} + LS method, although the algorithm finds a reduced Gaussian mixture of four functions despite being initialized with ten functions (K = 10). Table 10.9 shows the averaged measurements after 30 experiments. It is evident that the J + ABC and J + DE are the slowest to converge (iterations) and the J + AIS shows the highest computational cost (time elapsed) because it requires operators which demand long times. On the other hand, the Jnew + LS shows an acceptable trade-off between its convergence time and its computational cost. Therefore, although the implementation of J^{new} + LS in general requires more code than most of other evolution-based segmentators, such a fact is

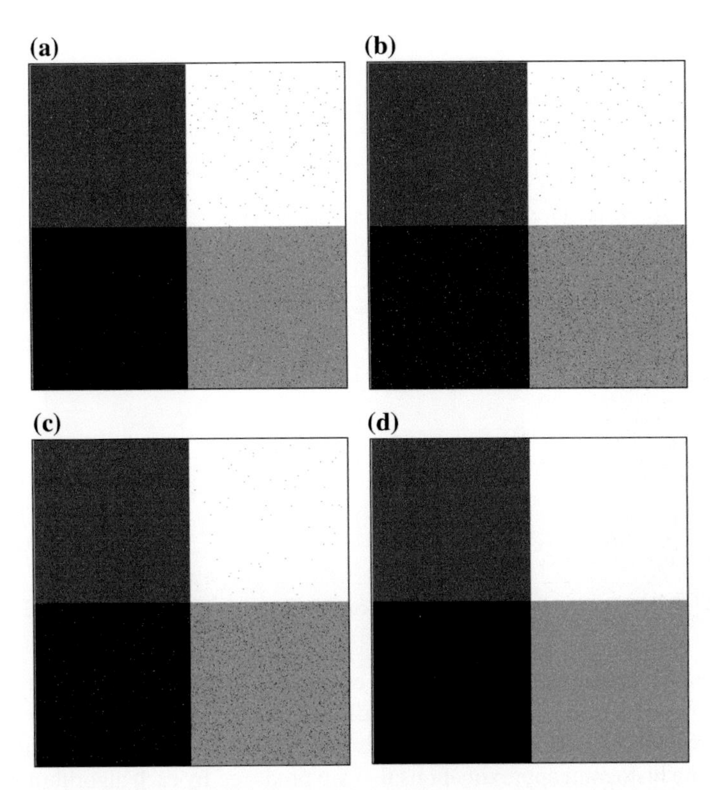

Fig. 10.18 Segmentation results in terms of accuracy. **a** J + ABC, **b** J + AIS, **c** J + DE and **d** the presented J^{new} + LS approach

Table 10.9 Iterations and time requirements of the J + ABC, the J + AIS, the J + DE and the J^{new} + LS algorithm as they are applied to segment benchmark images (see Fig. 10.17)

Iterations / Time elapsed	(a)	(b)	(c)	(d)
J + ABC	855	833	870	997
	2.72 s	2.70 s	2.73 s	3.1 s
J + AIS	725	704	754	812
	1.78 s	1.61 s	1.41 s	2.01 s
J + DE	657	627	694	742
	1.25 s	1.12 s	1.45 s	1.88 s
J^{new} + LS	314	298	307	402
	0.98 s	0.84 s	0.72 s	1.02 s

not reflected in the execution time. Finally, Fig. 10.19 shows the segmented images as they are generated by each algorithm. It can be seen that the presented approach generate more homogeneous regions whereas J + ABC, J + AIS and J + DE present several artifacts that are produced by an incorrect pixel classification.

Fig. 10.19 Images employed in the computational cost analysis

10.7.3 Performance Evaluation of the Segmentation Results

This section presents an objective evaluation of segmentation results that are produced by all algorithms in the comparisons. The ill-defined nature of the segmentation problem makes the evaluation of a candidate algorithm difficult [58]. Traditionally, the evaluation has been conducted by using some supervised criteria [59] which are based on the computation of a dissimilarity measure between a segmentation result and a ground truth image. Recently, the use of unsupervised measures has substituted supervised indexes for the objective evaluation of segmentation results [60]. They enable the quantification of the quality of a segmentation result without a priori knowledge (ground truth image).

10.7.3.1 Evaluation Criteria

In this chapter, the unsupervised index *ROS* proposed by Rosenberg [61] has been used to objectively evaluate the performance of each candidate algorithm. This index evaluates the segmentation quality in terms of the homogeneity within segmented regions and the heterogeneity among the different regions. *ROS* can be computed as follows:

$$ROS = \frac{\overline{D} - \underline{D}}{2}, \tag{10.24}$$

where $\underline{D}$ quantifies the homogeneity within segmented regions. Similarly, $\overline{D}$ measures the disparity among the regions. A segmentation result S_1 is considered better than S_2, if $ROS_{S_1} > ROS_{S_2}$. The inter-region homogeneity characterized by $\underline{D}$ is calculated considering the following formulation:

$$\underline{D} = \frac{1}{N_R} \sum_{c=1}^{N_R} \frac{R_c}{I} \cdot \frac{\sigma_c}{\sum_{l=1}^{N_R} \sigma_l}, \tag{10.25}$$

where N_R represents the number of partitions in which the image has been segmented. R_c symbolizes the number of pixels contained in the partition c whereas I considers the number of pixels that integrate the complete image. Similarly, σ_c represents the standard deviation from the partition c. On the other hand, disparity among the regions $\overline{D}$ is computed as follows:

$$\overline{D} = \frac{1}{N_R} \sum_{c=1}^{N_R} \frac{R_c}{I} \cdot \left[\frac{1}{(N_R - 1)} \sum_{l=1}^{N_R} \frac{|\mu_c - \mu_l|}{255} \right], \tag{10.26}$$

where μ_c is the average gray-level in the partition c.

10.7.3.2 Experimental Protocol

In the comparison of segmentation results, a set of four classical images has been chosen to integrate the experimental set (Fig. 10.20). The segmentation methods used in the comparison are J + ABC [19], J + AIS [20] and J + DE [21].

From all segmentation methods used in the comparison, the proposed J^{new} + LS algorithm is the only one that has the capacity to automatically detect the number of segmentation partitions (classes). In order to conduct a fair comparison, all algorithms have been proved by using the same number of partitions. Therefore, in the experiments, the J^{new} + LS segmentation algorithm is firstly applied to detect the best possible number of partitions N_R. Once obtained the number of partitions N_R, the rest of the algorithms were configured to approximate the image histogram with this number of classes.

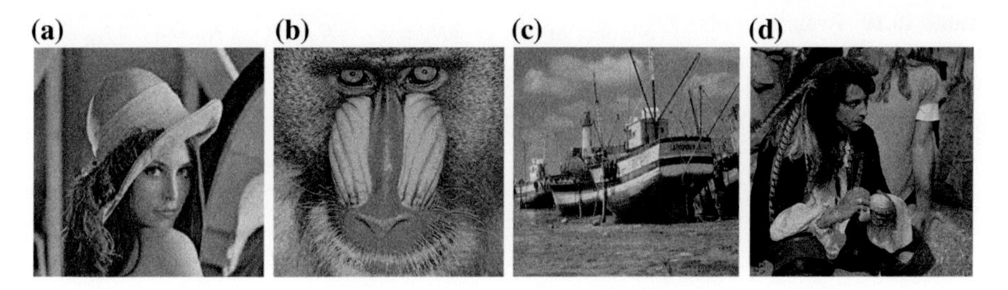

Fig. 10.20 Experimental set used in the evaluation of the segmentation results

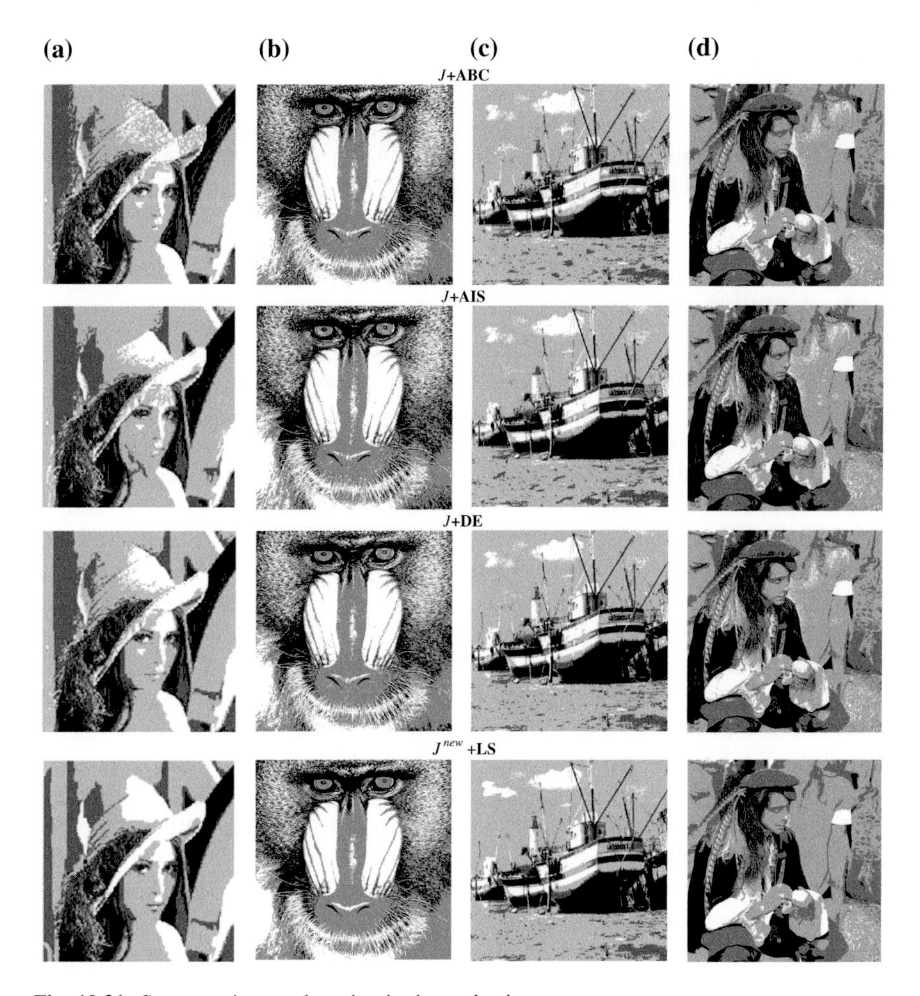

Fig. 10.21 Segmentation results using in the evaluation

Figure 10.21 presents the segmentation results obtained by each algorithm considering the experimental set from Fig. 10.20. On the other hand, Table 10.10 shows the evaluation of the segmentation results in terms of the *ROS* index. Such

Table 10.10 Evaluation of the segmentation results in terms of the ROS index

Number of classes	$N_R = 4$	$N_R = 3$	$N_R = 4$	$N_R = 4$
Image	**(a)**	**(b)**	**(c)**	**(d)**
J + ABC	0.534	0.328	0.411	0.457
J + AIS	0.522	0.321	0.427	0.437
J + DE	0.512	0.312	0.408	0.418
J^{new} + LS	0.674	0.401	0.514	0.527

values represent the averaged measurements after 30 executions. From them, it can be seen that the proposed J^{new} + LS method obtain the best *ROS* indexes. Such values indicate that the presented algorithm maintains the best balance between the homogeneity within segmented regions and the heterogeneity among the different regions. From Fig. 10.21, it can be seen that the presented approach generate more homogeneous regions whereas J + ABC, J + AIS and J + DE present several artifacts that are produced by an incorrect pixel classification.

10.8 Conclusions

Despite several evolutionary methods have been successfully applied to image segmentation with interesting results, most of them have exhibited two important limitations: (1) they frequently obtain sub-optimal results (misclassifications) as a consequence of an inappropriate balance between exploration and exploitation in their search strategies; (2) the number of classes is fixed and known in advance.

In this chapter, a new swarm algorithm for the automatic image segmentation, called the Locust Search (LS) has been presented. The presented method eliminates the typical flaws presented by previous evolutionary approaches by combining a novel evolutionary method with the definition of a new objective function that appropriately evaluates the segmentation quality with respect to the number of classes. In order to illustrate the proficiency and robustness of the presented approach, several numerical experiments have been conducted. Such experiments have been divided into two parts. First, the presented LS method has been compared to other well-known evolutionary techniques on a set of benchmark functions. In the second part, the performance of the presented segmentation algorithm has been compared to other segmentation methods based on evolutionary principles. The results in both cases validate the efficiency of the presented technique with regard to accuracy and robustness.

Several research directions will be considered for future work such as the inclusion of other indexes to evaluate similarity between a candidate solution and the image histogram, the consideration of spatial pixel characteristics in the objective function, the modification of the evolutionary LS operators to control the exploration-exploitation balance and the conversion of the segmentation procedure into a multi-objective problem.

References

1. Zhang, H., Fritts, J.E., Goldman, S.A.: Image segmentation evaluation: a survey of unsupervised methods. Comput. Vis. Image Underst. **110**(2), 260–280 (2008)
2. Uemura, T., Koutaki, G., Uchimura, K.: Image segmentation based on edge detection using boundary code. Int. J. Innovative Comput. Inf. Control **7**(10), 6073–6083 (2011)
3. Wang, L., Wu, H., Pan, C.: Region-based image segmentation with local signed difference energy. Pattern Recogn. Lett. **34**(6), 637–645 (2013)
4. Otsu, N.: A thresholding selection method from gray-level histogram. IEEE Trans. Syst. Man Cybern. **9**, 62–66 (1979)
5. Peng, R., Varshney, P.K.: On performance limits of image segmentation algorithms. Comput. Vis. Image Underst. **132**, 24–38 (2015)
6. Balafar, M.A.: Gaussian mixture model based segmentation methods for brain MRI images. Artif. Intell. Rev. **41**, 429–439 (2014)
7. McLachlan, G.J., Rathnayake, S.: On the number of components in a Gaussian mixture model. Data Mining Knowl. Discov. **4**(5), 341–355 (2014)
8. Oliva, D., Osuna-Enciso, V., Cuevas, E., Pajares, G., Pérez-Cisneros, M., Zaldívar, D.: Improving segmentation velocity using an evolutionary method. Expert Syst. Appl. **42**(14), 5874–5886 (2015)
9. Ye, Z.-W., Wang, M.-W., Liu, W., Chen, S.-B.: Fuzzy entropy based optimal thresholding using bat algorithm. Appl. Soft Comput. **31**, 381–395 (2015)
10. Sarkar, S., Das, S., Chaudhuri, S.S.: A multilevel color image thresholding scheme based on minimum cross entropy and differential evolution. Pattern Recogn. Lett. **54**, 27–35 (2015)
11. Bhandari, A.K., Kumar, A., Singh, G.K.: Modified artificial bee colony based computationally efficient multilevel thresholding for satellite image segmentation using Kapur's, Otsu and Tsallis functions. Expert Syst. Appl. **42**(3), 1573–1601 (2015)
12. Permutera, H., Francos, J., Jermyn, I.: A study of Gaussian mixture models of color and texture features for image classification and segmentation. Pattern Recogn. **39**, 695–706 (2006)
13. Dempster, A.P., Laird, A.P., Rubin, D.B.: Maximum likelihood from incomplete data via the EM algorithm. J. R. Stat. Soc. Ser. B **39**(1), 1–38 (1977)
14. Zhang, Z., Chen, C., Sun, J., Chan, L.: EM algorithms for Gaussian mixtures with split-and-merge operation. Pattern Recogn. **36**, 1973–1983 (2003)
15. Park, H., Amari, S., Fukumizu, K.: Adaptive natural gradient learning algorithms for various stochastic models. Neural Networks **13**, 755–764 (2000)
16. Park, H., Ozeki, T.: Singularity and slow convergence of the EM algorithm for Gaussian Mixtures. Neural Process. Lett. **29**, 45–59 (2009)
17. Gupta, L., Sortrakul, T.: A gaussian-mixture-based image segmentation algorithm. Pattern Recogn. **31**(3), 315–325 (1998)
18. Osuna-Enciso, V., Cuevas, E., Sossa, H.: A comparison of nature inspired algorithms for multi-threshold image segmentation. Expert Syst. Appl. **40**(4), 1213–1219 (2013)
19. Cuevas, E., Sención, F., Zaldivar, D., Pérez-Cisneros, M., Sossa, H.: A multi-threshold segmentation approach based on artificial bee colony optimization. Appl. Intell. **37**(3), 321–336 (2012)
20. Cuevas, E., Osuna-Enciso, V., Zaldivar, D., Pérez-Cisneros, M., Sossa, H.: Multithreshold segmentation based on artificial immune systems. Math. Prob. Eng., art. no. 874761 (2012)
21. Cuevas, E., Zaldivar, D., Pérez-Cisneros, M.: A novel multi-threshold segmentation approach based on differential evolution optimization. Expert Syst. Appl. **37**(7), 5265–5271 (2010)
22. Oliva, D., Cuevas, E., Pajares, G., Zaldivar, D., Osuna, V.: A multilevel thresholding algorithm using electromagnetism optimization. Neurocomputing **39**, 357–381 (2014)
23. Oliva, D., Cuevas, E., Pajares, G., Zaldivar, D., Perez-Cisneros, M.: Multilevel thresholding segmentation based on harmony search optimization. J. Appl. Math., art. no. 575414 (2013)

24. Cuevas, E., Zaldivar, D., Pérez-Cisneros, M.: Seeking multi-thresholds for image segmentation with Learning Automata. Mach. Vis. Appl. **22**(5), 805–818 (2011)
25. Tan, K.C., Chiam, S.C., Mamun, A.A., Goh, C.K.: Balancing exploration and exploitation with adaptive variation for evolutionary multi-objective optimization. Eur. J. Oper. Res. **197**, 701–713 (2009)
26. Chen, G., Low, C.P., Yang, Z.: Preserving and exploiting genetic diversity in evolutionary programming algorithms. IEEE Trans. Evol. Comput. **13**(3), 661–673 (2009)
27. Kennedy, J., Eberhart, R.: Particle swarm optimization. In: Proceedings of the 1995 IEEE International Conference on Neural Networks, vol. 4, pp. 1942–1948, Dec 1995
28. Storn, R., Price, K.: Differential evolution—a simple and efficient adaptive scheme for global optimisation over continuous spaces. Technical Report TR-95–012, ICSI, Berkeley, CA (1995)
29. Wang, Y., Li, B., Weise, T., Wang, J., Yuan, B., Tian, Q.: Self-adaptive learning based particle swarm optimization. Inf. Sci. **181**(20), 4515–4538 (2011)
30. Tvrdík, J.: Adaptation in differential evolution: a numerical comparison. Appl. Soft Comput. **9** (3), 1149–1155 (2009)
31. Wang, H., Sun, H., Li, C., Rahnamayan, S., Jeng-shyang, P.: Diversity enhanced particle swarm optimization with neighborhood. Inf. Sci. **223**, 119–135 (2013)
32. Gong, W., Fialho, Á., Cai, Z., Li, H.: Adaptive strategy selection in differential evolution for numerical optimization: an empirical study. Inf. Sci. **181**(24), 5364–5386 (2011)
33. Gordon, D.: The organization of work in social insect colonies. Complexity **8**(1), 43–46 (2003)
34. Kizaki, S., Katori, M.: A stochastic lattice model for locust outbreak. Phys. A **266**, 339–342 (1999)
35. Rogers, S.M., Cullen, D.A., Anstey, M.L., Burrows, M., Dodgson, T., Matheson, T., Ott, S.R., Stettin, K., Sword, G.A., Despland, E., Simpson, S.J.: Rapid behavioural gregarization in the desert locust, Schistocerca gregaria entails synchronous changes in both activity and attraction to conspecifics. J. Insect Physiol. **65**, 9–26 (2014)
36. Topaz, C.M., Bernoff, A.J., Logan, S., Toolson, W.: A model for rolling swarms of locusts. Eur. Phys. J. Spec. Top. **157**, 93–109 (2008)
37. Topaz, C.M., D'Orsogna, M.R., Edelstein-Keshet, L., Bernoff, A.J.: Locust dynamics: behavioral phase change and swarming. Plos Comput. Biol. **8**(8), 1–11 (2012)
38. Oster, G., Wilson, E.: Caste and Ecology in the Social Insects. Princeton University Press, New Jersey (1978)
39. Hölldobler, B., Wilson, E.O.: Journey to the Ants: A Story of Scientific Exploration (1994). ISBN 0-674-48525-4
40. Hölldobler, B., Wilson, E.O.: The Ants. Harvard University Press, Cambridge (1990). ISBN 0-674-04075-9
41. Tanaka, S., Nishide, Y.: Behavioral phase shift in nymphs of the desert locust, Schistocerca gregaria: Special attention to attraction/avoidance behaviors and the role of serotonin. J. Insect Physiol. **59**, 101–112 (2013)
42. Gaten, E., Huston, S.J., Dowse, H.B., Matheson, T.: Solitary and gregarious locusts differ in circadian rhythmicity of a visual output neuron. J. Biol. Rhythms **27**(3), 196–205 (2012)
43. Topaz, C.M., Bernoff, A.J., Logan, S., Toolson, W.: A model for rolling swarms of locusts. Eur. Phys. J. Spec. Top. **157**, 93–109 (2008)
44. Benaragama, I., Gray, J.R.: Responses of a pair of flying locusts to lateral looming visual stimuli. J. Comp. Physiol. A **200**(8), 723–738 (2014)
45. Sergeev, M.G.: Distribution patterns of grasshoppers and their kin in the boreal zone. Psyche **2011**, p. 9, Article ID 324130 (2011)
46. Ely, S.O., Njagi, P.G.N. , Bashir. M.O., El-Amin, S.E.-T., Hassanali1, A.: Diel behavioral activity patterns in adult solitarious desert locust, Schistocerca gregaria (Forskål). Psyche **2011**, 9, Article ID 459315 (2011)
47. Yang, X.-S.: Nature-inspired metaheuristic algorithms. Luniver Press, Beckington (2008)

48. Cuevas, E., Echavarría, A., Ramírez-Ortegón, M.A.: An optimization algorithm inspired by the States of Matter that improves the balance between exploration and exploitation. Appl. Intell. **40**(2), 256–272 (2014)
49. Ali, M.M., Khompatraporn, C., Zabinsky, Z.B.: A numerical evaluation of several stochastic algorithms on selected continuous global optimization test problems. J. Global Optim. **31**(4), 635–672 (2005)
50. Chelouah, R., Siarry, P.: A continuous genetic algorithm designed for the global optimization of multimodal functions. J. Heuristics **6**(2), 191–213 (2000)
51. Herrera, F., Lozano, M., Sánchez, A.M.: A taxonomy for the crossover operator for real-coded genetic algorithms: an experimental study. Int. J. Intell. Syst. **18**(3), 309–338 (2003)
52. Cuevas, E., Zaldivar, D., Pérez-Cisneros, M., Ramírez-Ortegón, M.: Circle detection using discrete differential evolution optimization. Pattern Anal. Appl. **14**(1), 93–107 (2011)
53. Sadollah, A., Eskandar, H., Bahreininejad, A., Kim, J.H.: Water cycle algorithm with evaporation rate for solving constrained and unconstrained optimization problems. Appl. Soft Comput. **30**, 58–71 (2015)
54. Kiran, M.S., Hakli, H., Gunduz, M., Uguz, H.: Artificial bee colony algorithm with variable search strategy for continuous optimization. Inf. Sci. **300**, 140–157 (2015)
55. Li, F.K.J., Ma, Z.: Rosenbrock artificial bee colony algorithm for accurate global optimization of numerical functions. Inf. Sci. **181**, 3508–3531 (2011)
56. Wilcoxon, F.: Individual comparisons by ranking methods. Biometrics **1**, 80–83 (1945)
57. Garcia, S., Molina, D., Lozano, M., Herrera, F.: A study on the use of non-parametric tests for analyzing the evolutionary algorithms' behaviour: a case study on the CEC'2005 Special session on real parameter optimization. J Heurist. (2008). doi:10.1007/s10732-008-9080-4
58. Unnikrishnan, R., Pantofaru, C., Hebert, M.: Toward objective evaluation of image segmentation algorithms. IEEE Trans. Pattern Anal. Mach. Intell. **29**(6), 929–944 (2007)
59. Unnikrishnan, R., Pantofaru, C., Hebert, M.: A measure for objective evaluation of image segmentation algorithms. In: Proceedings of the CVPR Workshop Empirical Evaluation Methods in Computer Vision, 2005
60. Zhang, Y.J.: A survey on evaluating methods for image segmentation. Pattern Recogn. **29**(8), 1335–1346 (1996)
61. Chabrier, S., Emile, B., Rosenberger, C., Laurent, H.: Unsupervised performance evaluation of image segmentation. EURASIP J. Appl. Sign. Process. **2006**, 1–12, Article ID 96306 (2006)

Appendix A
RANSAC Algorithm

RANSAC solves the problem of model parameters estimation by finding the best hypothesis h^B among the set of all possible hypotheses H generated by the source data. Such source data are typically contaminated by noise. In order to build the hypothesis h_i about the unknown parameters, a sample $\mathbf{S}_i$ of the minimum size (s) required for model estimation is obtained (for example, a sample of only two points is sufficient to calculate a straight line, $s = 2$, seven to calculate a fundamental matrix $s = 7$ and four to obtain a homography $s = 4$). Under this consideration, the probability of finding an outlier is reduced. Considering that the number of elements contained in a sample is small, the amount of possible samples that can be generated from the complete source data $\mathbf{U}$ is enormous. Under such circumstances, the exhausting testing of all samples for a reasonable time is impossible. RANSAC face such problem because it only considers G samples which are randomly selected and evaluated. Algorithms of the RANSAC family consist of G iterations of the following cycle:

(1) Construct a sample $\mathbf{S}_i \subset \mathbf{U}$ consisting of s different elements.
(2) Build the hypothesis h_i based on the sample $\mathbf{S}_i$.
(3) Evaluate the degree of agreement A_i of the hypothesis h_i with the set of all source data $\mathbf{U}$.

After the construction and evaluation of all G hypotheses, the hypothesis h^B with the best degree of agreement is chosen among them. It is considered as a robust estimate of the model parameters. Such operation can be described as follows:

$$h^B = \underset{i=1,\ldots,G}{\arg\max}\, A_i(\mathbf{U}, h_i) \tag{A.1}$$

The maximization of the degree of agreement (number of inliers) is equivalent to the minimization of the penalty function whose value depends on the number of outliers. Therefore, the degree of agreement $A_i(\mathbf{U}, h_i)$ is computed as follows:

E. Cuevas et al., *Applications of Evolutionary Computation in Image Processing and Pattern Recognition*, Intelligent Systems Reference Library 100,
DOI 10.1007/978-3-319-26462-2

$$A_i(\mathbf{U},h_i) = \sum_{j=1}^{M} \theta\left(e_j^2(h_i)\right), \quad j = 1,\ldots,M$$
$$\theta\left(e_j^2(h_i)\right) = \begin{cases} 0 & e_j^2(h_i) > Th \\ 1 & e_j^2(h_i) \leq Th \end{cases}, \tag{A.2}$$

where Th is a permissible error, M the number of elements contained in the source data $\mathbf{U}$ and $e_j^2(h_i)$ is the quadratic error produced by the jth data considering the hypothesis h_i. In the context of this chapter, $e_j^2(h_i)$ corresponds to EF_i^2 or EH_i^2 which represent the errors produced by the ith correspondence, considering the fundamental matrix $\mathbf{F}$ or homography $\mathbf{H}$, respectively.

Appendix B
List of Benchmark Functions

In Table B.1, n is the dimension of function, f_{opt} is the minimum value of the function, and **S** is a subset of R^n. The optimum location ($\mathbf{x}_{opt}$) for functions in Table B.1 is in $[0]^n$, except for f_5 with $\mathbf{x}_{opt}$ in $[1]^n$.

The optimum location ($\mathbf{x}_{opt}$) for functions in Table B.2, are in $[0]^n$, except for f_8 in $[420.96]^n$ and $f_{12} - f_{13}$ in $[1]^n$.

Table B.1 Unimodal test functions

Test function	S	f_{opt}
$f_1(\mathbf{x}) = \sum_{i=1}^{n} x_i^2$	$[-100, 100]^n$	0
$f_2(\mathbf{x}) = \sum_{i=1}^{n} \lvert x_i\rvert + \prod_{i=1}^{n} \lvert x_i\rvert$	$[-10, 10]^n$	0
$f_3(\mathbf{x}) = \sum_{i=1}^{n} \left(\sum_{j=1}^{i} x_j\right)^2$	$[-100, 100]^n$	0
$f_4(\mathbf{x}) = \max_i\{\lvert x_i\rvert, 1 \le i \le n\}$	$[-100, 100]^n$	0
$f_5(\mathbf{x}) = \sum_{i=1}^{n-1} \left[100(x_{i+1} - x_i^2)^2 + (x_i - 1)^2\right]$	$[-30, 30]^n$	0
$f_6(\mathbf{x}) = \sum_{i=1}^{n} (x_i + 0.5)^2$	$[-100, 100]^n$	0
$f_7(\mathbf{x}) = \sum_{i=1}^{n} ix_i^4 + rand(0, 1)$	$[-1.28, 1.28]^n$	0

E. Cuevas et al., *Applications of Evolutionary Computation in Image Processing and Pattern Recognition*, Intelligent Systems Reference Library 100,
DOI 10.1007/978-3-319-26462-2

Table B.2 Multimodal test functions

Test function	S	f_{opt}
$f_8(\mathbf{x}) = \sum_{i=1}^{n} -x_i \sin\left(\sqrt{\lvert x_i \rvert}\right)$	$[-500, 500]^n$	$-418.98*n$
$f_9(\mathbf{x}) = \sum_{i=1}^{n} \left[x_i^2 - 10\cos(2\pi x_i) + 10\right]$	$[-5.12, 5.12]^n$	0
$f_{10}(\mathbf{x}) = -20\exp\left(-0.2\sqrt{\frac{1}{n}\sum_{i=1}^{n} x_i^2}\right) - \exp\left(\frac{1}{n}\sum_{i=1}^{n}\cos(2\pi x_i)\right) + 20$	$[-32, 32]^n$	0
$f_{11}(\mathbf{x}) = \frac{1}{4000}\sum_{i=1}^{n} x_i^2 - \prod_{i=1}^{n}\cos\left(\frac{x_i}{\sqrt{i}}\right) + 1$	$[-600, 600]^n$	0
$f_{12}(\mathbf{x}) = \frac{\pi}{n}\left\{10\sin(\pi y_1) + \sum_{i=1}^{n-1}(y_i - 1)\left[1 + 10\sin^2(\pi y_{i+1})\right] + (y_n - 1)^2\right\} + \sum_{i=1}^{n} u(x_i, 10, 100, 4)$ $y_i = 1 + \frac{x_i + 1}{4} \quad u(x_i, a, k, m) = \begin{cases} k(x_i - a)^m & x_i > a \\ 0 & -a < x_i < a \\ k(-x_i - a)^m & x_i < -a \end{cases}$	$[-50, 50]^n$	0
$f_{13}(\mathbf{x}) = 0.1\left\{\sin^2(3\pi x_1) + \sum_{i=1}^{n}(x_i - 1)^2\left[1 + \sin^2(3\pi x_i + 1)\right] + (x_n - 1)^2\left[1 + \sin^2(2\pi x_n)\right]\right\} + \sum_{i=1}^{n} u(x_i, 5, 100, 4)$	$[-50, 50]^n$	0

Printed in the United States
By Bookmasters